AF616219

Guides to Information Sources

Information Sources in

Metallic Materials

Guides to Information Sources

A series under the General Editorship of
D. J. Foskett, MA, FLA
and
M. W. Hill, MA, BSc, MRIC

This series was known previously as 'Butterworths Guides to Information Sources'.

Other titles available are:

Information Sources in the Earth Sciences (Second edition)
edited by David N. Wood, Joan E. Hardy and Anthony P. Harvey

Information Sources in Polymers and Plastics
edited by R. T. Adkins

Information Sources in Energy Technology
edited by L. J. Anthony

Information Sources in the Life Sciences
edited by H. V. Wyatt

Information Sources in Physics (Second edition)
edited by Dennis F. Shaw

Information Sources in Law
edited by R. G. Logan

Information Sources in Management and Business (Second edition)
edited by K. D. C. Vernon

Information Sources in Politics and Political Science: a survey worldwide
edited by Dermot Englefield and Gavin Drewry

Information Sources in Engineering (Second edition)
edited by L. J. Anthony

Information Sources in Economics (Second edition)
edited by John Fletcher

Information Sources in the Medical Sciences (Third edition)
edited by L. T. Morton and S. Godbolt

Guides to Information Sources

Information Sources in
Metallic Materials

1989

Editor
M. N. Patten

BOWKER-SAUR
London • Edinburgh • Munich • New York
Singapore • Sydney • Toronto • Wellington

British Library Cataloguing in Publication Data

Information sources in metallic materials
1. Metals. Information sources
I. Patten, M. N. Series
669′.007

ISBN 0-408-01491-1

Library of Congress Cataloging-in-Publication Data

Information sources in metallic materials / editor M. N. Patten.
p. cm.—(Guides to information sources)
ISBN 0-408-01491-1 (alk. paper) : £4.00
1. Metals—Bibliography. 2. Materials—Bibliography.
3. Metals—Information services. 4. Materials—Information services.
I. Patten, M. N. II. Series: Guides to Information sources (London, England)
Z6678.I48 1989
[TA459]
016.669—dc20 89-39752
CIP

Bowker-Saur is the library and information division of Butterworths, Borough Green, Sevenoaks, Kent TN15 8PH

Cover design by Calverts Press
Printed on acid-free paper
Printed and bound in Great Britain by
Biddles Ltd, Guildford and King's Lynn

Series Editors' Foreword

Daniel Bell has made it clear in his book *The Post-Industrial Society* that we now live in an age in which information has succeeded raw materials and energy as the primary commodity. We have also seen in recent years the growth of a new discipline, information science. This is in spite of the fact that skill in acquiring and using information has always been one of the distinguishing features of the educated person. As Dr Johnson observed, 'Knowledge is of two kinds. We know a subject ourselves, or we know where we can find information upon it.'

But a new problem faces the modern educated person. We now have an excess of information, and even an excess of sources of information. This is often called the 'information explosion', though it might be more accurately called the 'publication explosion'. Yet it is of a deeper nature than either. The totality of knowledge itself, let alone of theories and opinions about knowledge, seems to have increased to an unbelievable extent, so that the pieces one seeks in order to solve any problem appear to be but a relatively few small straws in a very large haystack. That analogy, however, implies that we are indeed seeking but a few straws. In fact, when information arrives on our desks, we often find those few straws are actually far too big and far too numerous for one person to grasp and use easily. In the jargon used in the information world, efficient retrieval of relevant information often results in information overkill.

Ever since writing was invented, it has been a common practice for men to record and store information; not only fact and figures, but also theories and opinions. The rate of recording accelerated after the invention of printing and moveable type, not because that in itself could increase the amount of recording but because, by making it easy to publish multiple copies of a document and sell them at a profit, recording and distributing information became very lucrative and hence attractive to more people. On the other hand, men and women in whose lives the discovery of the handling of information plays a large part usually devise ways of getting what they want from other people rather than from books in their efforts to avoid information overkill. Conferences, briefings, committee meetings are one means of this; personal contacts through the 'invisible college' and members of one's club are another. While such people do read, some of them voraciously, the reading of published literature, including in this category newspapers as well as books and journals and even watching television, may provide little more than 10% of the total information that they use.

Computers have increased the opportunities, not merely by acting as more efficient stores and providers of certain kinds of information than libraries, but also by manipulating the data they contain in order to synthesize new information. To give a simple illustration, a computer which holds data on commodity prices in the various trading capitals of the world, and also data on currency exchange rates, can be programmed to indicate comparative costs in different places in one single currency. Computerized data bases, i.e. stores of bibliographic information, are now well established and quite widely available for anyone to use. Also increasing are the number of data banks, i.e. stores of factual information, which are now generally accessible. Anyone who buys a suitable terminal may be able to arrange to draw information directly from these computer systems for their own purposes; the systems are normally linked to the subscriber by means of the telephone network. Equally, an alternative is now being provided by information supply services such as libraries, more and more of which are introducing terminals as part of their regular services.

The number of sources of information on any topic can therefore be very extensive indeed; publications (in the widest sense), people (experts), specialist organizations from research associations to chambers of commerce, and computer stores. The number of channels by which one can have access to these vast collections of information are also very numerous, ranging from professional literature searchers, via computer intermediaries, to Citizens' Advice Bureaux, information marketing services and information brokers.

The aim of the Guides to Information Sources (formerly Butterworths Guides to Information Sources) is to bring all these sources and channels together in a single convenient form and to present a picture of the international scene as it exists in each of the disciplines we plan to cover. Consideration is also being given to volumes that will cover major interdisciplinary areas of what are now sometimes called 'mission-oriented' fields of knowledge. The first stage of the whole project will give greater emphasis to publications and their exploitation, partly because they are so numerous, and partly because more detail is needed to guide them adequately. But it may be that in due course the balance will change, and certainly the balance in each volume will be that which is appropriate to its subject at the time.

The editor of each volume is a person of high standing, with substantial experience of the discipline and of the sources of information in it. With a team of authors of whom each one is a specialist in one aspect of the field, the total volume provides an integrated and highly expert account of the current sources, of all types, in its subject.

D. J. Foskett
Michael Hill

Preface

This new contribution to the guides to Information Sources Series covers information sources for metals and related materials, i.e. those which have a role in the processing of metals or which may combine with metals to form new materials. The only significant diversion from this will be found in the chapter on packaging materials where reference is made to alternative and competing materials. Rubber and plastics will be dealt with in a further volume which has been commissioned for the series.

This volume does not follow the usual pattern of identifying different forms of literature such as reports and patents in individual chapters as this has been very competently tackled in adjacent books in the Series such as L. J. Anthony's *Information Sources in Engineering*. Instead, it sets out to treat all information regardless of its form as being of equal value when one is examining what is available on a particular metal. Those who wish, therefore, to explore the literature forms in detail are recommended to examine the companion volumes in the series, as many of the sources are common to material, manufacture and application.

The intention was not to produce a guide which would merely adorn the shelves of academic, industrial and reference libraries but to attempt to fill a gap in the information tools of the metals researcher, manufacturer and user. The coverage of end-user interests could have been much greater although this would have resulted in considerable overlap with *Information Sources in Engineering*.

This book has no pretentions to comprehensiveness but seeks to set out a series of references which, although valuable in themselves, are often more important because they act as pointers to a much wider range of source material. Only recently, I came across a reference (in a book on university teaching) to the fact that between its editions of 1924 and 1958, Gmelin's *Handbook of Inorganic Chemistry* reported that 25,000 research investigations had taken place on zinc alone. Such statements no longer surprise us for we are conscious of the outpourings of *Metals Abstracts*, *Chemical Abstracts* and *Referativnyi Zhurnal* to name but a few. Such sources are all information stores but through their citations and bibliographic coupling capabilities they become very powerful pointers to more detailed information. In many ways, the contributors to this guide are themselves pointers to specialist information as one will see from their affiliations that they have all been (and many still are) concerned with the metals industry. A number of them are members of the Metals Information Group, an informal organization of metals information specialists set up to improve the information flow in the metals community within the UK. It was this body which was instrumental in persuading the British Library to set up the Metals Information Review Committee under the auspices of the Metals Society in 1978 to examine the availability of metals information in the UK and the extent to which user needs were being met. The findings of the committee were published in British Library Research and Development Report No.5717 'Scientific and technical information in the metals industry: report of the Metals Information Review Committee, July 1982'. Many of the needs expressed by those who participated in the review have yet to be met but the need for guides to the literature was an early recommendation of the Committee and this has provided much of the impetus for this volume, coming as it does with its companion volumes in engineering, physics and science and technology.

The emergence of Materials Information in recent years as the non-profit-making arm of the American Society for Metals and the Metals Society (which has since merged with the Institution of Metallurgists and reverted to a former title – the Institute of Metals) is a healthy sign for the future and it is encouraging to see that it is paying attention to other needs expressed by the Committee mentioned above – notably in the areas of numerical data and techno-commercial information.

Some comments on perceived trends in the literature pertaining to metals may be appropriate. In recent years, there has been a noticeable decline in the number of books being published in the field although this has been more than offset by the huge volume of periodical, conference and report literature which has appeared. Thus, for a number of metals, the book references may appear to be quite old but they nevertheless remain important in many cases. The Metals Information Review Committee recommended the establishment of a metallurgical database which represented at least fifty years of research in partial recognition of this fact. Jevons' *Metallurgy of Deep Drawing and Pressing* is still a classic in its field despite appearing forty-five years ago. A number of contributors also make the point, and one can only hope that our current fascination with on-line databases and the like will not blind us to the need for good well-referenced texts on processing, properties and applications of materials.

The contents of individual chapters have been determined largely by the contributors themselves and they have identified those sources which they, from their own experience, have found to be useful. The contributors have sought to identify not only printed sources of information but also machine-readable databases of both textual and numeric data and a wide variety of companies and organizations which either publish and/or provide information on specific materials. Part I begins with a chapter on general sources of metallurgical information in an attempt to avoid repeating titles and their contents too frequently – although titles are subsequently emphasized in other chapters where this is felt to be necessary. This is followed by extraction metallurgy and the ferrous metals, followed by the main non-ferrous metals in alphabetical order with the miscellany in the Other Metals (Ch. 10) bringing up the rear. This section of the book is rounded off with powder metallurgy and refractories. Part II is devoted to applications of metals with chapters on design, welding, corrosion and specific use areas such as aeronautics, construction and packaging. For convenience, a list of relevant organizations with their addresses and acronyms where they are known by these has been included as an appendix.

This volume owes something perhaps to the recognition by its contributors of the need to help to fill an ever-widening gap in the literature. It owes considerably more to the fact that a number of very busy people have given their time and effort to help to achieve something that they all believe to be necessary. For some, the task has not been made any easier by the economic pressures which have seen many well-known metals information services reduced in scope and operation or even closed down altogether. It seems sad that there is often no true recognition, either at a local or at a national level, of the value of such services until it is too late. My grateful thanks go, therefore, to all the contributors for their helpful advice and suggestions, for their dedication to the task and for their patience and understanding.

Contributors

V. A. Callcut
Copper Development Association, UK

Dr D. S. Coleman
Institute of Polymer Technology and Materials Engineering, Loughborough University of Technology, UK

M. J. Conway
Lead/Zinc/Cadmium Development Associations, London, UK

R. L. Davies
British Steel Welsh Laboratory, Port Talbot, UK

L. J. Dumper
The Welding Institute, Cambridge, UK

R. F. Flint
Fulmer Research Institute, Slough, UK

Dr D. R. Gabe
Institute of Polymer Technology and Materials Engineering, Loughborough University of Technology, UK

Susan Harris
Technica Ltd, UK

B. A. Hicks
Steel Sheet Information Service, British Steel Port Talbot, UK

C. B. Hubbard
Esso Engineering Ltd, UK

D. Keevil
Aluminium Federation, Birmingham, UK

P. R. Lawrance
British Steel Grangetown Laboratory, Middlesborough, UK

J. Loader
The Welding Institute, Cambridge, UK

Dr D. Murfin
British Ceramic Research Ltd, Stoke-on-Trent, UK

M. N. Patten
Institute Librarian, Southampton Institute of Higher Education, UK

Dr D. K. Thomas
Director, University of Durham Industrial Research Laboratories, UK

Susan Tupholme
Formerly with British Steel Swinden Laboratories, Rotherham, UK

Contents

PART ONE

Materials

CHAPTER ONE

General sources in metallurgy

S. HARRIS, P. R. LAWRANCE and M. N. PATTEN

In the United Kingdom, the Institute of Metals and the Iron and Steel Institute published the bulk of the historical material supported by regional institutions such as the West of Scotland Iron and Steel Institution. In the United States, the American Institute of Mining, Metallurgical and Petroleum Engineers (AIME), the Association of Iron and Steel Engineers (AISE) and the American Society for Metals (ASM) and their regional publications were and still are the major contributors to the history of metals. A recent work is that by R. F. Tylecote entitled *The Prehistory of Metallurgy in the British Isles* (Institute of Metals, 1986). It is a revised edition of a work which first appeared in 1962. Across the Atlantic, P. M. Molloy's *The History of Metal Mining and Metallurgy* (Garland Publications, New York, 1986) has a very good annotated bibliography and is an ideal starting point for those seeking to research the background to metals.

General guides to the literature are largely out of date but are still useful. Those to note are E. G. Gibson and E. W. Tapia *Guide to Metallurgical Information* (SLA Bibliography No. 3, 2nd edn, Special Libraries Association, New York, 1965); V. L. Wilcox, *Guide to Literature for Metals and Metallurgical Engineering* (American Society for Engineering Education, Washington, 1970); M. R. Hyslop, *A Brief Guide to Sources of Metals Information* (Information Resources Press, Washington, 1973); K. Boodson, *Nonferrous Metals: A Bibliographical Guide* (Macdonald and Evans, 1972); and *Sources of Technical Information – Materials* (Federation of Materials Societies, Washington, 1975). Gmelin's *Handbuch der anorganischen Chemie* should not be overlooked and the reader will find this described in the section dealing with Handbooks (p. 7). A very comprehensive work entitled *Encyclopedia of Materials Science and Engineering*, edited by M. B. Bever and published in eight volumes by Pergamon in 1986, contains a very good chapter by H. H. Westbrook, 'Materials information sources', which is particularly valuable as a pointer to sources of raw data. Publications such as the National Bureau of Standards Special Publication 396–1: *Critical Surveys of Data Sources: Mechanical Properties of Metals*, by R. B. Gavert, R. L. More and J. H. Westbrook (NBS, 1974), are still relevant sources.

General and process metallurgy

Anyone with little or no knowledge of metals and wishing to acquire rapidly some basic information about metals and their characteristics could well begin by reading *Metals in the Service of Man* by A. Street and W. Alexander (Penguin Books, 1944); now in its 9th edn (1989). A very good volume with which to follow would be A. Cottrell's well-referenced *An Introduction to Metallurgy* (Crane, Russak, New York, 1975). From here onwards, the choice becomes difficult but mention should be made of J. J. Moore, *Chemical Metallurgy* (Butterworths, 1981); R. Higgins, *Metallurgical Process Metallurgy* (Krieger, Florida, 1983); D. Brown, *Metallurgy Basics* (Van Nostrand, 1983); and J. E. Neely, *Practical Metallurgy and Materials of Industry* (Wiley, 1984). Vladimir Sedlacek, *Nonferrous Metals and Alloys* has recently been translated in English (Elsevier, 1986) and includes some good bibliographies. Another work worth mentioning here is V. A. Krivandin and B. L. Markov, *Metallurgical Furnaces* (Central Books, 1980) which also has a good bibliography. A more recent work is L. Coudurier *et al.*, *Fundamentals of Metallurgical Processes* (Pergamon, 2nd edn, 1986).

Turning to engineering metallurgy, an important work to note is R. Higgins, *Engineering Metallurgy* (R. E. Krieger, Florida, 1983). This two-volume work covering physical and process metallurgy is well-referenced, as is M. C. Nutt's *Metallurgy and Plastics for Engineers* (Pergamon, 1977) and G. Dieter *Mechanical Metallurgy* (McGraw-Hill, 1986). Other works of note include E. C. Rollason, *Metallurgy for Engineers* (Arnold, 1973) and F. W. J. Bailey, *Fundamentals of Engineering Metallurgy and Materials* (Cash, 1972). Mention could also be made here of *Properties and Selection of Tool Materials* (ASM, 1975) which originally appeared in the 'Tool materials' section of the 8th edition of ASM's *Metals Handbook*.

Physical metallurgy

It is useful to remember that many older works are not only good starting points but that they often contain material difficult to locate elsewhere. R. E. Smallman, *Modern Physical Metallurgy* (Butterworths, 1970; 4th edn 1985) is a good starting point. The literature in this area is considerable and one can do little more than direct the reader towards a number of well-referenced works such as E. R. Petty, *Physical Metallurgy of Engineering Materials* (Allen & Unwin, 1970); K. W. Andrews, *Physical Metallurgy: Techniques and Applications* (Wiley, 1973); R. E. Reed-Hill, *Physical Metallurgy Principles* (Van Nostrand, 1972); S. H. Avner, *Introduction to Physical Metallurgy* (McGraw-Hill, 1974); and J. D. Verhoeven, *Fundamentals of Physical Metallurgy* (Wiley, 1975). More recent publications include *Physical Metallurgy*, edited by R. W. Calm and I Haasen (North-Holland, 1983); R. Higgins, *Applied Physical Metallurgy* (Krieger, 1985); and P. Haasen, *Physical Metallurgy*, translated by J. Mordike (Cambridge University Press, 1986). Reverting to earlier works, one should note W. A. Wood, *The Study of Metal Structures and Their Mechanical Properties* (Pergamon, 1971); P. Cotterill and P. R. Mould, *Recrystallization and Grain Growth in Metals* (Surrey University Press, 1976); and C. S. Barrett, *Structure of Metals*, which is now in its 3rd edition (Pergamon, 1980).

Turning to phase transformations, specific works of note include D. A. Porter, *Phase Transformations in Metals and Alloys* (Van Nostrand Reinhold, 1981) and J. W. Christian, *The Theory of Transformations in Metals and Alloys: An Advanced Textbook in Physical Metallurgy*, which first appeared in 1965 and subsequently was published in an enlarged two-volume edition in 1975. In 1986, the proceedings of a symposium sponsored by the Metallurgical Society in New Orleans in March 1986 was published by TMS under the title of *Mechanical Properties and Phase Transformations in Engineering Materials*, edited by S. D. Antolovich *et al.* Two recent conferences of interest whose proceedings have been published are *Materials at Their Limits* (Institute of Metals, 1986) and *Strength of Metals and Alloys*, 7th International Conference on the Strength of Materials and Alloys (Montreal, 1985), edited by H. J. McQueen and others (Pergamon, 1986). Finally, mention should be made of the large volume of periodical literature dealing with aspects of physical metallurgy such as the cover-to-cover translation of *Metallofizika* as *Physics of Metals*.

Metallography needs to be mentioned and a recent work is *Applied Metallography*, edted by G. F. Van der Voort (Van Nostrand Reinhold, 1986).

Alloys of metals

Some general works on alloys should be noted such as W. F. Smith, *Structure and Properties of Alloys* (McGraw-Hill, 1981); *The Metallurgy of the Light Alloys* (Institution of Metallurgists, 1983); and *Light Alloys: Metallurgy of the Light Metals* (ASM, 1981). Important reference points are the *Atlas of Binary Alloys: A Periodic Index* by K. P. Standhamnar and L. E. Murr (Dekker, 1973) and *The Alloy Phase Diagrams Symposium* held in Boston in 1982 and subsequently edited by L. H. Bennett and others and published by North Holland (1983). More recently, the AIME published the proceedings of a number of committees with interests in alloy phase diagrams under the title of *Noble Metals, Alloys, Phase Diagrams, Alloy Phase Stability, Thermodynamic Aspects, Properties and Special Features*, edited by T. B. Massalski *et al.*, (AIME, 1986). This work contains some good bibliographies. In 1987, the ASM published a new two-volume edition of *Binary Alloy Phase Diagrams*, again edited by T. B. Massalski *et al.*, and covering more than 1,850 alloy systems and containing key references for some 700 systems.

Advanced High-temperature Alloys: Processing and Properties, edited by S. M. Allen *et al.* (ASM, 1986) arose from the proceedings of a 1985 Symposium sponsored jointly by MIT and ASM, and discusses topics such as hot isostatic pressing, advances in superplastic materials, creep damage and rapidly cast crystalline thin sheet materials.

Toxicology of metals

The toxicology of metals has long been a topic of interest to man and a number of relevant works have been published recently: *Toxic Metals in the Atmosphere*, edited by J. O. Nriagu and C. I. Davidson (Wiley, 1986), is well-referenced as is G. Mance, *Pollution Threat of Heavy Metals in Aquatic Environments* (Elsevier, 1987). *Toxicology of Metals: Clinical and Experimental Research* – Proceedings of the VIIth UDEH International symposium held in Kitakyushu, Japan, 1986 – was published for the International Union of Pure and Applied Chemistry by Ellis Horwood (1987) and was edited by S. S. Brown and Y. Kodama. Mention should also be made of B. L. Carson, *Toxicology and Biological Monitoring of Metals in Humans including Feasibility and Need* (Lewis Publishers, Chelsea, Michigan, 1986).

Recycling of metals

Shortages of particular metals for a variety of reasons, both political and economic, have intensified interest in recycling certain metals, notably tin and precious metals among the non-ferrous group, while the use of steel scrap in substantial proportions in steel manufacture, both by electric arc and oxygen processes, is likely to remain a feature of the industry for the foreseeable future. A good introduction to this subject is *Recycled Metals in the 1980s* (National Association of Recycling Industries, New York, 1982), containing a useful bibliography. This Association also produces guides to the recycling of specific metals such as *Recycling Copper and Brass* (1980), and *Recycling Aluminium* (1981). Statistics and prices of scrap are dealt with in the technocommercial section of this chapter but a useful technical source to mention here is R. Newell *et al.*, *A Review of Methods for Identifying Scrap Metals* (US Bureau of Mines, 1982). M. E. Henstock, *Design for Recyclability* was published by the Institute of Metals on behalf of the Materials Forum in 1988.

Reference works

Dictionaries

Those dictionaries which deal exclusively with a particular metal will be best pursued in the chapter or section dealing with that metal. Two guides to dictionaries which may be consulted for particular subjects or languages are: *Bibliography of Scientific, Technical and Specialised Dictionaries – Polyglot, Bilingual, Unilingual*, compiled by C. W. Rechenbach and E. R. Garnett (Catholic University of America, Washington, 1969) and *Bibliography of Interlingual Scientific and Technical Dictionaries* (UNESCO, Paris, 1969).

A good general dictionary source was A. D. Merriman's *A Dictionary of Metallurgy* which first appeared in 1958 and was subsequently published as *A Concise Encyclopaedic Dictionary of Metallurgy* by Macdonald and Evans (1965) with nearly 7,000 definitions. Over 4,000 terms with an emphasis on mechanical properties and uses will be found in D. Birchon's *Dictionary of Metallurgy* (Newnes, 1965). More recently, C. R. Tottle produced his *Encyclopaedia of Metallurgy and Materials* based on Merriman's *Dictionary* and published jointly by Macdonald and Evans and The Institute of Metals (1984). This volume contains extensive factual information, including explanations of trade names and mechanical and physical properties and is well illustrated. While discussing terminology, it would be useful to mention ASM's *Thesaurus of Metallurgical Terms: A Vocabulary Listing for Use in Indexing, Storage and Retrieval of Technical Information in Metallurgy* (1986).

Multilingual dictionaries

Elsevier's *Dictionary of Metallurgy and Metal Working* (in six languages) last appeared in 1978 with definitions in English and synonyms and indexes in French, Spanish, Italian, Dutch and German. In the previous year, Elsevier had published E. F. Tyrkiel's *Dictionary of Physical Metallurgy* (in English, German, French, Polish and Russian). More recently, Elsevier have produced a *Dictionary of Physical Metallurgy* (in five languages – English, German, French, Russian and Spanish) by R. Freiwillig *et al.* in 1987, containing about 100,000 terms. An earlier work worthy of note is the *International Dictionary of Metallurgy, Mineralogy, Geology, Mining and Oil Industries* (in four languages, English, French, German, Italian), compiled by A. C. Schwicker (McGraw-Hill, 1968). *A Multilingual Glossary of Heat Treatment Terminology* was produced by the Institute of Metals (1986).

Bilingual dictionaries

Beginning with French, J. O. Kettridge's *Dictionary of Technical Terms and Phrases* covers metallurgy, metal processing and materials extensively and appeared in a new edition in 1983. Published by Routledge and Kegan Paul, this dictionary is a long-standing favourite with translators. Two specialist sources to note are H. Piraux *Dictionnaire français-anglais de l'electrotechnique* (Editions Eyrolles, 1973) and R. Ernst, *Dictionnaire général de la Technique Industrielle, vol. IX, French–English* (Oscar Brandstetter Verlag, 1982).

An important German source containing some 40,000 terms is K. Stolzel, *Dictionary of Metallurgy and Foundry Technology* (Elsevier: English–German, 1984; German–English, 1986). Other general sources favoured by translators include R. Ernst, *Worterbuch der Industriellen Technik* (Oscar Brandstetter Verlag, 4th edn, 1981) and the *German–English Dictionary of Metallurgy*, with related material on ores, mining and minerals, crystallography, welding, metal working, tools, metal products and metal chemistry (McGraw-Hill, 1945), still considered useful.

Italian is represented by G. Marolli, *Technical Dictionary* (Le Monnier, Florence, 1972) and a useful although dated Spanish source with some 70,000 terms is L. L. Sell, *Comprehensive*

Technical Dictionary, Spanish–English (including Metallurgy) (McGraw-Hill, 1949). *An English–Portuguese Metallurgical Dictionary* by J. L. Taylor was published by Stanford University in 1963.

An English–Chinese Dictionary of Metals and their Heat Treatment by Tand Longen and Ni Zenglian was published by Industrial Press, New York (1981). *Gakujutsu yogushu* (Japanese scientific terms) was published by the Japanese Ministry of Education in Tokyo between 1954 and 1964 in twelve volumes. The Metallurgy volume was published early in the series and is now rather dated.

A. A. Rybarzh's *English–Russian Metallurgy Terms* (Moscow, Gosudarstvennoe Izdat Technik-teoretischeskoi Literatury, 1950) was soon followed by A. R. Macandrew, *Glossary of Russian Technical Terms used in Metallurgy* (Varangian Press, New York, 1953). A later publication is T. Demguine, *Russian–English Dictionary of Metallurgy and Allied Science* (Ungar, New York, 1962). A more recent general dictionary is the *English–Russian Polytechnical Dictionary*, edited by A. E. Chernukhin (Pergamon, 1977).

Handbooks

The term 'handbook' covers a range of works from pocket-size compilations of facts through to large multi-volume treatises; it is used here merely in the sense of a digest of information presented in a concise form with a minimum of elaboration, and often with tabulated data and references to the literature. Handbooks grade into textbooks on the one hand and encyclopaedias on the other. Some of the more general handbooks are given below, but the list is necessarily selective, particularly in the case of general technical handbooks which deal with metals:

CRC Handbook of Chemistry and Physics 69th edn, Boca Raton (Florida: CRC Publishing, 1988). Known as 'the Rubber Book'.

Encyclopaedia of Chemical Technology 3rd edn, R. E. Kirk and D. F. Othmer (eds) (New York: Wiley Interscience, 1984, 26 vols). A one-volume concise version of this work was published in 1985, and the full edition is available in full text form in machine-readable form via the online hosts, Datastar, DIALOG, BRS, and Tech-Data.

Gmelins Handbuch der anorganischen Chemie 8th edn (Berlin: Springer, 1924–). This is a massive work; all aspects of the pure and applied chemistry of each (*inter alia*) metallic element is covered in great depth: manganese, for example, has 16 volumes devoted to it. Earlier volumes are in German, later ones in English. Extraction technology and theory is covered in the volume on the 'element'; a literature review and process summaries are provided. For iron and steel there is a supplement: *Gmelin Durrer Metallurgie des Eisens*, 4th edn, 1964– , with 8 two-part volumes published to 1985.

Handbook of Solvent Extraction, T. C. Lo, M. H. I. Baird and C. L. Hilton (eds) (New York, Wiley Interscience, 1983). This covers general principles, industrial equipment and industrial processes for organic and inorganic products (including metals), together with costs and general aspects of process design and use. There are individual chapters covering the metals to which this extraction method is applicable.

Handbook on Material and Energy Balance Calculations in Metallurgical Processes, H. A. Fine and G. H. Gerger (Warrendale, Pennsylvania: Metallurgical Society of AIME, 1979).

Inorganic Chemical and Metallurgical Process Encyclopaedia, M. Sittig (Park Ridge, N.J.: Noyes Data. Corp., 1968).

Materials and Technology: A Systematic Encyclopaedia of the Materials used in Industry and Commerce Vol. 3, Metals and Ores (London: Longman; Amsterdam: de Bussy, 1970).

Metals Handbook 8th edn (Metals Park, Ohio: American Society for Metals, 1960–77) 11 vols (9th edn in preparation; 7 vols published to date). The work is written primarily for the industrial user of metals and is concerned with physical metallurgy. Of the 8th edition, vols 1 (properties and selection) and 8 (metallography, structures and phase diagrams) are of most interest, and of the volumes so far published of the 9th edition, vols 1 and 2 (properties and selection of irons and steels and of non-ferrous alloys and pure metals respectively). A one-volume desk edition is also available.

Perry's Chemical Engineer's Handbook 6th edn, Robert H. Perry and Don W. Green (eds) (New York: McGraw-Hill, 1984).

Selected Values of the Thermodynamic Properties of the Elements, R. Hultgren *et al.* (Metals Park, Ohio: American Society for Metals, 1973). The 'most reliable published and hitherto unpublished data, critically evaluated and presented in tabular form'.

Smithell's Metals Reference Book 6th edn, E. A. Brandes (ed.) (London: Butterworths, 1983). A comprehensive compilation of metals data – the sections of most interest to the extraction metallurgist will be the general physical and chemical contents, crystallographic and crystal chemical data, metallurgically important minerals, thermochemical data, phase diagrams, and physical properties of molten salts.

Thermodynamic Data for Mineral Technology, L. B. Pankratz, J. M. Stuve, N. A. Gokcen (*US Bureau of Mines Bulletin*, no. 677, 1984).

Thermodynamic Properties of 65 Elements – Their Oxides, Halides, Carbides and Nitrides, C. E. Wicks and F. E. Block (*US Bureau of Mines Bulletin*, no. 605, 1963); together with an ongoing series to revise and expand the data therein, the first of which is *Thermodynamic Properties of Elements and Oxides*, B. L. Pankratz (*US Bureau of Mines Bulletin*, no. 672, 1982).

The ASM *Metals Handbook* is now in its ninth edition and the volumes produced so far deal with the following topics:

(1) properties and selection: irons and steels;
(2) properties and selection: non-ferrous alloys and pure metals;
(3) properties and selection: stainless steels, tool materials and special-purpose metals;
(4) heat treating;
(5) surface cleaning, finishing and coating;
(6) welding, brazing and soldering;

(7) powder metallurgy;
(8) mechanical testing;
(9) metallography and microstructures;
(10) materials characterization;
(11) failure analysis and prevention;
(12) fractography.

A one-volume index as well as a one-volume desk version of the *Handbook* are available from ASM and owners of the earlier editions should remember that these often contain data and references which do not appear in the current edition.

The 12th edition of G. S. Brady and H. R. Clauser, *Materials Handbook* (McGraw-Hill, 1986) is, by contrast, a much more modest publication, but still manages to give composition, methods of production, major properties and characteristics, uses and commercial designations or trade names of some 13,000 substances. Edited by C. Wick and R. Veilleux, *The Tool and Manufacturing Engineers Handbook* from the Society of Manufacturing Engineers appeared in a four volume edition in 1985. Volume 3 deals with materials, finishing and coating and, as well as listing and classifying many materials and their properties, aspects of engineering materials such as heat treatment and surface and edge preparation are discussed.

ASM's *Guide to Materials Engineering Data and Information* appeared in 1986 and contains summary tables of properties of elements and major industrial alloys, a detailed directory of the major manufacturers of metals, composites and other engineered materials. Sources of standards are well covered and AISI steels are cross-referenced to most key equivalents. There is a good bibliography of standards reference books and an extensive list of journals dealing with materials, plus a glossary of over 40,000 materials engineering terms. *The Heat Treating Source Book*, edited by P. S. Gupton (ASM, 1986) is a collection of some seventy articles from the literature of heat treating and covers all aspects of the current technology. The second edition of R. C. McMaster, *Nondestructive Testing* (ASM, 1982) is in two volumes, the first covering leak testing and the second dealing with a wide range of liquid penetrant tests. Pearson's *Handbook of Crystallographic Data for Intermetallic Phases* by P. Villars and L. D. Calvert (ASM, 1986) is a reprinted edition of the 1967 compilation with minor amendments and contains critically evaluated data from over 33,000 references, covering the international literature from 1913 to 1983.

R. W. Evans and B. Wilshire, *Creep of Metals and Alloys* (Institute of Metals, 1984) assesses factors affecting the accuracy and reproducibility of creep and fracture properties in relation to the design and operation of tensile and compressive creep equipment. All the software used in the calculations is available from the Institute of Metals.

Journals

Journals, defined loosely, include newspapers and newsletters and range through the trade and industry magazines and commercial organizations' house publications to the learned scientific periodicals which provide the primary means of formal dissemination of original work. A number of computerized databases, often full text, cover the more ephemeral general interest material (weeklies, business publications) and are considered more fully in the section on secondary sources. References to journal articles from industry and learned publications form the bulk of many subject-based or general engineering abstract services and bibliographic databases.

A useful source of information on journals, providing subject access as well as bibliographic details is *Ulrich's International Periodicals Directory*, which is published annually by R. R. Bowker, with quarterly updates. More specifically, a comprehensive listing of metallurgy journals (though minerals industry publications are not so well covered) may be found in *Source Journals in Metals and Materials*, published in 1986 by the American Society for Metals on behalf of Materials Information. This lists over 1,300 journals scanned for inclusion in the *Metadex* database, *Metals Abstracts* and the other secondary sources produced by Metals Information. Similar lists, even if not formally published, are often available from other database compilers or abstracting services. It would not be sensible to reproduce extensive lists of titles here but the reader seeking to glean some knowledge of the key sources available could gain from a study of the titles discussed in some detail in Chapters 2 and 3.

Trade and industry journals

These cover the international and domestic scene in a particular industry by means of news, technical articles and editorial comment, and are typically produced by editorial staff expert in the field, and published either by a commercial technical publisher or a professional organization. News items cover technical developments and equipment, industrial, legislative and company news, together with announcements of courses and meetings and new publications. The longer articles vary in technical depth; some are little more than advertising material while others more closely approach the substantive items found in 'learned' journals. Most publications of this type include a considerable amount of advertising material; the brief digests on new products can usually be supplemented by sending off for information on a product or service via the editorial office – a convenient way of acquiring trade literature.

Many of these journals are international in their coverage though those produced in countries with a high level of indigenous metal production (Australia, South Africa, USA) naturally tend to emphasize domestic activities.

Also noteworthy under this heading are the business journals devoted to the mineral and metals industries: these are concerned more with metals and minerals as commodities and with economic and political factors affecting the industry than with technical developments.

Mention should also be made of the newsletters (usually monthly) of the various professional bodies in the field, for example *IMM Bulletin*, which aim not only to inform the membership of the activities of the parent body, but contain in addition news of a more general nature.

On-line databases

The term 'secondary source' means no more than the tools available for tracking down relevant literature and therefore includes library catalogues, abstracts journals, bibliographies and on-line bibliographic databases.

This section will concentrate on machine-readable databases (although printed sources will also be covered) since for the majority of users they are the most cost-effective means of accessing literature not only in central areas of interest but also in peripheral fields; most (though not all) are available on a pay-as-you-go basis and all that is required for searching is a terminal, or microcomputer or word processor with communications software, telephone and modem, together with a password from the administrators of the host computer. The individual or small company therefore need not invest in expensive printed compilations in very limited areas and for would-be users without access either to the technology or to a willing searcher many libraries (including public libraries) will carry out searches on behalf of their clients and some database producers will do so. Firms or individuals who will carry out searches, 'information brokers', are listed in *UK Online Search Services*, 3rd edn, Geraldine Turpie (ed.) (London, Aslib, 1987), and for the US: *Online Databases Search Services Directory*, 2nd edn (Detroit, Michigan, Gale Research, 1988). Most of these will undertake one-off searches and produce regular updates on topics of interest.

In the UK Aslib – the Association for Information Management – is an invaluable source of information for anyone contemplating using these services who is uncertain how to go about it. They produce a guide for the beginner entitled *Going Online* (most recent edition 1988), and a monthly newsletter *Online Notes* which reports new developments and databases. To track down databases of interest in unfamiliar fields two works will be found useful: *Online Bibliographic Databases: A Directory and Sourcebook*, by James L. Hall, 4th edn (London: Aslib, 1986; updated every few years) and a more comprehensive work which appears every six months (each issue also has a supplement): *Directory of Online Databases* (New York: Cuadra/Elsevier). *Brit-Line: A Directory of British Databases* (Lingfield, Surrey: EDI Publishing, 1987) lists British-produced databases.

Useful databases and their printed equivalents

Brief details are given of the coverage, producer and 'hosts' on whose computers the databases are mounted. The hosts are those to contact for access passwords and instructions on searching; the databases producers for information on database coverage, subject indexing and other database-specific matters. Many databases are mounted on more than one host; there may be slight differences in coverage, search capabilities and costs among them. Printed equivalents of the databases are noted and their coverage in time is often more extensive than that of the online version back in time, though the online version may be more up to date with new additions. All databases are bibliographic (contain citations to the literature) and are in English unless otherwise stated.

Databases covering the general metallurgical/minerals industry literature

Note: sources dealing exclusively with one metal only are excluded; they are covered elsewhere in this book.

IMMAGE Institution of Mining and Metallurgy, UK. 1979– . Worldwide technical literature on minerals exploration, mining, processing, non-ferrous extractive metallurgy and related industry matters. Online access is available via Pergamon Financial Data Services as a private file; potential users should consult the IMM initially. Printed equivalent: *IMM Abstracts*.

Materials Business File Metals Information (joint venture of American Society of

Metals and the Institute of Metals, UK). 1985– . Summaries of techno-commercial literature covering new developments and markets in metals and non-metallic materials. Available via CAN/OLE, CEDOCAR, Orbit. Printed equivalents: *Non-ferrous Alert, Steels Alert, Polymers/Ceramics/Composites Alert.*

Metadex Metals Information. 1966– , depending on host. Worldwide technical literature on metallurgy and metals. Available via CAN/OLE, CEDOCAR, CISTI, DIALOG, ESA-IRS, INKA, Orbit, STN. Printed equivalent: *Metals Abstracts* (1968–), *Review of Metal Literature* (1966–67), *Alloys Index* (1974–) and other Metals Information publications. *Metadex* and *Materials Business File* and their respective printed equivalents are mutually exclusive.

Minproc CANMET, Canada. 1978– . Literature on mineral processing and metal recovery which has relevance to the Canadian industry. In French and English. Available via QL Systems. Printed equivalent: *Minproc Abstracts.*

NODULES Bureau National des Données Océaniques (BNDO) and others, France. 1876– . Literature on extraction of mineral values from ocean floor deposits, especially polymetallic nodules. In French and English. Available via BNDO in Brest.

Non-ferrous Metals Abstracts BNF Metals Technology Centre, UK. 1961–83. Note that this is a closed file – no more information will be added to it. Literature on all aspects of non-ferrous metallurgy, mostly English language. Available via DIALOG, ESA-IRS. Printed equivalent: *BNF Metals Abstracts.*

PASCAL: Métaux – Métallurgie Centre National de la Recherche Scientifique and Centre de Documentation Scientifique et Technique, France. 1973– . Worldwide technical literature on metals and metallurgy. In French; titles and descriptors in English. Available via ESA-IRS, Télésytèmes-Questel. Printed equivalent: *Pascal thema 240: métaux, métallurgie.*

SDIM 1/SDIM 2 Fachinformationszentrum Werkstoffe, FRG. SDIM 1: 1972–79; SDIM 2, 1979– . Worldwide literature on metallurgy. In English, German and French. Available via INKA, STN.

Other relevant science and engineering databases

CA Search Chemical Abstracts Service (CAS). 1967– , depending on host. Covers the world literature on pure and applied chemistry; there are also related databases which cover source publications, substance registry numbers, etc. Available via BRS, CAN/OLE, CISTI, Datastar, DIALOG , ESA-IRS, Orbit, STN, Tech-Data, Télésytèmes-Questel; most comprehensive service is from STN. Printed equivalent: *Chemical Abstracts.* Useful for theory of reactions and processes, hydrometallurgy, mineral and surface chemistry and anything 'chemical' in emphasis.

Chemical Enginering Abstracts Royal Society of Chemistry, UK. 1970– . World literature on chemical engineering. Available via Datastar, ESA-IRS, Orbit. Printed equivalent: *Chemical Engineering Abstracts.* Useful for process and reaction engineering, unit processes and operations, particularly hydrometallurgy.

Compendex Engineering Information Inc., USA. 1969– , depending on host. Covers the world literature on all aspects of engineering and general matters such as maintenance and testing, tribology, pollution, waste disposal, process control etc; substantial coverage of minerals and metals extraction. Available via: BRS, CEDOCAR, CISTI, CAN/OLE, Datastar, DIALOG , ESA-IRS, Knowledge Index, Pergamon Infoline, SDC, STN, Tech-Data. Printed equivalent: *Engineering Index Monthly.*

DECHEMA Deutsche Gesellschaft fuer Chemisches Apparatewesen

(DECHEMA), FRG. 1976– . World literature on chemical engineering. In German; titles in English. Available via STN.

Engineering Materials Abstracts Now on-line via ESA. Produced by Materials Information and covers technical developments in polymers, ceramic and composite materials. The coverage of the Japanese literature is extensive and the file currently holds some 22,500 records and covers material from 1982.

Fluidex BHRA – the Fluid Engineering Centre, UK. 1973– . Covers world literature on behaviour of fluids in engineering. Useful for mixing and separation technology, fluid control, flow measurement, solid/liquid-, two- and multiphase flow. Available via DIALOG , ESA-IRS. Printed equivalent: *BHRA Abstracts* journals (various titles).

Inspec Institution of Electrical Engineers, UK. 1969– , depending on host. World literature on physics, electrical engineering and electronics, computing, and information technology. Useful for measurement and process control engineering. Available via BRS, CEDOCAR, CISTI, CAN/OLE, Datastar, DIALOG, ESA-IRS, INKA, JICST, Knowledge Index, Orbit, STN, Tech-Data. Printed equivalents: *Physics Abstracts, Electrical and Electronics Abstracts, Computer and Control Abstracts, IT Focus*.

Kirk-Othmer Online John Wiley and Sons, UK. 1977– . Full text of the 3rd edition of the *Kirk-Othmer encyclopaedia of chemical technology*. Useful for basic reference for all aspects of chemical technology, including extraction of metals. Available via BRS, Datastar, DIALOG , Tech-Data.

NTIS US National Technical Information Service. 1964– , depending on host. Unrestricted reports of US (mostly) government-sponsored research and development in all areas of science and technology. Available via BRS, CEDOCAR, CISTI, CAN/OLE, Datastar, DIALOG, ESA-IRS, INKA, JICST, Knowledge Index, Mead Data Central, Reference Service, Orbit, STB, Tech-Data. Printed equivalent: *Government Reports Announcements*.

Scisearch Institute for Scientific Information Inc., USA. 1974– . General science and technology; source journals are those most cited. Allows searching for cited references from an article. Available via DIALOG and (a more limited version) DIMDI and Datastar. Printed equivalents: *Science Citation Index, Current Contents*.

Minerals industry business databases

Certain minerals industry publications are available in full-text form on line as part of the general *NEXIS* service operated by Mead Data Central. This is one of a number of general full-text news services available on a subscription basis and is noteworthy because it covers – as well as general and business news publications, wire services and broadcast transcripts – *Metals Week*, and the three Mining Journal publications *Mining Journal, Mining Magazine* and *Mining Annual Review*, all from 1981.

Recent advances in information technology have allowed means of access other than online to large remote computers: some smaller 'source' (rather than bibliographic) databases are available in diskette form for manipulation on personal computers, and new high-density storage media such as CD-ROM (compact disc – read only memory) can now be used to make complete databases locally available to the customer for use on his/her own computer.

Tape lease of some larger databases is possible and economic for large users. Many databases and hosts now permit downloading of data from the host to the user's computer for storage and further processing, though there are usually certain contractual restrictions.

Conferences

Conferences, symposia and formal meetings of various kinds are important fora in their own right for the dissemination of new ideas and practice, regardless of whether the results are distributed to a wider audience via the issue of related publications. New approaches to problem areas are often taken as themes and the cross-fertilization of ideas from separate but related fields is undoubtedly of value to the participants; for maximum benefit to the wider community, however, it is obviously desirable that the proceedings are published more widely. The significance of the conference paper as a permanent information source is reflected in the increasing extent to which presentations are subsequently quoted in the literature. Publishing practices and editorial and bibliographic control are in general not good: papers are often only available to registrants or by writing to the organization or authors, are usually unrefereed, and if formally published are often delayed until some time after the event, thus reducing their value which in part lies in their currency. The smaller local meetings in particular often remain unrecorded in secondary sources.

Sources of information on conferences and conference proceedings

Most trade and industry magazines and the newsletters of professional institutions carry notices of forthcoming events organized by their parent bodies and others and include both local meetings and larger scale events. The Institute of Metals publishes *World Calendar of Forthcoming Meetings: Metallurgical and Related Fields* (quarterly, 1965–). There are in addition a number of general guides to forthcoming scientific and technical meetings produced; these are listed in *Walford's Guide to Reference Material* (op. cit.).

General secondary sources (abstracting/indexing services, databases, bibliographies) covering the literature of the field will include conference material where possible and efforts are often made by the producers of such services to obtain it, but coverage may well be patchy for the more obscure meetings, and individual papers are not always recorded – one entry for the entire proceedings often sufficing. As well as the general subject sources, specialist services covering conference publications are found. *Conference Papers Index* (1973–), produced by Cambridge Scientific Abstracts in Washington, covers about 800 scientific and technical meetings worldwide annually via programmes and abstracts, and is available both in printed form (monthly) and online via IRS/Dialtech and DIALOG. *EI Meetings Index* (1979–), produced by Engineering Information in New York indexes papers from about 2,000 'significant' published proceedings a year in the engineering field. It is available online via IRS/Dialtech, Datastar and DIALOG. With some database hosts it has recently been incorporated in *Compendex*, the database of engineering journal and report literature. From 1983 a printed version, *Engineering Conference Index*, is available as an annual index in 6 sections; the section *Mining, Metals and Fuel and Nuclear Engineering* for 1983–4 will cover 260 conferences and 12 700 papers. The *Index of Conference Proceedings Received* produced monthly with annual and 10-yearly (1964–73) cumulations by the British Library Document Supply Centre (Boston Spa: 1964–), is arranged by keyword taken from the title, and is available as *Conference Proceedings Index* via Blaise-Line for the same period. Further printed sources of information on conference proceedings include the following:

Current Index to Conference Papers in Engineering World Meetings Information Center (New York: CCM Information Corp., 1969–). (Monthly, with twice yearly cumulations; subject and author indexes.)

Directory of Published Proceedings: Series SEMT (science/engineering/medicine/technology) Harrison, N.Y.: Interdok, 1965–). (10 per year, with indexed cumulative annual volume.)

Index to Scientific and Technical Information Proceedings (Philadelphia: Institute for Scientific Information, 1978–). (Monthly, with twice-yearly cumulations; various indexes. Also available online via DIMDI.)

Where appropriate, references are made to specific conferences throughout this volume.

The next chapter gives a good indication of the range of conferences organized by metallurgical organizations and the publications catalogues of these should be regularly examined in order to keep abreast of published proceedings. Most organizations regard conferences as very profitable ventures and the number of these has grown significantly in recent years.

Patents

Patents undoubtedly form an underexploited reservoir of technical information; their principal advantages as a source of knowledge about new processes are that, once an application is accepted, the user knows the technology described is by definition novel and also that the theory on which it is based has already been translated into practice. Their disadvantage is that the terminology and classifications used sometimes seem designed to conceal as much as they reveal and are not intended to be transparent to the non-specialist and, in the past at least, searching through patent indexes has been a time-consuming exercise not always thought worthy of the return.

An exhaustive treatment of the patent literature is beyond the scope of this section. There are a number of works dealing with patents and their exploitation, of which some of the most recent and useful for the non-specialist are A. Wittman, R. Schiffels and M. Hill, *Patent Documentation* (London: Sweet and Maxwell, 1979), *Patents: An Information Resource* (Advisory Group on Aerospace Research and Development report, 1980) and a paper by Charles Oppenheim: *Patent Information Online – a review* (pp. 91–9 in: *Proceedings of the 5th Online Information Meeting, London, 1981*, Learned Information, Oxford). The most cost-effective method to access the patents literature for the casual user is to search the various on-line databases available on an ad-hoc dial-up basis. Table 1.1 gives a brief summary of the coverage of those available which give bibliographic information on patent applications and patents granted. It does not include those covering patent law and patent law decisions. Some bibliographic subject databases include patents in their coverage, the best known of these being CA Search.

Most developed countries have some kind of classified official patents journal for formal publication and these provide a regular current awareness. The *UK Official Journal of Patents* and *US Patents Gazette* are examples. The scanning of such publications is often time-consuming and the user may wish to avail him/herself of the various official abridgement or commercial abstract services available. These are often classified by country or subject and the *World Patent Information* and *World Patent Abstracts* published by Derwent Publications (UK) are good examples. The Derwent service is available online and can be searched by subject group, patentee etc.

Table 1.1. General patents databases available online

Name	*Producer*	*Country/date coverage*	*Hosts*
BREV	Ministry of Economic Affairs, Belgium	Belgium; 1973–	Belindis
CHINAPATS	International Patent Documentation Center	China; 1985–	Dialog, Orbit
CIBERPAT	Registro de la Propiedad Industrial, Spain	Spain; 1968–	Registro de la Propiedad Industrial Spain
CLAIMS – various databases	IPI/Plenum	USA; date is database-dependent	Dialog
EUROPEAN PATENTS REGISTER	European Patents Organisation/Institut National de la Proprieté Industrielle	European patents; 1978–	European Patents Organisation; Télésytèmes-Questel
INPADOC/INPANEW (new patents)	International Patent Documentation Center	European, WIPO, 51 countries; date is database- and country-dependent	INKA, STN, Orbit
INPADOC PATENTE	International Patent Documentation Center	Austria, West Germany, Switzerland; 1978–	INKA, STN
INPI-1	Institute National de la Proprieté Industrielle	France, 1969–	Télésytèmes-Questel
INPI-2	European Patents Organisation/Institut National de la Proprieté Industrielle	European, 1978–	Télésytèmes-Questel
INPI-3	European Patents Organisation/Institut National de la Proprieté Industrielle	Cross refers among patents from various countries/authorities 1969–	Télésytèmes-Questel
INPI-4 (International patent classification)	World Intellectual Property Organisation	International classification for patents	STN, Télésytèmes-Questel, Registro de la Propiedad Industrial (Spain)
JAPIO	Japanese Patent Information Organisation	Japan; 1977–	Orbit
LEXPAT	Mead Data Central	USA; 1975– (full text)	Mead Data Central

Table 1.1. *Continued*

Name	*Producer*	*Country/date coverage*	*Hosts*
PATDATA	BRS	USA; 1971–	BRS, Techdata
PATDATA	Deutsches Patentamt	West Germany; 1973–	STN
PATOS – various databases	Bertelsmann Information Service GmbH, Deutsches Patentamt	Germany; date is database-dependent	Bertelsmann Information Service GmbH
SITADEX	Registro de la Propiedad Industrial, Spain	Spain, 1979–	Registro de la Propiedad Industrial, Spain
USCLASS	Derwent Inc	US patent classification	Orbit
USPATENTS/US PATENTS ALERT (new patents)	Derwent Inc	USA; 1970–	Orbit
WPI (World Patents Index)	Derwent Inc	Chemical, electrical and mechanical patents issued by 29 authorities; 1963– (subject-dependent)	Orbit, DIALOG , Télésytèmes-Questel

Reports

In the metals industry, as elsewhere, there are a number of organizations whose work is of obvious and immediate relevance and such bodies, as well as emphasizing the technical, are also interested in the social and environmental milieu in which the industry operates. There are of course organizations whose principal concern is solely with these matters, such as the Health and Safety Executive in the UK or the Environmental Protection Agency in the USA.

The work of these various bodies is disseminated principally via the report literature. The problem for the would-be user of this information is twofold; identifying the organizations producing the information and tracking down the documents in their often semi-published form. The difficulties in the identification and bibliographic control of the technical report literature are well documented in almost any guide to scientific and technical literature. Subject- or industry-based databases and other secondary sources cover what is available to them, and for the USA in particular there are long-established secondary sources dealing exclusively with government-sponsored research reports; the latter are dealt with more fully in the section on US report literature. Recent efforts to improve the coverage of semi-published material of this type has included the database SIGLE (System for Information on Grey Literature in Europe) available via the hosts BLAISE-LINE and INKA. In practice, personal contact with the producing organizations remains an important means of obtaining such information; informal newsletters and research reports can help.

Documents produced by the Commission of the European Community (EEC) and the European Coal and Steel Community (ECSC) are generally restricted to member organizations but unclassified or 'Open Reports' are available from the Commission's headquarters in Brussels. One should not overlook the many statistical, survey and technical report format publications of the United Nations and its offspring such as UNIDO.

The *Aslib Index to Theses* has a number of subject headings which enable the source of theses to be located, e.g. Chemical metallurgy; Corrosion; Physical metallurgy; Steel structures. In addition to this list of subject headings, there is an adequate subject index of 'keywords'. The following subsections cover some of the principal sources, mostly government-funded, of this kind of information in the main metal-producing English-speaking countries. The principal public source within the UK for obtaining loan or microform copies of technical reports is the British Library Document Supply Centre.

Australia

Federal government-sponsored research in Australia is performed chiefly under the auspices of CSIRO – the Commonwealth Scientific and Industrial Research Organisation. Two divisions within its Institute of Energy and Earth Resources are responsible for research in extractive processing of metals: the Division of Mineral Chemistry and the Division of Mineral Engineering. Annual research reports of each division give informative extended summaries of project work under way and external publications are listed. The divisions produce separate lists of their reports and outside publications of their staff and an annual list of publications and reports issued and available is published as *Serial Publications, Monographs and Pamphlets Issued by CSIRO*.

The Australian Mineral Industries Research Association (AMIRA) is a limited company which coordinates and administers contract research (by CSIRO, educational institutions, private consultants etc.) on behalf of its members who are companies involved in this industry. Summaries of research undertaken by the member companies themselves are given in the annual *Non-confidential Research Information*.

The Australian Mineral Development Laboratories (AMDEL) is managed jointly by AMIRA, the Commonwealth (national) government and the government of South Australia and is an independent commercial contracting organization engaged in R. and D. and consultancy within the minerals industry. Its *Annual Report* and *AMDEL Bulletin* (annual) describe its work, and a full publications list is given in the former.

Canada

Canada's mineral and metal production is large, accounting for more than 10 per cent of its GNP. There are thus considerable resources devoted to government-sponsored R. and D. in this field, largely undertaken on behalf of the Department of Energy, Mines and Resources by CANMET – the Canada Centre for Mineral and Energy Technology. 500–600 reports a year are produced by CANMET or organizations working on its behalf. The *Catalogue of CANMET Publications*, produced annually, lists publications or presentations by CANMET staff or reports of contracted work, and includes work published by staff. The main series is *CANMET reports*; there are in addition 'investigation' and 'technical' reports which are made availabe if unclassified or declassified. Other branches of the Department of Energy, Mines and Resources producing reports on metals- and mineral-related matters include the Mineral Policy Sector and the Office of Environmental Affairs.

South Africa

The Republic of South Africa is the world's largest producer of gold, platinum group metals, vanadium and chrome ore and a significant producer of other metals including iron, steel and ferroalloys, uranium, antimony and manganese. Contribution to minerals and metals technology is thus substantial; research and development being carried out principally by the Chamber of Mines Research Organisation (funded by the industry) and Mintek – the Council for Mineral Technology (formerly the National Institute for Metallurgy – NIM). The Chamber of Mines Research Organisation lists its *Research Reports* and the external publications of its staff in its *Annual Report*. Much of its effort on the metallurgical front is devoted to gold extraction. The Council for Mineral Technology, whose work embraces mineral processing and ferrous and non-ferrous extractive metallurgy publishes its work via its *Reports* series (about 60 per year), formerly *NIM Reports*, together with occasional bulletins and special publications. Cumulative lists of publications are issued from 1966 to date; the most recent being *Special Publication* no. 6, listing all unrestricted NIM/Mintek publications, externally published work by staff, and South African and foreign patents granted, all to 1984. The bimonthly *Research digest* summarizes work under way and announces new reports.

Both the Chamber of Mines and the Minerals Bureau of the Department of Mines produce reports and regular bulletins on economic and policy matters of interest to the mineral industry.

UK

The Department of Trade and Industry's Warren Spring Laboratory undertakes research and development, including contract work for private firms, in mineral processing, metals extraction from primary and secondary raw materials, waste recovery, process automation and control, bulk materials handling and pollution. The scale of activity is small compared with some of the larger metal-producing countries; some 20 reports covering all aspects of the laboratory's work per year are issued in the main reports series; these are listed in quarterly publications announcements. The various divisions of the laboratory also produce occasional newsletters and bibliographies of material published by staff. Relevant divisions include the Mineral Processing Division and the Metals Extraction Division.

Other government-run establishments carrying out work of possible relevance to the minerals industry include the Laboratory of the Government Chemist (analysis), the Atomic Energy Research Establishment at Harwell (separation technology, hazardous wastes), the National Engineering Laboratory (engineering measurements and fluid technology), the National Physical Laboratory (physical metrology, high-temperature properties of metals and melts, materials testing), British Geological Survey (economic evaluation of ores and mineral deposits), Health and Safety Executive (occupational and industrial health and safety).

The Department of Industry's guide *Technical Services for Industry* is now somewhat outdated (1981) but is still a useful guide to sources of technical expertise within government departments and quangos. The *Civil Service Yearbook* produced by the Cabinet Office (HMSO) is a further source of information.

The British Library Document Supply Centre's monthly list of *British Reports, Translations and Theses* provides a current awareness service for new UK-produced reports, classified by subject and with a monthly and annually-cumulated keyword index.

As far as government reports and the like are concerned, the reader can consult the HMSO daily, weekly or monthly lists, use the DTI Technical Reports Centre service which lists and provides subject searches on government-sponsored work in the UK and overseas – a service incidentally also online via DIALTECH – and arrange to see specialist report lists produced by issuing organizations.

USA

The USA is by far the world's largest producer of report literature. The National Technical Information Service (NTIS) is the body responsible for the collection and dissemination of unclassified, publicly available, unlimited distribution reports produced by federal agencies or their contractors. The NTIS database contains over 1 million records from 1964 to date – this in itself demonstrating the size and importance of the US technical report literature – and is available online via many of the large database hosts. The chief printed source is NTIS's *Government Reports Announcements and Index* (semimonthly) and its predecessors. In the UK the principal agent for US technical reports is Microinfo, who, as well as obtaining the publications themselves distribute a variety of announcement and alerting services.

The principal federal organization involved in the minerals and metals extraction industry is the US Bureau of Mines. The chief report series issued are the *Information Circulars* and the *Reports of Investigations*. The former are surveys of technical and economic activity rather than accounts of original research; between 100 and 200 a year are issued covering all aspects of the minerals industry. The latter, a similar number of which are issued, document original research. Preliminary accounts of research are produced as *Technical Progress Reports* and also *Open File Reports* (unpublished), together with various other series concerned with the economics of the minerals industry, mostly in the US. All Bureau publications are announced in a monthly list, with annual and five-yearly indexed cumulations from 1960 and a fifty year cumulative list from 1910–60: *List of Publications Produced by the US Bureau of Mines*. They are in addition included in the NTIS database and printed documentation.

Libraries and information services

To locate comprehensive information relevant to a typical industry-based topic it is desirable to consult a number of sources. Experience shows that even those claiming to cover the same area often apparently show surprisingly little overlap. Problems of scatter remain, even in the large subject-based databases. To answer fully the question, for example, 'What are the economic and technical prospects for metal recovery from manganese nodules?', ideally one would search databases on oceanography, international law, and marine engineering as well as the more obvious commodity- and minerals/metals-related ones. With limited resources this will often be impracticable, in which case a search which aims to cover the literature relevant to all aspects of the industry, including the economic, historical, environmental etc., will be more cost-effective to exploit than those which cover a specific academic-based discipline.

In the UK, Metals Information and the Institution of Mining and Metallurgy are the principal providers of a general industry-based service. There are also the commodity-based services which are considered elsewhere in this book.

Metals Information is a joint venture between the American Society of Metals and the Institute of Metals in the UK. Although the services provided are aimed squarely at the metals – and increasingly the materials – technologist, extraction is also covered, as are all branches of ferrous metallurgy. Services provided of interest to the extraction metallurgist are the online databases already mentioned: *Metadex* and *Materials Business*

File, of use for retrospective searching, with the printed *Metals Abstracts* (subject indexes are also produced) and *Steels Alert* and *Non-ferrous Alert* (online as part of *Materials Business File*) as current awareness publications. A literature search service undertakes both on-off and regular current awareness searches on customer-defined topics. Bibliographies compiled from *Metadex* are also published at intervals on, *inter alia*, extraction-related topics; examples are direct reduction, electric arc melting, gold and silver extraction.

The Institution of Mining and Metallurgy provides services which are to a large extent complementary to those of Metals Information, concentrating on the 'upstream' end of the industry; its coverage of mineral technology and non-ferrous extraction metallurgy is unrivalled, but ferrous metallurgy is largely excluded. Services include *IMM Abstracts* (which has printed indexes only from 1985); prior to that the Library's card indexes must be used for manual searching); the online equivalent *IMMAGE* – also available via tape lease or on diskette – covers from 1979. A unique feature is the historical range of the collection; detailed indexing of the contents of industry journals started in 1894. This is of some relevance in the minerals industry where the vagaries of metal prices and demand and of the political situation of potential metal-producing countries, when combined with improved technology of extraction may render deposits once regarded as defunct once more economic! Retrospective literature searches are carried out using these various sources and selective current awareness profiles run for users on topics of their choice. The coverage of the secondary sources is the significant technical literature in economic geology, mining, minerals technology and non-ferrous metals recovery from primary and secondary sources, with considerable emphasis also on related topics (safety, environment, economics) particularly since the mid-1970s.

Some useful general guides to sources of information in the UK:

Aslib Directory of Information Sources in the United Kingdom, vol. 1: Science, Technology and Commerce. 5th edn, E. M. Codlin (ed.) (London: Aslib, 1982). Provides a listing, with indexes, of organizations providing information in documentary form, as statistics or other media, or direct advice.

Councils, Committees and Boards (Beckenham: CBD Research). Lists, with indexes, UK advisory, non-governmental regulatory, investigatory and other appointed bodies acting nationally or regionally.

Directory of British Associations and Associations in Ireland (Beckenham: CBD Research). Lists national and regional bodies with voluntary membership.

Guide to Government Department and Other Libraries (London: British Library Science Reference and Information Service). Includes specialist information services outside government that are willing to assist outside enquirers.

Industrial Research in the United Kingdom: A Guide to Organisations and Programmes (Harlow: Longman).

BSI

British Standards Institution
Information Department
Linford Wood
Milton Keynes
MK14 6LE

British and foreign standards, quality assurance, 'Technical Help to Exporters' service.

Various High Commission and Embassy Libraries can often help with enquiries about a country and its official publications. Although not set up to deal with outside enquiries, many specialist publishers of industry-related books and journals (e.g. *Metal Bulletin* and *Mining Journal*) are sources of expertise in their areas and willing to assist, even if they do not have formal libraries or information services.

From this chapter it will have been apparent that the mining and metallurgical institutes of the English-speaking countries are substantial providers of published information in the English language, and for all countries these kinds of organization will be a valuable first point of call for enquiries about the industry in that country. The addresses of the principal ones are given below.

Both Metals Information and the IMM Information Services are able to provide document delivery from their libraries (the Institute of Metals for Materials Information in the UK with backup where required from the American Society for Metals) which are valuable resources in their own right for seekers of information, though potential users should check any restrictions of use or charges for non-members of the Institutions concerned. The directories quoted later in this chapter provide a guide to organizations with libraries or information services – with the same proviso concerning use by non-members – and a brief listing, not intended to be exhaustive, of some libraries in the UK which have been found useful for industry-related, but not necessarily industry-specific, information, is appropriate. General sources rather than those dedicated to one commodity are listed. Many will lend items via the British Library Document Supply Centre's interlibrary loan scheme; the BLDSC listed below will provide details of this scheme.

National Libraries

British Library – Document Supply Centre
Boston Spa
Wetherby
West Yorkshire LS23 7BQ

Centre for UK library interlending. Journals, monographs, technical reports, UK theses – all subjects.

British Library – Science Reference and Information Service
25 Southampton Buildings
Chancery Lane
London WC2A 1AW

Physical and chemical sciences and technologies, engineering, business information, UK and foreign patents, trade literature. Reference/enquiry service.

British Library – Newspaper Library
Colindale Avenue
London NW9 5HE

Local and regional newspapers and weeklies. Reference service.

British Library – Official Publications Library
Great Russell Street
London WC1B 3DG

UK government and legislative publications, EEC and intergovernmental bodies' publications. Reference/enquiry service.

Statistics and Market Intelligence Library
1 Victoria Street
London SW1H 0ET

Part of the Department of Trade and Industry's library services. UK and overseas official statistics. Enquiry/reference service.

United Kingdom Atomic Energy Authority
Atomic Energy Research Establishment
The Library
Harwell
Oxfordshire OX11 0RB

Nuclear technology but also metallurgy, chemical engineering, waste treatment and disposal, pollution. Technical reports. Enquiry/reference service.

Warren Spring Laboratory
Gunnels Wood Road
Stevenage
Herts SG1 2BX

Minerals technology, waste treatment, pollution, reclamation from wastes, chemical engineering, biotechnology. General enquiry service.

Institution libraries

Institute of Metals Library
1 Carlton House Terrace
London SW1Y 5DB

Metals and materials technology. Reference/enquiry service.

Institution of Chemical Engineers
165–171 Railway Terrace
Rugby CV21 3HQ

Chemical technology. Reference/enquiry service.

Institution of Mechanical Engineers
1 Birdcage Walk
London SW1H 9JJ

Engineering, especially mechanical engineering. Reference/enquiry service.

Institution of Mining and Metallurgy
Library and Information Services
44 Portland Place
London W1N 4BR

All aspects of the minerals industry. Reference/enquiry service.

Public libraries

Most large provincial centres have central libraries with substantial resources in technical, business and commercial information often including patents, standards and trade literature collections. Worthy of note in London are the City Business Library and Westminster Central Reference Library. Outside London, and not to be confused with the Metals Society/ASM Metals Information Service, is **WMI (World Metals Index)** sponsored by Sheffield City Libraries. It provides documentary information on standards, advice on materials selection and performance and referral service related to metals testing and research facilities.

Techno-commercial information sources for metals

The aim of this section is to bring to the reader's attention the wide variety of sources which are waiting to be tapped – including a number which do not appear to have much connection with the metals industry. The Metals Information Review Committee which was set up in 1978 by the Metals Society under the sponsorship of the British Library's Research and Development Department devoted considerable energies to examining the need for 'techno-commercial information' before it finally reported in 1982 (British Library R. & D. Report no. 5717), 1,000 questionnaires were sent to firms and individuals in the UK and answers to questions revealed that both large and small companies had difficulty in obtaining information in a number of areas. The report looked at a whole range of information provision such as databanks, in-house data, information for management and information for small businesses and concluded that much needed to be done to meet the needs of the metals community in certain areas. Progress has been made in the intervening years but it has been spasmodic and uncoordinated. One of the problems with techno-commercial information is that it is scattered across a wide variety of information sources and the reader's attention is drawn to four general guides to business information sources: P. Wasserman *et al.*, *Encyclopedia of Business Information Sources* (Dale, Michigan) (US entries only but very good on databases); D. M. Brownstone and G. Carruth, *Where to Find Business Information* (Wiley) (International English language coverage); *Aslib Directory, Vol. 1: Information Sources in Science, Technology and Commerce* (specialist information sources in UK organizations); and *Current British Directories* (CBD Research Ltd.).

Directories

Titles to be sought out are those published by *Metal Bulletin* such as *Iron and Steel Works of the World*; *Non-Ferrous Metal Works of the World*; *Metallurgical Plantmakers of the World*; *Metal Traders of the World*; *European and North American Scrap Directory*; *Metal Bulletin Handbook* (Vol. 2 is devoted to metals prices); *Ferro-Alloy Directory* and a host of directories dealing with specific metals. Dun & Bradstreet publish *The Metalworking Directory*, covering manufacturing plants in the US. An essential UK source is *Ryland's Directory* (Fuel and Metallurgical Journals Ltd).

All of the directories mentioned above are updated every 2–3 years or more frequently, with the exception of *Aslib Directory*.

Trading, trade marks, names, literature and associations

A recent guide to the art of trading is *Trading in Metals* by T. Tarring and P. Robbins (*Metal Bulletin*, 1983) and this covers trading in specific metals and scrap trading technique. Brenda Rimmer of the British Library's Science Reference Library has published *Trade Marks: A Guide to the Literature and Directory of Lists of Tradenames* (1976) which is international in its coverage. Specific lists of trade names can be found in sources such as *Trade Names Directory* (Gale Research Corp., Michigan) and Ryland's *Directory*. The British Library has produced two other useful guides in this context in *Trade Literature: A Review and Survey* by Martin Thomson and *Trade Directory Information in Journals*.

Sources of information on trade and research organizations include the *Aslib Directory* already mentioned; *European Research Centres: A Directory of Organisations in Science, Technology, Agriculture and Medicine* (Longmans); *European Sources of Scientific and Technical Information* (Longman) (with details of information services available); *Industrial Research in the United Kingdom* (contains a useful section on trade and development associations) (11th edn, Longman, 1985); *Industrial Research Laboratories in the United States* (Bowker); and *World Guide to Scientific Associations* (Saur, N.Y.) (contains a useful list of organizations involved in metallurgy).

Prices

In addition to historical sources such as *Metals Handbook*, current prices for metals can be found in *The Wall Street Journal* and *The Financial Times* and in publications appearing several times a week such as *American Metal Market*, *Metal Bulletin* and *Japan Metal Bulletin*. *Metals Week* produces a global newsletter and also runs a price notification service and produces a substantial annual price handbook. It also runs an online database. Chase Econometrics Associates Inc. run three databanks – Non-Ferrous Metals Forecast; Iron and Steel Forecast; and Steel Forecast. *The Financial Times* runs a World Commodity Report service which covers metals.

Markets and competition

Some of the key sources to be monitored have been mentioned under Prices but it will be apparent that apart from a number of general sources of a periodical or review nature there are journals dealing with specific metals and aspects of metal manufacture and use. These are too numerous to mention and the reader is advised to consult *Ulrich's International Periodicals Directory* and *Ulrich's Directory of Irregular Serials* – both published by Bowker. Until recently, commercial information has been largely ignored by the database producers with the exception of files such as Predicast which had been abstracting commercial information for its printed form – *F & S. Index* – for a number of years. This is a very comprehensive source with US, international and European editions of the printed form of the database. *Research Index* which covers UK daily newspapers and key weeklies is a very current source as are the databases of the *Financial Times* and the *New York Times* to name but two.

A number of stockbrokers are active analysts of metal markets and trends and produce regular detailed reports, as do specialist sources which provide information on a subscription basis such as the Stamford Research Institute Long Range Planning Service. Frost & Sullivan and Chase Econometrics are obviously filling a need. Metals Information (ASM and the Institute of Metals) has recently launched *Steel Alert*, *Non-ferrous Alert* and *Polymers/Ceramics/Composites Alert* – all designed to keep management informed about metals and materials developments.

An online version of these files exists in the *Materials Business File*.

Innovation and technology transfer

The patent literature has been dealt with earlier but it cannot be stressed too often that the filing of a patent may be the only evidence that a product or process exists. Three useful intelligence sources are *Product Licensing Index* (announcing products available for license and licensees looking for products); *Research Disclosure* (allowing low-cost disclosure of innovations without the expense of patenting); and *Planned Innovation* listing selected innovations and covering licensing, marketing and business aspects. A number of periodicals carry details of technologies sought and offered including *UNIDO Newsletter, International Licensing, Technology Transfer Action* and *Information for Innovators*. Three computer databases are the French Transinove which matches an individual's interest profile against the technology available on a regular basis; Technotec which is a US database listing submitted profiles of technology sought and technology offered from worldwide sources. Management, distribution and marketing material is also included; and World Bank of Licensable Technology, giving details of products and processes available for licensing. Other useful sources are the UK's Technology Reports Centre, the US National Technical Information Service (NTIS), Battelle Memorial Institute and the Stanford Research Institute. The latter maintains extensive backup files on microfilm of raw data and evidence gathered in support of each report produced. Conference sources have received detailed mention elsewhere.

Market research

The Predicast database has proved to be particularly useful because of its indexing structure and the emphasis placed on companies, materials and products. Many trade and research organizations undertake market research on behalf of their members. The range and output of some of these can be seen in *Market Research and Industry Surveys: A List of Reports Held by the Science Reference Library* (British Library, 1984). The British Overseas Trade Board issues the *International Directory of Published Market Research* and Industrial Aids Ltd. produces *Published Data on European Industrial Markets*. *Sources of UK Marketing Information* by E. Tupper and G. Wills was published by Benn in 1975 but is still useful.

Product information

Predicasts and its hard-copy version *F & S. Index* are good general sources but this is the sort of information which is spread over sources ranging from standard specification handbooks to advertisements in trade journals. A new work which has tackled the problem of identifying product information sources extremely well is *Finding and Using Product Information – from Trade Catalogues to Computer Systems*, edited by Raymond Wall (Gower, 1986).

Statistical and economic surveys

This section attempts to provide an overview of general sources of information on supply, production etc. of raw metals; many of the sources also cover ores, concentrates and finished or semi-finished products. General sources covering more than one commodity are emphasized as the more specific information on individual metals can be found in the chapters devoted to them elsewhere in this book.

Information on metals production is gathered and published by government agencies – often Departments of Mines or Mineral Resources – and yearbooks or more frequent cumulations are produced; too numerous to list individually here. Supranational and international bodies such as ASEAN, the UN, the OECD and the EEC also produce statistics for their areas of influence and commodity associations (such as the International Tin Institute, the Lead Development Association, International Iron and Steel Institute) maintain extensive databases of statistical information on their metal and its sources and products.

As well as the various sourcebooks for organizations mentioned earlier in this section, one work which will be found useful is the compilation by Roskill Information Services, updated every few years: *Roskill's Directory of Sources for Metals and Minerals Data* (London: Roskill Information Services, 5th edn, 1982) which lists published sources by metal and country. In the UK the Statistics and Market Intelligence Library of the Department of Trade and Industry – a useful repository of this kind of information in its own right – produces short guides to the statistical and marketing information available in various countries.

Several commercial enterprises produce regular publications containing statistical information ranging from the daily and weekly prices, for which the most detailed sources are *Metal Bulletin* and *Metals Week*, to annual or five-yearly cumulations. Regular summaries such as these latter often include brief reviews of technical and commercial developments and prognostications for the industrial sectors concerned, but for more detailed analysis one must turn to commodity or country/regional surveys. These are produced principally by specialist commodity analysts; they may be multi-client studies available only to subscribers initially; more often they are unrestricted but command a high price particularly if much forecasting and market analysis is involved. Some are little more than regurgitations, albeit in a useful and convenient form, of material already published and available. Firms in the UK producing commodity and regional reports include the Commodities Research Unit, Metals and Minerals Research Services, *Metal Bulletin*, Mining Journal Research Services, Roskill Information Services, Anthony Bird Associates, Shearson Lehmann/Hutton, and Rudolf Wolff and Company. New reports are announced in the business press, particularly *Metal Bulletin*, *Mining Journal*, and in *Minline*, a useful newsletter reporting on mineral industry information developments which is itself produced by one of the above-mentioned firms – Metals and Minerals Research Services. The publications of Roskill Information Services in particular are noteworthy because they cover the less common metals for which information is otherwise scarce; titles in their series *The Economics of. . . .*, each of which is revised at intervals, include for example *Caesium and Rubidium* (2nd edn, 1984), *Scandium* (1st edn, 1974), *Selenium* (4th edn, 1983), as well as the more usual high-volume and strategic metals. Notable overseas examples of enterprises overseas producing commodity and regional studies include Metals Economics Group in the USA and Australian Mineral Economics Pty. in Australia. Mention should also be made of *American Metal Market* and *Iron Age*, typical of the range of periodicals with good statistical information. Several business and economics sources: newsletters, statistics and journals, are now available online. These are covered in the section on databases (see p. 12).

Government agencies also produce publications of this type, perhaps with less analysis than the commercially produced variety, but with the possible advantage to the user of knowing that the information is made available on a statutory basis to the producers of the report and thus is as complete as possible within its scope. In the UK the Minerals Strategy and Economics Group of the British Geological Survey produce on behalf of the Mineral Resources Consultative Committee the *Mineral Dossier* series (HMSO), and also a number of other regional and commodity reviews. Other examples of major series include the *Untersuchungen ueber Angebot und Nachfrage mineralische Roehstoffe (Investigations into supply and demand of mineral raw materials)* and *Roehstoffwirtschaftliche Laenderberichte* (Raw materials economics country reports) produced by the West German Bundesanstalt fuer Geowissenschaften und Roehstoffe, and the *Mineral Commodity Profiles*, *Mineral Issues* and *Mineral Perspectives* of the US Bureau of Mines, and the report series of the Mineral Policy Sector of the Canadian Department of Energy, Mines and Resources. Naturally the emphasis of these reports is on those commodities or countries of special interest to the producing countries for strategic or economic reasons at the time of writing.

General statistical compilations

This list includes the major, regularly produced, statistical compilations which cover a number of commodities on an international basis. Annual statistics are generally available two years after the year to which they apply, but the year given in the title of the work may be the year of its publication or the year the statistics refer to, depending on the publisher's policy.

Annuaire Minemet: statistiques (Paris: Minemet – Groupe Imetal, annual). Contains data for 4 years.

Annual Review of the Metal Markets (London: Shearson/American Express).

Metal Bulletin Handbook, Vol. 1: Prices; Vol. 2: Statistics and Memoranda (Worcester Park: Metal Bulletin, annual, 1968). Contains data for 5 years. (Succeeds *Quin's Metal Handbook*.)

Metal Statistics: The Purchasing Guide of the Metals Industries. American Metal Market (New York: Fairchild Publications, annual, 1908–). Span of coverage varies by commodity.

Metalli non-ferrosi: statistiche (Rome: SAMIM, annual, 1948–). Contains data for 7 years.

Metallstatistik (Frankfurt-am-Main: Metallgesellschaft, annual, 1889–). Contains data for 10 years.

Metals Analysis and Outlook (London: Metals and Minerals Publications, quarterly, 1976–).

Metals Week Price Handbook (New York: McGraw-Hill, annual, 1973–). Contains data for year of edition.

Mineral Commodity Summaries (Washington: US Bureau of Mines, annual, 1982–). Contains data for 5 years.

Mineral Facts and Problems (Washington: US Bureau of Mines, every 4–5 years). Published as part of the Bureau's *Bulletin* series; the 1985 edition is no. 675 of that series. Chapters comprise commodity surveys which are printed individually as *Mineral Commodity Profiles*.

Minerals and Materials (Washington: US Bureau of Mines, bimonthly, 1977–).

Minerals Yearbook, Vol. 1: Metals, Minerals and Fuels; Vol. 2: Area Reports – Domestic [i.e. US]; Vol. 3: Area Reports – International (Washington: US Bureau of Mines, annual, 1933–). Preprints of individual chapters are published up to 18 months before the compilation.

Non-ferrous Metal Data (New York: American Bureau of Metal Statistics, annual, 1921–). Formerly *Yearbook of ABMS*. Contains data for 5 years.

The Non-ferrous Metal Industry (Paris: OECD, annual, 1954–). Contains data for 3 years.

Roskill's Metals Databook (London: Roskill Information Services, annual, 1979–). Span of coverage varies by commodity.

World Metal Statistics (monthly) and *World Metal Statistics Yearbook* (annual cumulation) (London: World Bureau of Metal Statistics, 1948–). Yearbook contains data for 10 years.

World Mineral Statistics: Production, Exports, Imports (Minerals Strategy and Economics Research Group, British Geological Survey. London: HMSO, annual (previously every two years), 1978–). Contains data for 5 years. Succeeds: *Statistical Summary of the Mineral Industry: World Production, Imports and Exports*, 1913–77.

Significant reviews of economic developments over the previous year are to be found, along with the technical reviews, in the

Mining Annual Review (London: Mining Journal), with coverage both by commodity and by country. *Engineering and Mining Journal's* annual review issue also contains economic reviews but with emphasis on the US. Neither contains significant tabulated statistical data. Most of the industry journals contain occasional 'one-off' regional or commodity surveys; *Metal Bulletin Monthly* in particular concentrates on this kind of item.

Brief mention must be made of the government-produced yearbooks in the large mineral and raw metal producing countries. In Canada, US and Australia these may be issued also on a state or province level as well as federally; it is not possible to list them all. Regardless of their geographical scope they usually cover the field on a commodity basis and include significant statistical information, often in considerable detail. Publication in some cases is often erratic, particularly for the African and South American countries, and even in the so-called developed countries is often delayed. Of these publications, the following are typical:

Australian Mineral Industry Annual Review (Canberra: Australian Government Publishing Service for the Bureau of Mineral Resources, Geology and Geophysics).
Canadian Minerals Yearbook (Ottawa: Energy, Mines and Resources Canada).
Indian Minerals Yearbook (Nagpur: Indian Bureau of Mines).
Minerals Yearbook (US) – see previous page.

Translating services

For iron and steel, the best known source is the BISITS collection in the Institute of Metals which produces several hundred translations each year. The Institute cooperates with ASELT (Association dans le domaine de la Siderurgie) which funds translations for ECSC member countries and publishes details of these in World Transindex and in Euro Abstracts. The British Library produces *British Reports, Translations and Theses* monthly and a further source of metallurgical translations is the monthly Translations Register Index published by the National Translations Center of the John Crerar Library, University of Chicago. Other organizations active in this field include Aslib, standards bodies, commercial translating agencies such as Ricanski, embassies and trade and research associations.

CHAPTER TWO

Extraction Metallurgy

S. HARRIS

Introduction

The subject areas covered by this chapter are the processing of as-mined minerals and secondary raw materials for the recovery of metals, i.e. those areas conventionally designated 'mineral processing' and 'extractive metallurgy'.

The boundary between these two fields remains a real one in terms of both the organization of metal production and the sources of information about it, though the distinction is becoming increasingly blurred with the advent of hydrometallurgical extraction techniques for the base, precious and less common metals. A further distinction is found between the organizations and publications concerned with the production of iron and steel on the one hand and non-ferrous metals on the other. A note on terminology is appropriate: 'non-ferrous' is taken to embrace all metals apart from iron and steel, and to include the ferroalloying elements such as vanadium, chromium and manganese. Sources dealing with extractive metallurgy as a whole, rather than those whose scope is restricted to one metal, are emphasized.

Reference material: handbooks

Extraction metallurgy interacts with and draws on a number of other disciplines – chemistry, physics, chemical engineering – and many works of reference used in the chemical and process industries are standard also in the extractive processing of metals. There is a trend away from the publication of textbooks and handbooks – an inspection of the Institution of Mining and Metallurgy's holdings reveals a number of such works published prior to 1960 and relatively few since. The rapid advance of technology and the increasing volume of literature produced renders large compilations slow to produce and as 'state-of-the-art reviews' they date quickly. There is still however a need for compilations of basic data in all branches of the process industries which ensures a continuing demand for the large chemical data sources. Handbooks of general interest to the metals user have been discussed in Chapter 1.

A particular source to note here is: *Handbook of Solvent Extraction*, T. C. Lo, M. H. I. Baird and C. L. Hilton (eds) (New York: Wiley Interscience, 1983). This covers general principles, industrial processes for organic and inorganic products (including metals), together with costs and general aspects of process design and use. There are individual chapters covering the metals to which this extraction method is applicable.

There are few handbooks on extractive processing. A long-awaited update to Arthur F. Taggart, *Handbook of Mineral Dressing: Ores and Industrial Minerals* (New York: Wiley; London: Chapman and Hall, 1945) has recently been produced and covers not only mineral processing but also pyrometallurgical and hydrometallurgical extraction: *SME Mineral Processing Handbook*, edited by N. L. Weiss (New York: Society of Mining Engineers of AIME, 1985). An introduction covers economic minerals (including non-metallic) and their properties; part 1, physical and chemical extraction techniques; part 2, examples and descriptions of flowsheets and plant operations for individual metals; part 3, general matters such as process design, control, automation, sampling, waste treatment.

Reference material: dictionaries and glossaries

Dictionaries covering extraction metallurgy in detail tend either to cover metallurgy as a whole, thus including metal-forming, fabrication, metal products, finishing and testing, or to deal with the minerals industry, including mining, mineral processing and, sometimes, economic geology. The emphasis of the coverage is usually obvious from the title.

There is a shortage of an up-to-date English language glossary or encyclopaedic dictionary of mineral processing and extraction metallurgy. The most recent and complete is the *Dictionary of Mining, Mineral and Related Terms*, compiled by Paul W. Thrush and the staff of the US Bureau of Mines and published in 1968 in

Washington by the US Department of the Interior. It is detailed in its coverage, containing some 55,000 entries (including economic geology and mining) for English-language terms from all over the world, with variants of uses and sources of information quoted; however it invariably reflects the state of technology at the time of its compilation, and advances since then have rendered its once comprehensive coverage incomplete. An earlier work of more restricted scope is the *Dictionary of Mineral Technology*, by E. J. Pryor (London: Mining Publications, 1963). Over 4,000 terms covering the processing of ores and their subsequent metallurgical treatment, along with relevant physico- and electrochemical terms, are included. Useful for historical terms, and the source of inspiration for Thrush's 1968 work (op. cit.) is A. H. Fay, *A Glossary of the Mining and Mineral Industry* (*US Bureau of Mines Bulletin*, no. 95, 1920) with some 18,000 definitions.

Dictionaries covering metallurgy as a whole are better represented and can be found in the introductory chapter.

There are a few multilingual dictionaries covering mineral technology:

Gruvteknisk ordlista (Mining technology glossary) (Kiruna: Luosssavaara-Kiirunavaara Aktiebolag (LKAB), 1961). English, Swedish, German, French, with master listing in Swedish.

International Dictionary of Metallurgy, Mineralogy, Geology, A. C. Schwicker (Milan: Technoprint in association with McGraw-Hill, 1968). English, French, German, Italian with master listing in English.

Lexique quadrilingue de la préparation des minerais (Four language dictionary of mineral processing) (St Etienne: Société de l'Industrie Minérale, 1963). German, English, French, Russian.

Trilingual Dictionary for Comminution (St Etienne: Société de l'Industrie Minérale, 1975). German, French, English.

World Mining Glossary of Mining, Mineral Processing and Geological Terms. R. J. M. Wyllie and George O. Argall (2nd edn, San Francisco: Miller Freeman, 1975). English, Swedish, German, French, Spanish, with master listing in English).

There are in addition a number of bilingual dictionaries. The coverage is patchy; German, for example, being better served than French. For some languages a general scientific-technical bilingual dictionary will be all that is available. These may be located using the guides to dictionaries and to scientific and engineering literature mentioned at the beginning of this book. Two more specific in their scope are:

Fachworterbuch Bergbau (Mining technical dictionary). I. H. Goergen *et al.* (Essen: Verlag Gluckauf, 1981). 2 vols: English–German/German–English.

Technik-Worterbuch Bergbautechnik und Aufbereitung (Technical dictionary – mining technology and processing). H. Schmidt (Berlin: VEB Verlag Technik, 1980). English–German/German–English.

Textbooks

Books and monographs are produced by professional bodies and learned societies as well as by commercial publishers and with the reputation of the parent organization behind them the former are generally highly regarded. Foremost among the organizations with a publishing programme in the fields of extractive metallurgy and mineral technology are the constituent bodies of AIME (the American Institute of Mining, Metallurgical and Petroleum Engineers). AIME's Society of Mining Engineers publishes work on mineral processing, hydrometallurgy – particularly that of uranium, copper and precious metals – and on general aspects of the minerals industry. The Metallurgical Society of AIME is concerned with extractive metallurgy as a whole and also for specific metals other than iron, as well as other areas of metallurgical science and technology, including some aspects of steels. The Iron and Steel Society of AIME deals with iron and steelmaking.

Similar bodies in other countries have a more limited publishing programme in the same areas; it is generally the organizations who are concerned with the minerals industry incorporating the 'upstream' end of metallurgical technology, rather than those involved in the fabrication and use of metals, that are important in this respect. The major English language representatives are the Canadian Institute of Mining and Metallurgy (CIM), the South African Institute of Mining and Metallurgy (SAIMM), the Australasian Institute of Mining and Metallurgy (Aus. IMM) and in the UK the Institution of Mining and Metallurgy (IMM).

Of the textbooks produced by the commercial publishers, four series are noteworthy. The first is the *Ellis Horwood Series in Industrial Metals* (Chichester: Wiley under the Ellis Horwood imprint), intended to provide concise summaries on the properties, occurrence, extraction, alloys, processing, products and economics of the particular metal, aimed at both students and practising engineers; titles so far published have covered cobalt, copper, nickel, tin and hydrometallurgical extraction, and, in an earlier part of the series when it was published by Macdonald and Evans, zinc. Elsevier's *Process Metallurgy* series comprises five titles to date: on solvent extraction, the extractive metallurgy of tin, polarization measurements in metal deposition, biohydrometallurgy, and tailing disposal. Also produced by Elsevier is the *Developments in Mineral Processing* series, with nine titles so far; these are mostly textbook-style treatments or conference proceedings, rather than Elsevier's more usual review format. The NATO Advanced Study Institutes proceedings form up-to-date reviews of the 'state of the art' and several titles have been published covering mineral processing, including *E75 – The Scientific Basis of Flotation; E117 – Mineral Processing at a Crossroads: Problems and Prospects; E122 – Mineral Processing Design*. This series is published by Martinus Nijhoff (Dordrecht, Manchester, New York).

At an introductory level, intended for undergraduate students, are two pairs of companion volumes. *Mineral Processing Technology: An Introduction to the Practical Aspects of Ore Treatment and Mineral Recovery* (4th edn, 1988) by B. A. Wills, and *Extraction Metallurgy*; 2nd edn, by J. D. Gilchrist, nos 29 and 30 respectively in the *International Series on Materials Science and Technology*, are published by Pergamon (1980); *Introduction to Mineral Processing*, by E. G. Kelly and D. J. Spottiswood, and *Nonferrous Extractive Metallurgy*, by C. B. Gill are published by Wiley Interscience (1980 and 1982).

A further general text on mineral processing is *Unit Operations in Mineral Processing*, by J. M. Carne (Colorado School of Mines Press, 1978), again aimed at students. A number of older general works at a more advanced level include those by notable practitioners in mineral technology such as E. J. Pryor, R. H. Richards and A. F. Taggart (including Taggart's *Handbook* (op. cit., p. 37)) but are now dated. The most recent of these is E. J. Pryor's *Mineral Processing*, 3rd edn (London: Applied Science, 1965). Not textbooks *sensu stricto* but compilations of articles on plant practice and new processes reprinted from the journal and collected under appropriate subject headings are *Engineering and Mining Journal's E. & MJ Operating Handbook of Mineral Processing* (New York: McGraw-Hill) of which two volumes, published in 1977 and 1980, and along the same lines, the *E & MJ Book of Flowsheets*, have appeared. Extraction metallurgy processes are included in the latter. A similar idea is used in the *Minerals Engineering Handbook*: a series of articles published in the journal *Mine and Quarry* covering the physical processing of ore minerals and coal which the reader collects and updates to form the handbook (Derby: Minerals Engineering Society, 1984–). One further general work in mineral processing is *Mineral Processing Plant Design*, 2nd edn, edited by A. L. Mular and R. B. Bhappu (New York: Society of Mining Engineers of AIME, 1980).

Mention should also be made of the work by G. Tarjan and published under the title *Mineral Processing* by the Akademiai Kiado in Budapest. Volume 1 (1981), entitled *Fundamentals, Comminution, Sizing and Classification*, and Volume 2 (1987), *Concentration, Flotation, Separation and Back-up Processes*, have appeared to date. *Design and Installation of Concentration and Dewatering Circuits*, edited by A. L. Mular and M. A. Anderson (Littleton, Co.: Society of Mining Engineers of AIME, 1986) comprises contributions by specialists on aspects of flotation, solid-liquid and solid-solid separation and related design and operational topics.

Considering mineral processing further, the monographs which deal with the initial stages of extractive processing, such as comminution, gravity concentration, flotation, dewatering, should be briefly mentioned. The approach in these can vary widely; books on flotation, for example, can be highly theoretical treatments of surface chemistry or can take a more practical approach. Many works of more general applicability are relevant in mineral processing – for example those on solid–liquid separation, and powder technology. Books and monographs in mineral technology generally tend to be conventional in their approach; new areas of concern, such as the integrated processing of fine-grained, complex or difficult-to-treat ores, or approaches addressing topics of wider concern but with special reference to the mineral industry – such as energy conservation, process control, environmental problems – are more usually treated as topics for conferences. Some representative textbooks in mineral technology published in the last ten years are cited below.

Comminution

Design and Installation of Comminution Circuits, K. L. Mular and G. V. Jergensen (eds) (New York: Society of Mining Engineers of AIME, 1982).

The Process of Fine Grinding (Developments in Mineral Science and Engineering, Vol. 1), B. Beke (The Hague; London: Martinus Nijhoff/Dr W. Junk, 1981).

Process Engineering of Size Reduction: Ball Milling, L. G. Austin *et al.* (New York: Society of Mining Engineers of AIME, 1984).

Flotation

Flotation (A. M. Gaudin memorial volume), M. C. Fuerstenau (ed.) (New York: AIME, 1976, 2 vols.)

Principles of Flotation, R. P. King (ed.) (*South African Institute of Mining and Metallurgy monograph series*, no. 3, 1982).

Gravity concentration

Gravity Concentration Technology (*Developments in mineral processing*, no. 5), R. O. Burt (Amsterdam: Elsevier, 1984).

Handbook of Mineral Jigs (San Francisco: Consolidated Placer Dredging Inc., 1983).

Hydrocyclones, L. Svarovsky (London: Holt, Rinehart and Winston, 1984).

Dewatering

Advances in Drying, Vols 1 and 2, A. S. Mujumdar (ed) (Washington; London: Hempisphere Publishing, 1980 and 1983).
Solid-Liquid Separation (*Butterworths Monographs in Chemistry and Chemical Engineering*), 2nd edn, L. Svarovsky (ed.) (London: Butterworths, 1981).

Magnetic and Electric Separation

Electrostatic Mineral Separation, I. I. Inculet (Letchworth: Research Studies Press; New York, Chichester: Wiley, 1984).
Magnetohydrodynamic and Magnetohydrostatic Methods of Mineral Separation, U. Andres (New York: Wiley; Jerusalem: Israel Universities Press, 1976).

Pelletization

Particle Size Enlargement, C. E. Capes (Amsterdam, Elsevier Scientific, 1980).
Pelletising of Iron Ores, K. Meyer (Berlin: Springer; Düsseldorf: Verlag Stahleisen, 1980).

Turning to extraction metallurgy proper, similar patterns of publication are evident. There are a number of recent works taking a generalized approach which may be more or less theoretical or practical in their approach. Again, the treatment is usually a conventional one; individual metals or groups of metals come in for scrutiny, or particular facets – pyrometallurgical practice, chemical thermodynamics, or the various hydrometallurgical processes, conventionally applied to gold and silver ores (cyanidation) but now applied increasingly to other metals, has gone some way to breaking down the distinction between mineral processing and extraction metallurgy, or even, in the case of in-situ, leaching, between mining and extraction metallurgy. Hydrometallurgical processing also embraces marginal areas such as the recovery of metals from dilute process streams and wastes, and also the techniques of biological leaching, as well as the more conventional areas of dissolution, solvent extraction and ion exchange which are also widely applicable in the process industries generally. Works broader in scope than extractive metallurgy therefore may also be relevant.

A selection of some of the better known earlier works also exemplifies the 'textbook' approach. Applications of chemical theory and techniques to metallurgical systems are treated in works such as *Chemical Analysis in Extractive Metallurgy* by R. S. Young (London: Charles Griffin, 1971) and *Metallurgical Thermochemistry*, 4th edn, by O. Kubachewski, E. L. Evans and C. B. Alcock (Oxford: Pergamon, 1967). A chemical approach to the general theory of metallurgical extraction processes is given by

A. R. Burkin's *Chemistry of Hydrometallurgical Processes* (London: Spon, 1966), and F. Habashi's multi-volume work of the same title – *Principles of Extractive Metallurgy* (New York; London: Gordon and Breach, 1969–). Of this last work, 2 volumes have so far been published: Vol. 1: *General Principles* (1969) and Vol. 2: *Hydrometallurgy* (1970). More applied in flavour, describing examples of equipment and industrial processes are *Extractive Metallurgy: Principles and Application*, by W. H. Dennis (London: Pitman, 1965) and *Unit Processes of Extractive Metallurgy*, by R. D. Pehlke (London: American Elsevier, 1973). Finally, there are works describing the principles and practice of particular types of equipment or process, for example *Electric Smelting Processes*, by A. G. E. Robiette (London: Griffin, 1973).

Some recent (past ten years) examples of general extraction metallurgy textbooks are listed below.

Chemical Metallurgy, J. J. Moore (London: Butterworths, 1981).

Extractive Metallurgy Laboratory Exercises, H. A. Fine (ed.) (Warrendale, Pennsylvania: Metallurgical Society of AIME, 1982). (*Instructor's Guide and Solution Manual* also available.)

Flash Smelting: Analysis, Control and Optimisation, W. G. Davenport and E. H. Partelpoeg (Oxford: Pergamon, 1987).

Fundamentals of Metallurgical Processes 2nd edn, L. Coudivier *et al.* (Oxford: Pergamon, 1985).

Hydrometallurgical Extraction and Reclamation (*Ellis Horwood Series in Industrial Metals*), E. Jackson (Chichester: Ellis Horwood, 1986).

International Technology for the Non-ferrous Smelting Industry, T. K. Corwin *et al.* (Park Ridge: Noyes Data Corporation, 1982).

'An Introduction to Chemical Metallurgy' (*International Series on Materials Science and Technology*, no. 26), 2nd edn, R. H. Parker (Oxford: Pergamon, 1978).

Ion Exchange Technology, D. Naden and M. Streat (eds) (Chicester: Ellis Horwood for the Society of Chemical Industry, 1984).

Mineralogy for Metallurgists: An Illustrated Guide, H. W. Fander (ed.) (London: Institution of Mining and Metallurgy, 1985).

Mining and Metallurgical Practices in Australia (*Aus. IMM Monograph Series*, no. 10), J. T. Woodcock (ed.) (Parkville: Australasian Institute of Mining and Metallurgy, 1980).

Molten Salt Technology, D. G. Lovering (ed.) (New York; London: Plenum, 1982).

Nonferrous Extractive Metallurgy, C. B. Gill (Chichester: Wiley Interscience, 1982).

Principles of Extractive Metallurgy, T. Rosenqvist (New York; London: McGraw-Hill, 1983).

Principles of Pyrometallurgy, C. B. Alcock (London: Academic Press, 1976).

Process and Fundamental Considerations of Selected Hydrometallurgical Systems, M. C. Kuhn (ed.) (New York: Society of Mining Engineers of AIME, 1981).

Rate Processes of Extractive Metallurgy, Hong Yong Sohn and M. E. Wadsworth (New York; London: Plenum, 1979).

Separation Processes in Hydrometallurgy, G. A. Davies (ed.) (Chichester: Ellis Horwood for the Society of Chemical Industry, 1987).

Stoichiometry and Thermodynamics of Metallurgical Processes, Y. K. Rao (Cambridge, UK: Cambridge University Press, 1985).
Topics in Non-ferrous Extractive Metallurgy (*Critical Reports on Applied Chemistry*, no. 1), R. Burkin (ed.) (Oxford: Blackwell Scientific for the Society of Chemical Industry, 1980).

Minerals industry and metallurgy conferences

Regularly held and one-off conferences are arranged by the major professional and scientific organizations within the minerals and metallurgical industries; there are also a number of standing councils and committees that exist solely for the purpose of organizing larger-scale events every few years. Many commercial and trade and industry organizations run meetings on particular metals, often covering the whole range of economic, technical and commercial aspects of its exploitation. Examples are the World Conference on Tin, the ILAFA (Institute Latinoamericano del Fierro y Acero) iron and steel conferences, the ICSOBA (International Committee for the Study of Bauxite, Alumina and Aluminium) conferences on aluminium, International Iron and Steel Congress, and those series on individual metals organized by Metal Bulletin Congresses on aluminium, copper and iron ore. These usually include a substantial proportion of papers on extraction technology, but are not considered further here, being covered elsewhere in this book. Some recent general conferences on extraction metallurgy and mineral technology are given below followed by a list of the more important regularly held ones. All those listed have proceedings published by the organizers indicated.

Advances in Mineral Processing: A Half Century of Progress. . . . (New Orleans: Society of Mining Engineers of AIME, 1986).
Advances in Sulphide Smelting. Part of the 1983 Metallurgical Society of AIME extraction and process metallurgy fall meeting.
Automation in Mining, Mineral and Metal Processing. Selected papers from the 5th IFAC Symposium. (Tokyo 1986; Pergamon 1987).
Chloride Hydrometallurgy (Brussels: Benelux Metallurgie, 1977).
Complex Sulphide Ores (Rome: Institution of Mining and Metallurgy, 1978).
Complex Sulfides: Processing of Ores, Concentrates and By-products (San Diego, California: Metallurgical Society of AIME, 1985).
Control 84 – Mineral/Metallurgical Processing (Los Angeles: Society of Mining Engineers of AIME, 1984).
Design and Construction of Tailings Dams. (Golden, Colorado: Colorado School of Mines, 1980).
Energy Considerations in Electrolytic Processes. (Newcastle-upon-Tyne: Society of Chemical Industry, 1980).

Hydrometallurgical Process Fundamentals (NATO Advanced Research Institute, Cambridge, UK, 1982. Proceedings published by Plenum Press, New York, 1984).

Indigenous Raw Materials for Industry (Institute of Metals, 1984) is the published proceedings of the conference sponsored by the Materials Forum and the Fellowship of Engineering, co-sponsored by the (then) Metals Society in November 1983. The book covers primary resources, recycling and waste as valuable sources in their own right and future requirements and perceived trends.

Innovative Technology and Reactor Design in Extractive Metallurgy (*Reinhardt Schumann International Symposium*) (Colorado Springs, Metallurgical Society of AIME, 1986).

International Blast Furnace Hearth and Raceway Symposium (Newcastle, N.S.W., Australian Institute of Mining and Metallurgy, 1981).

Physical Chemistry of Extractive Metallurgy (Metallurgical Society of AIME, 1983).

Raffinationsverfahren in der Metallurgie (*Refining processes in metallurgy*) (Hamburg, Gesellshaft Deutscher Metallhutten- und Bergleute, 1983).

Reagents in the Mineral Industry (Rome, Institution of Mining and Metallurgy, 1984).

Recycle and Secondary Recovery of Metals (Fort Lauderdale, Florida, Metallurgical Society of AIME, 1985).

Sulphur Dioxide Control in Pyrometallurgy (Chicago, Metallurgical Society of AIME, 1981).

Regularly held minerals technology/extraction metallurgy conferences

Advances in Extractive Metallurgy/Extraction Metallurgy (Institution of Mining and Metallurgy). Every 4–5 years, last held 1985 (6th).

Applications of Computers and Mathematics in the Mineral Industries (APCOM). Organized in turn by various bodies – e.g. IMM, Aus. IMM, Colorado School of Mines, Society of Mining Engineers of AIME. Every 2–3 years, last held 1986 (19th).

Applied Mineralogy in the Minerals Industry. Organized in turn by various bodies – e.g. AIME, National Institute for Metallurgy (South Africa). Every 3 years?, held 1984 (2nd), no details of 3rd available at going to press.

Council of Mining and Metallurgical Institutions Congress. (Formerly *Commonwealth Mining and Metallurgical Congress.*) Organized on behalf of the Council by one of the constituent bodies. Every 4 years, last held 1986 (13th).

IFAC Symposium on Automation in Mining, Mineral and Metal Processing. Organized by various bodies for the International Federation of Automatic Control. Every 3 years, last held 1986 (5th).

Instrumentation in the Mining and Metallurgical Industries (Annual Mining and Metallurgy Industries Symposium and Exhibit) (Instrument Society of America).

International Colloquium on Refractories (Verein Deutscher Eisenhuttenleute, Institut für Gesteinhuttenkunde). Annual.

International Mineral Processing Congress. Organized in turn by various bodies on behalf of the International Committee, recent congresses held in Warsaw, Toronto, Rio de Janeiro, Cannes. Every 2–3 years, last held 1988 (16th).

International Solvent Extraction Conference (....) ISEC. Organized by various bodies, e.g. University of Liege, CIM, American Institute of Chemical Engineers. Every 3 years, last held 1986 (8th).

International Symposium on Biohydrometallury. Organized by various bodies for the international advisory board for the symposium. Every 2–3 years, last held 1986 (6th).

Pyrometallurgy. Organized by the Institution of Mining and Metallurgy in conjunction with the Institute of Metals. 1st in proposed series held 1987.

Professional institutions organize regular meetings at both local and national level. Large scale meetings whose proceedings are published, and notes on the organizations' publishing policy for such meetings, are briefly described below.

American Institute of Mining, Metallurgical and Petroleum Engineers.

Annual and fall meetings The Metallurgical Society (TMS) and the Society of Mining Engineers (SME) both issue preprints. Selected papers are published for the SME sessions in the journals *Mining Engineering*, in *Minerals and Metallurgical Processing* and in AIME *Transactions*, and for TMS sessions in *Journal of Metals*, which also publishes programmes with abstracts. Special symposia, some jointly organized by SME and TMS, may be published separately, e.g. *Chloride Electrometallurgy*, part of the 1982 annual meeting and published by TMS, and *Hydrometallurgy of Copper, Its Byproducts and Rarer Metals*, part of the 1981 annual general meeting and published by SME. The *Extractive and Process Metallurgy Fall Meeting* and *International Symposium on Hydrometallurgy* are of particular interest.

Other conferences Many are formally published by the constituent society.

Australasian Institute of Mining and Metallurgy

Annual Conference Located each year in different mining areas, the conference tends to reflect the interests of the mineral industry in that area. Examples: *Broken Hill Conference 1983*, *New Zealand Conference 1980*. Proceedings are published by the Institute in its *Conference series*.

Other conferences Published in the *Symposia series*. Recent titles on extraction include *Extraction Metallurgy Symposium* 1984, *Carbon-in-pulp Technology for the Extraction of Gold* 1982, *Mill Operators' Conference* 1982, *Australia-Japan Extractive Metallurgy Symposium* 1980, *Principles of Mineral Flotation* 1984, *Scientific and Technical Developments in Extraction Metallurgy* 1985, *Research and Development in Extractive Metallurgy* 1987.

Canadian Institute of Mining and Metallurgy

Annual meeting Programmes with abstracts of the general meeting and the meetings of the divisions, including the Canadian Mineral Processors and the Metallurgical Society, are usually published in *CIM Bulletin*. Some papers may appear in *CIM Bulletin* or *Canadian Metallurgical Quarterly*; occasionally the metallurgical symposia that form part of the annual meetings are published as special issues of the latter. The Canadian Mineral Processors, a division of the Institute, issue volumes of papers presented at their own annual meeting. The *Annual Hydrometallurgical Meeting* of CIM's Metallurgical Society is published separately, usually in loose-leaf form; these take a specific metal or topic as the theme, e.g. *Liquid-solid Separation* (1981), *Uranium* (1982), *Zinc* (1983). Other proceedings of the annual meeting of the CIM Metallurgical Society are similarly available, e.g. *Quality Control in Non-ferrous Hydrometallurgical Processes* 1985, *Copper* 1984. The 1987 proceedings was published by Pergamon Press in Oxford. It is not the practice to publish the collected proceedings of the general annual meeting of CIM.

Other conferences These are occasionally published as *Special volumes* of the Institute. Individual papers may incidentally be published in the Institute's journals.

Technical reviews

Review articles are useful concise summaries of the state of knowledge on a topic or recent advances in it, and appear on a regular or occasional basis in journals alongside articles describing original work, or in publications devoted to such an approach. Traditionally, the 'technical review' condenses and summarizes the work of others and usually includes an extensive reference list of published work, but the concept is often extended to include reviews of products, or developments in a class of equipment. Less academic in tone, the latter appear in industry magazines rather than learned journals; *Mining Magazine*, for example, in its 1984 volume (vol. 151) contained several features reviewing theory, practice and major types of equipment in a particular field, and followed this with a list of suppliers. Flotation reagents (September 1984, pp. 202–19), gravity separation (October 1984, pp. 235–41, 29 refs) and classifiers (July 1984, pp. 27–44, 24 refs) are examples of the topics covered.

Regularly published reviews include the following:

Annual review of extractive metallurgy. *Journal of Metals*, April issue (no. 4). Reviews the previous years's developments in physical chemistry and the basic principles of extractive and process metallurgy, electrometallurgy, and hydrometallurgy. Technical and economic reviews of selected metals follow, the metals varying from year to year. The 1984 issue (vol. 36, no. 4) covered copper, titanium, aluminium, iron and steel, molybdenum, cobalt, tungsten and tin.

Annual review of mineral processing. *Mining Engineering*, May issue (no. 5). The previous year's technical developments are covered. The annual review issue of the journal also includes exploration and mining activity, the coal industry and economic reviews of non-metallic commodities, with reference principally to the US.

Mining annual review. London: *Mining Journal*. This comprises two sections, one reviewing the previous year's developments in extraction technology (covering mineral processing and extraction metallurgy) and the other giving an overview of economic developments in the minerals industry by commodity and country.

A fairly new journal in the field is *Mineral Processing and Technology Review* (quarterly, 1983–), published by Gordon and Breach. This aims to review developments in mineral processing and extractive metallurgy, emphasizing the engineering aspects, together with economic and environmental concerns. The same publisher in 1973 issued Volume 1 of *Progress in Extractive Metallurgy* primarily as a review publication, but only the one volume has so far appeared.

Journals

Much of the information on extraction is found in the minerals industry rather than the metal products coverage. The exceptions here, presumably because of the more intimate association in ferrous metallurgy between extraction, alloying and further processing and also the relative importance of ferrous metals in general, are iron and steel, which have a number of publications devoted exclusively to them. These are dealt with more fully in Chapter 3 on iron and steel.

Weekly or more frequent business publications:

Latin American Mining Letter (fortnightly) (London: Metals and Minerals Publications, 1982–).

Mining Journal (twice weekly) (London: Mining Journal, 1835–).

Skilling's Mining Review (weekly) (Duluth: Skilling's Mining Review, 1912–).

Monthly journals (unless otherwise specified)

Arab countries

Arab Mining Journal (Amman: Arab Mining Company, 1980–).

Australia

Australian Mining (Chippendale, N.S.W.: Thomson Publications (Australia), 1980–).
Mining Monthly (Leederville, W.A.: Mining Monthly, 1980–).
Mining Review (Dickson, A.C.T.: Australian Mining Industry Council, 1970–).

Austria

Berg-u. Huttenmannische Monatshefte. (Vienna, Springer, 1841–).

Canada

Canadian Mining Journal (Don Mills, Ontario: Southam Communications, 1879–).
CIM Bulletin (Montreal: Canadian Institute of Mining and Metallurgy, 1898–).
Northern Miner (weekly) (Toronto: Northern Miner Press, 1915–). Incorporates *Western Miner* from 1984.

France

Industrie minérale (St Etienne: Société de l'Industrie Minérale, 1855–).
Industrie minérale: les techniques (St Etienne: Société de l'Industrie Minérale, 1972–).
Journal du four électrique (Paris: PYC Edition, 1895–). Incorporating *Mines et métallurgie*.

Germany – Federal Republic

Aufbereitungstechnik (Wiesbaden: Verlag für Aufbereitung, 1960–).
Erzmetall (Clausthal-Zellerfeld: Gesellschaft Deutscher Metallhutten- u. Bergleute, 1912–).

India

Indian Mining and Engineering Journal (Bombay: Mining Engineering Association, 1962–).
Journal of Mines, Metals and Fuels (Calcutta: Books and Journals Pvt., 1953–).

Italy

Industria Mineraria (Rome: Associazione Mineraria Italiana, 1927–).

Latin America

De re metallica de la mineria y los metales, revista (Lima: Instituto Geologico Minero y Metalurgico, 1984–).
Geomimet (Mexico City: Asociacion de Ingenieros de Minas, Metalurgistas y Geologicos de Mexico, 1973–).
Mineracao metalurgia (Sao Paulo: Editorio Scorpio, 1936–).

Scandinavia

Vuoriteollisus/Bergshanteringen (Espoo: Vuoriteollisus Bergsmannaforeningen, 1943–).

South Africa

South African Mining and Engineering Journal (Johannesburg: Thomson Publications, 1891–).
South African Mining World (Sandown: Phase 4 (Pty), 1982–).

UK

International Mining (London: International Mining, 1984–).
Metal Bulletin Monthly (London: Metal Bulletin, 1972–).
Mining Magazine (London: Mining Journal, 1909–).

USA

Engineering and Mining Journal (New York: McGraw-Hill, 1866–).
Mining Engineering (Littleton, Colorado: Society of Mining Engineers of AIME, 1949–).

Zimbabwe

Mining and Engineering (Harare: Thomson Publications (Zimbabwe), 1934–).
Chamber of Mines Journal (Harare: Argosy Press for the Chamber of Mines of Zimbabwe, 1958–).

Learned journals

Commercially published journals in the field include the following:

Hydrometallurgy (quarterly) (Amsterdam: Elsevier, 1975–). Subtitle: An international journal devoted to all aspects of the aqueous processing of metals.
International Journal of Mineral Processing (quarterly) (Amsterdam: Elsevier, 1974–).
Mineral Processing and Technology Review (quarterly) (New York; London: Gordon and Breach, 1983–).
Minerals Engineering (quarterly) (Oxford: Pergamon, 1988–).

The principal English language journals produced by professional bodies are listed below, together with an indication of coverage if it is not apparent from the title.

Canadian Metallurgical Quarterly (Toronto; Oxford: Pergamon for the Metallurgical Society of the Canadian Institute of Mining and Metallurgy, 1962–). '. . . all aspects of metallurgy and materials science' (from the instructions to authors, but a third to a half of the contributions are on extractive metallurgy).
Journal of Metals (monthly) (Warrendale, Pennsylvania: Metallurgical Society of AIME, 1949–). 4 or 5 articles per issue on extractive and process metallurgy. Annual review in April issue. As a house journal of TMS/AIME it includes industry and institute news also.
Journal of the South African Institute of Mining and Metallurgy (monthly) (Marshalltown: the Institute, 1894–). Covers mining, mineral processing, extraction, process and physical metallurgy, with emphasis on South African practice and the metals produced there.

Metallurgical Transactions B. (quarterly) (Warrendale, Pennsylvania: Metallurgical Society of AIME; Metals Park, Ohio: American Society for Metals, 1969–). Extraction and process metallurgy.

Minerals and Metallurgical Processing (quarterly) (Littleton, Colorado: Society of Mining Engineers of AIME, 1984–). Mineral processing, hydrometallurgy.

NML Technical Journal (quarterly) (Jamshedpur, India: National Metallurgical Laboratory, 1959–). Metallurgical research and development, including extraction technology.

Proceedings. Australasian Institute of Mining and Metallurgy (monthly) (Parkville, Victoria: 1898–). Combined with Aus. *IMM Bulletin* – the Institute's newsletter – from 1984, appearing as *Proceedings* section within it. Covers economic geology, mining, mineral processing and extraction metallurgy, with emphasis on Australian practice.

Transactions. Indian Institute of Metals (bimonthly) (Calcutta, 1946–).

Transactions of the Institution of Mining and Metallurgy, section C: Mineral Processing and Extractive Metallurgy (quarterly) (London: 1892–).

USSR, China and Japan

Some notes on the Chinese, Russian and Japanese journal literature are appropriate in view of the relative size and importance of the contributions to the literature from those languages.

USSR

The USSR contribution to the metallurgical literature is substantial. Some of the principal journals are translated cover to cover; most are included in one or other of the major databases (Metadex, Compendex), which provide titles, index terms and often an abstract which will assist in selecting items worthy of translation. USSR and Eastern European scientific and technical literature are comprehensively covered in *Referativnyi Zhurnal*, the Russian language abstracting and indexing service produced monthly by the Institut Nauchnoi Informatsii, Akademiya Nauk SSSR (Institute of Scientific Information of the USSR Academy of Sciences) (Moscow, VINITI, 1960–). It covers the international literature in a number of sections; the principal ones of interest are *Gornoe delo* (mining), which includes mineral processing, and *Metallurgiya*. Further information is contained in *A Guide to Referativnyi Zhurnal* (London: Science Reference Library, 1975). *Selected Abstracts from the Abstract Journal 'Metallurgy', Part A: Science of Metallurgy* and *Part B: Technology of Metals* (Oxford: Pergamon, 1961–) is a monthly publication giving translations of abstracts of periodical articles and conference papers included in the *Metallurgiya* section of *Referativnyi Zhurnal*. An online database in English, *Soviet Science and Technology* (produced by IFI/Plenum in the USA and available via the host DIALOG,

1975–) is also available which includes metallurgy in its coverage.
The USSR journals issued in translation are:

Russian Metallurgy (*Izvestiya Akademii Nauk SSSR, Metally*) (bimonthly) (London: Scientific Information Consultants, original 1959– , translation 1960–); formerly *Russian Metallurgy and Mining, Russian Metallurgy and Fuels*. Cumulative indexes are published.
Soviet Journal of Non-ferrous Metals (*Tsvetnye metally*) (monthly) (New York: Primary Sources, original 1926– , translation 1960–).
Soviet Non-ferrous Metals Research (*Izvestiya Vysshikh Uchebnykh Zavedenii: Tsvetnye metally*) (bimonthly) (Stonehouse, Glos.: Technicopy, original 1959– , translation 1973–).
Steel in the USSR (monthly) (London: Institute of Metals, 1971–). Contains a selection of translated material from *Stal'* and *Izvestiya Vysshikh Uchebnykh Zavedenii: Chernaya metallurgiya*.

China

China's desire to cooperate economically with the West and the occurrence in the country of significant resources of metallic ores – tin, tungsten and rare metals in particular – together with the development of the technology to exploit them have led to an increase in the importance of the Chinese metallurgical literature. Most of the available journals contain informative English abstracts and are covered by the appropriate major secondary sources. Significant titles in extractive metallurgy are:

Acta Metallurgica Sinica (quarterly) (Beijing: Chinese Society of Metals, 1965–).
Journal of the Central South Institute of Metallurgy (quarterly) (Changsha, 1956–).
Iron and Steel (bimonthly) (Beijing: Chinese Society of Metals, 1966–).
Non-ferrous Metals (quarterly) (Beijing: Chinese Society of Metals, 1949–).
Rare Metals (biannual) (Beijing: Chinese Society of Metals, 1982–).

Japan

All the following journals have at least titles and summaries in English.

Journal of the Iron and Steel Institute of Japan (monthly) (Tokyo, 1915–).
Journal of the Japan Institute of Metals (monthly) (Sendai, 1937–).
Journal of the Mining and Metallurgical Institute of Japan (monthly) (Tokyo, 1885–).
Metallurgical Review of MMIJ (biannual) (Tokyo: Mining and Metallurgical Institute of Japan, 1984–). English text.

Reports and government-sponsored research

In the extraction of metals, as in other industries, there are a number of organizations whose work is of obvious and immediate relevance and such bodies, as well as emphasizing the techniques of extraction itself, are interested also in the social and environmental milieu in which the industry operates. There are of course, organizations whose principal concern is solely with these matters, such as the Health and Safety Executive in the UK or the Environmental Protection Agency in the USA. To avoid repetition, research organizations have been discussed in the introductory chapter.

Patents

Specific sources of information on patents in the field of extraction metallurgy include the patent digest published every few months in *Mining Magazine*, which covers UK and European patent applications of interest to the minerals industry some 18 months after their original submission, by which time a preliminary check has been carried out to establish their validity. Noyes Data Corporation, publishers in the USA, have produced useful summaries of US patents in extraction metallurgy in two titles in their series *Chemical Technology Reviews*: no. 93 is entitled *Extractive Metallurgy; Recent Advances* (edited by E. J. Stevenson, Park Ridge, New Jersey, 1977) and no. 227 is *Extractive Metallurgy: Developments since 1980* (edited by M. J. Collie, 1984). Chapter 1 discusses patents in general and has a useful table of sources.

Sources of information on organizations, companies and other institutions

Companies – producers and suppliers of metals

The directories produced by Metal Bulletin Books cover primary and secondary production and trading of ferrous and non-ferrous metals. Two companion volumes: *Iron and Steel Works of the World*, 8th edn (1983), and *Non-ferrous Metal Works of the World*, 4th edn (1985), and also the *Ferro-alloy Directory* (1984), list by country major producers, giving brief details of plants and products, with indexes to company and commodity and buyers' guides. Trading is covered in *Metal Traders of the World*, 3rd edn (1986), and *Steel Traders of the World*, 3rd edn (1984), also the *Ferro-alloy Directory* mentioned above. The *European and North American Scrap Directory* (1981), lists companies involved in the various aspects of scrap trading and processing.

Directories covering the mining and minerals industry are usually also concerned with metals extraction. Covering the international scene and including smelting and refining are the

Financial Times Mining International Yearbook (Harlow: Longman) and the *EMJ International Directory of Mining* (annual; New York: McGraw-Hill). Many others cover single countries or even particular regions; they often have some statistical information and/or economic reviews of production and may include guides to local associations and government agencies with an interest in the industry. Some examples of these are:

Annuaire de la Fédération des Chambres Syndicales de Minérais et des Métaux Non-ferreux (France).
Canadian Mines Handbook (Toronto: Northern Miner Press, annual).
Jahrbuch für Bergbau, Energie, Mineraloel und Chemie (Essen: Verlag Gluckauf, West Germany).
Jobson's Mining Yearbook (Melbourne: Dun and Bradstreet, Australia).

Companies – equipment suppliers and trade literature

Metal Bulletin's *Metallurgical Plantmakers of the World*, 2nd edn (1981), is the most comprehensive single source, listing by alphabetical order of company their activities and products, with indexes by company and type of equipment. Companies engaged in plant design and plant and project engineering and consultancy are listed also. Other general trade directories such as *Ryland's – The Directory of the Engineering Industry* (annual; Birmingham: Guardian Communications) and *Kompass* (annual – editions for most European countries are published by Kompass Publishers) cover metallurgical plant, metal products and consumables such as reagents though the product classification used is often fairly general.

The UK firm Technical Indexes provides a subscription service on microfilm of UK trade catalogues, together with printed indexes, in various areas of engineering; relevant sections are *Process Engineering Index, Manufacturing and Materials Handling Index*, and *Laboratory Equipment Index*. Files of US, Canadian and Australian trade catalogues are also available from them.

For mineral processing plant recourse may be made to mining equipment directories: *Mining Directory – Mines and Mining Equipment Companies Worldwide*, 4th edn (Don Nelson Publications, 1988 (published in the UK; no place of publication given)) and the buyers' guide of the Association of British Mining Equipment Companies (ABMEC) (Sheffield: every 2–3 years) are two useful examples.

The industry press should not be neglected as a source of information on new products, services and equipment, though, apart from the reviews of equipment in particular areas or applications published as editorial matter, in their raw state the items are more useful as current awareness tools than as reference sources.

Chemical industry directories such as the *Chemical Industry Directory and Who's Who* (Tonbridge: Benn, annual) or the *Chemical Industries' Association Directory of Chemicals on the UK Market* (updated every few years) are valuable for tracking down process reagents.

Consultancy services

The services available for process or project development in the minerals/metals industry range from those of the individual consultant to divisions of large multinational corporations undertaking project construction and engineering work. In addition many university departments or government-sponsored bodies will undertake, consultancy work in more specialized areas. A survey of the private sector services available in the UK to the minerals industry has been carried out by N. J. B. Pocock (see below) describing the services and their main markets; most of those listed in this section, though UK-based, operate worldwide and may be subsidiaries of foreign-owned companies. *Mining Magazine* periodically reviews the services available in this field, and there are a number of more general directories of consultants produced by their various UK umbrella organizations. Most mineral industry directories, both international and local, also carry lists of consultants and laboratories; further listings will be found in the 'professional directory' section of appropriate industry journals such as those listed below.

Association of Consulting Scientists Members and Services (Buntingford, Herts., annual). UK-based services.

British Consultants' Bureau Directory and Index of Specialisations. Every two years. UK-based services.

Chemical Engineers' Directory and Who's Who (Tonbridge: Benn, annual).

Consultant Chemical Engineers' Bureau directory (London, 1983).

Consultants and Contractors. Mining Magazine, vol. 147, no. 5 (Nov. 1982), 457–63. UK-based services.

Consulting Engineers' Who's Who and Yearbook (London: Municipal Publications for the Association of Consulting Engineers, annual). UK-based services.

Directory of Metallurgical Consultants and Translators (London: Metals Information, 1984). International coverage.

EMJ International Directory of Mining (New York: McGraw-Hill, annual). International coverage.

List of Consultants (Rugby: Institution of Chemical Engineers, 1987).

Metallurgical Plantmakers of the World, 2nd edn (Worcester Park: Metal Bulletin, 1981). Includes consultancy services.

NAMAS Concise Directory (Teddington: National Physical Laboratory, 1985). Testing and analytical services.

Private Consultancy Services for Mineral Processing and Extractive Metallurgy, N. J. B. Pocock. Paper presented at the seminar *Finding and Funding Technology for the Mineral Industry*, 9 Nov. 1983, organized by the Institution of Mining and Metallurgy and the Minerals Industry Research Organisation.

'The project engineers'. Mining Magazine, vol. 148, no. 3 (Mar. 1983), 207–31. International coverage.

Register of Consulting Scientists, Contract Research Organisations and Other Scientific and Technical Services, D. J. B. Copp (ed.) (Bristol: Adam Hilger, 1984). UK-based services.

Technical Services for Industry: Technical Information and Other Services Available from Governmental Departments and Associated Organisations (London: Department of Industry, 1981).

Educational establishments

Dated with regard to the information on research topics and personnel but nevertheless useful if only because it is the only source of its kind available is the *Worldwide Directory of Mineral Industry Education and Research*, edited by H. Wohlbier (Houston: Gulf Publishing Co., 1968). *Current Research in Britain* (formerly *Research in British Universities, Polytechnics and Colleges*), *vol. 1 – Physical Sciences* (Boston Spa: British Library, annual) lists doctoral and postdoctoral work in progress. There are in addition a number of more general directories of educational institutions of more limited use because research interests are not listed; these include the *World of Learning* (London: Europa Publications, annual) and the *Commonwealth Universities Yearbook* (London: Association of Commonwealth Universities, annual).

An interesting development is the appearance of *Annales Metallurgicae*, an annual compilation of current and planned research projects in metallurgy at universities and research institutes throughout the world, produced by Intag Publishing in Espoo, Finland.

Turning to choice of course rather than identification of research expertise the principal source for the UK is the *CRAC Degree Course Guide: Technology* (Cambridge: Hobson's Press for the Careers Research and Advisory Centre, every two years) whose sections include metallurgy, materials science and mining engineering. The courses obviously vary in their emphasis on extractive rather than physical metallurgy and not all may satisfy the requirements of the appropriate engineering institution for corporate (chartered engineer) membership – in the UK this would normally be the Institution of Mining and Metallurgy or the Institute of Metals. Interested employers have set up the Mineral Industry Manpower and Careers Unit to advise on career matters. The Institution of Mining and Metallurgy has recently published (1988) the proceedings of a seminar entitled *Careers in the Mineral Industry*.

Elsewhere the appropriate professional institution can assist; the Society of Mining Engineers of AIME for example produces a *Guide to Minerals Schools*. Many of the professional and educational institutions concerned also run programmes of short courses and seminars for continuing education of professionals within the industry.

Professional and industry associations and other organizations

There are a number of aids to finding out about institutes, organizations, trade associations and special interest groups available and it is beyond the scope of this chapter to mention all but a few of those of specific interest to the extraction metallurgist. It is similarly not possible to list even the British bodies individually due to space restrictions. A selection of the more general reference works listing these should be available in any large public or research library. The industry directories already mentioned in this chapter will be of use in this context. Associations and others with strong links with the metals industry have been included in an appendix at the end of this volume.

The organizations referred to in the following sources provide primary technical or economic information, advice or knowhow, and/or a library and information or economic intelligence service drawing on both in-house and external expertise – more usually some combination of these. Advice and information may only be provided to members or employees and in many cases may be chargeable. Reference is made elsewhere in this chapter to educational establishments and private firms.

Some directories listing metals and minerals organizations on an international basis (already cited elsewhere in this section) are:

EMJ International Directory of Mining
Financial Times Mining International Yearbook
Metal Bulletin Handbook, Vol. 2: Statistics and Memoranda.

Minerals industry business databases

Certain minerals industry publications are available in full-text form on line as part of the general NEXIS service operated by Mead Data Central. This is one of a number of general full-text news services available on a subscription basis and is noteworthy because it covers – as well as general and business news publications, wire services and broadcast transcripts – *Engineering and Mining Journal*, *Metals Week*, and the three Mining Journal publications *Mining Journal*, *Mining Magazine* and *Mining Annual Review*, all from 1981.

The US firm Chase Econometrics produces a number of databases containing time series analyses of past performance and forecasts in a number of sectors of the metals industry internationally: *Aluminium Forecast*, *Base Metals Forecast* (copper and brass, lead, zinc, tin and silver), *Ferroalloys and Strategic Metals Forecast* (chromium, cobalt, manganese, molybdenum, nickel, niobium, silicon and tungsten), *International Lead and Zinc Forecast*, *Iron and Steel*, *Metals Week* (prices for nonferrous metals), *World Steel Forecast*. A subscription is required.

Other subscription services include *Daily Metal Prices* (produced by Australian Bureau of Statistics; available via I. P. Sharp and Associates – 'typical' prices are given, from various sources) and *Daily Metals Report* (News-a-Tron; available via NewsNet – analysis and forecast based on LME prices). *Energy and Mineral Resources* is a newsletter also available as part of the NewsNet service.

Ecomine (1984–) is a database of summaries of minerals industry business and economic news produced by the French Bureau de Recherches Géologiques et Minières and Observatoires des Matières Premières, with a corresponding printed press review. The database is available via Télésytèmes-Questel. The techno-commercial information section in Chapter 1 should also be examined for likely sources in this context.

CHAPTER THREE

Iron and Steel

P. R. LAWRANCE

Introduction

In view of the vast amount of information of potential significance to the user of ferrous materials intelligence it is sensible to consider the various demands on the literature in order to be clear about the nature of the sources available. If users are aware of the different information requirements it will enable them to concentrate on those sources most relevant to their particular needs and not be overwhelmed by the volume and variety of information available. Equally important, it will enable the reader to impose some form of quality standard on the particular sources of information studied.

Once users have decided on the type and level of information required, they are faced straight away in the case of ferrous metals with a difficulty related to the fact that iron and steel are such versatile materials in that they pervade virtually the whole of technology from nuclear reactors to razor blades, brittle fracture to thermal fatigue, explosive forming to plasma welding. Indeed, the social and commercial aspects of iron and steel production mean that sources in the fields of economics and politics are also potentially relevant to the information user in this branch of metals and materials science.

It follows, therefore, that information on steel will be found in the literature relating to a number of industries, e.g.:

iron and steel manufacture;
engineering;
materials handling;
foundry technology;

welding and allied industries;
construction industry;
automobile industry;
domestic appliance manufacture;
chemical engineering; and
off-shore industry.

In addition to these and similar directly related industries, there are closely related topics such as mining and minerals, coal and coke manufacture; fuels and furnaces; corrosion and protection; instrumentation; and pollution control. When it is realized that all these subject fields have their own specialist sources of information, the iron and steel information user is faced with a daunting task in ensuring that any search is effective.

A further indication of the wide-ranging nature of the iron and steel related topics can be obtained by examining the breakdown of 800 information seekers in one particular iron and steel manufacturing company by subject discipline. Table 3.1 illustrates this.

Table 3.1. Example of iron and steel information users by discipline

Discipline	Number
Engineers	134
Senior managers	63
Ironmaking specialists	61
Rolling specialists	60
Administrative personnel	58
Steelmaking experts	56
Materials scientists	41
Safety officers	39
Medical staff	38
Computer specialists	34
Materials handling and transport personnel	30
Metallurgists	30
Refractory specialists	21
Chemists	21
Fuel and power specialists	19
Instrument and measurement personnel	15
Sales and marketing staff	15
Physicists	14
Environmentalists	11
Planning experts	8
Raw materials technologists	7
Others (training, work study, standards and information specialists)	25
TOTAL	800

In view of this therefore, there is a need for some discipline in approach by the user to the source material in iron and steel to achieve the maximum benefit. Although subject classification of the source material can go some way towards this, because of the great degree of overlap that exists, the basic format of the source is the chief manner in which sources have been grouped. However, where possible, for example, in the description of the reference sources and literature, some attempt to group by subject discipline is made. In the case of abstracts and standard specifications those considered most useful by the writer in providing information in the past are recommended; in the case of dictionaries, practical advice from an experienced specialist translator has assisted in the selection and, in the case of company and specialist literature, the writer's own experience in the industry played an important part in its evaluation and selection.

Such a compilation of sources can never be comprehensive. However, it is hoped that the selection of 'core' material mentioned will form the basis of a balanced guide to the sources of iron and steel, both metal science and engineering aspects, elementary and advanced, retrospective and current, specialized and general, well-known and less well-known.

Reference sources

Publications classed as reference sources include general dictionaries and glossaries, handbooks and encyclopaediae and a selection of key text books. The range of subjects covered include the general broad topics and specialized ones but it must be borne in mind that many of the general works are compilations of specialist data and, as such, may contain vital data on a specific aspect of iron and steel materials. Some of the publications mentioned, although published several years previously, still contain useful information relvant today and even if the data refer to materials no longer current, the investigation of older products requires that modern information should be regarded as an addition to, rather than a replacement of, earlier data. An additional point to remember is that a mass of new information, excellent and valuable though it may be, can often serve to obscure a vital piece of knowledge more easily retrievable from an older edition of a reference source.

Subject dictionaries

The Iron and Steel Industry – A Dictionary of Terms by W. K. V. Gale (David and Charles, 1971) is an elementary general Iron and Steel glossary as its name implies, but for the enquirer it gives brief, simple definitions of such terms as acicular cast iron, deep drawing (DD) steel, spheroidal graphite (s.g.) iron etc. and, in some cases, more than one definition, e.g. white iron. The terms included are a mixture of new and obsolete. A. K. Osborne, *Encyclopaedia of the Iron and Steel Industry* (Technical Press, 1967) goes a little further by providing more enhanced definitions supported by literature references, property tables and flow sheets for the production of iron and steel. Although rather dated, this is still an extremely useful encyclopaedia for specialists and non-specialists alike. Another less general dictionary is that by Eric N. Simons, *Dictionary of Ferrous Metals* (Frederick Muller, 1971) which contains many factual definitions of ferrous material arranged alphabetically and includes many proprietary alloys. It also contains a physical and mechanical property table, but it remains basically a dictionary of ferrous materials and as such is more specialized than the above-mentioned more general industrial dictionaries. Another general dictionary is D. Birchon, *Dictionary of Metallurgy* (Butterworths, 1965).

Process and properties handbooks

The 7th 1948 edition of the ASM *Metals Handbook* was one of the major comprehensive reference books ever published at the time with more than 500 individual contributors and it is still an excellent source of ferrous metallurgical knowledge of its time with many references to the literature of the day. It deals with all aspects of iron and steel from the manufacturing processes to application, and includes chapters on analysis, properties, treatment, working applications etc. Similarly, when it came out some years later, the 8th edition was the most extensive handbook available for the selection, interpretation and application of metals information by the general and specialist reader and ran to eleven volumes. For the authoritative coverage of iron and steel in particular, this work is excellent and represents the efforts of numerous experts in the field.

The 9th edition incorporates new technology, e.g. data on compacted graphite cast irons, maraging steels and HSLA steels. It also incorporates improvements in the presentation of information in the form of graphs, tables etc. with the usual excellent index. The twelve volumes published to date are fully described in Chapter 1.

Process metallurgy

From the point of view of iron and steel process metallurgy specification i.e. including actual plant and manufacturing theory, there was nothing to compare with the 9th (1971) edition of that excellent textbook edited by H. E. McGannon on behalf of United States Steel Corporation, *The Making, Shaping and Treating of Steel*. It covers everything from reaction kinetics to machinability tests, from planning steel plant process technology to liquid carburizing of steel components. A 'bible', if ever there was one, of some 1,420 pages and well over a 100 page index of high quality. Again, rather dated now, but a must for any iron and steel source collection as is the new 10th edition with 1,600 pages, 200 tables and 1,200 figures. *Blast Furnace Technology Science and Practice*, J. Szekely (ed.) (Marcel Dekker, 1972) is a reasonable attempt to cover the field of ironmaking technology and is the proceedings of a conference in 1970 at Buffalo, New York. USA. A much better textbook is *Blast Furnace – Theory and Practice*, Vols. 1 and 2 by J. H. Strassburger (Gordon and Breach, 1969). Vol. 1 covers the history of BF technology, the BF process, benefication, sintering, pelletizing coal instrumentation and operation with a chapter devoted to special grades of iron and ferro-alloys. Process analysis, heat balance and transfer are also dealt with. *Cast Iron Technology* by R. Elliott was published by Butterworths in 1988 and *Modern Ironmaking Methods*, by R. D. Walker (Institute of Metals, 1986) covers principles, raw materials, alternative processes and the future of ironmaking. A much more elementary book is *Practical Ironmaking* by G. D. Elliot and J. A. Bond (United Steel Companies, 1959). In its time, this was a useful students' guide and is still a good basic textbook on ironmaking. Pergamon Press published *The Iron Blast Furnace Theory and Practice* as part of its series on materials science some twenty years later. This was written by J. G. Peacey and W. G. Davenport and comprises a much more theoretical approach to the ironmaking process. Another useful series of papers is put together under the title of *An Intensive Course on Blast Furnace Ironmaking* and is in effect 17 lectures on all aspects of ironmaking given at McMaster University, Hamilton, Ontario, Canada in June 1977. The authors are predominantly American. The Process Technology Division of the Iron and Steel Society of AIME published a number of reference books relating to iron and steelmaking processes; in 1944 (revised 1951) *Basic Open Hearth Steelmaking*, in 1962 the 2-volume *Electric Furnace Steelmaking* and in 1969 *Blast Furnace Theory and Practice*. The latest in the series *BOF Steelmaking* is in 5 volumes: *Introduction, Theory, Design, Operation* and *Special Topics*. It is an excellent monograph series edited by R. D. Pehlke, W. F. Porter, R. F. Urbon and the late J. M. Gaines.

With regard to steelmaking, *Ferrous Production Metallurgy* by A. T. Peters (J Wiley, 1982) is a modern practical textbook on steel production covering raw materials, production of iron and steel, vacuum degassing, pitside practice, steel types, continuous casting, primary rolling etc. It is simply, but informatively, written and an excellent review and also covers pollution, energy and economic aspects. *Continuous Casting*, the proceedings of the Metallurgical Society of AIME technical sessions (Interscience, 1962), covers the specialized area of continuous casting practice, in particular conticasting of slab, billets, blooms, the effects of forging and rolling on conticast steels. The papers are edited by *D. L. McBride* and *T. E. Dancy*. As far as mini-mills are concerned, the major activities associated with steel works are covered in *Small-Scale Steelmaking* edited by R. D. Walker (Applied Science, 1983) whereas *Steel Production: Process, Products and Residuals* by C. S. Russell and W. J. Vaughan (Johns Hopkins UP, 1976) discusses residuals management as it applies to integrated iron and steel production by means of a linear programming model.

Three rather dated but useful little illustrated guides to iron and steel processes are *A Simple Guide to Basic Processes in the Iron and Steel Industry* published in the 1950s by British Iron and Steel Federation (BISF); *The Making of Steel Tubes* – a simple description which is a Stewarts and Lloyds booklet describing the history of steel tubes and the various processes involved, and *Guide to Shaping Processes in the Steel Industry* published by the British Steel Corporation.

Properties and physical metallurgy handbooks

The *Metals Reference Book* by C. J. Smithells, 6th edn edited by E. A. Brandes (Butterworths, 1983) is a reference tool well-known as a reliable source of metallurgical data. It is particularly valuable for Fe-alloy equilibrium diagrams and the physical properties of steels e.g. specific heats, specific gravity, coefficient of thermal expansion, thermal conductivity and electrical resistivity. Hardness value conversion tables for ASTM E140 steels are given. Nearly 100 pages of data on mechanical properties, e.g. UTS, YS elongation and impact strength values, are presented for steels, cast irons and cast steels. Pattern contraction allowances data for cast iron, malleable iron and cast steel are given in the foundry data section. The *ASM Metals Reference Book* contains valuable data on cast irons, carbon and alloys steels, tool steels and wrought stainless steels. The book is particularly strong on tabulated physical and mechanical property data relating to the various AISI-SAE grades, ASTM specifications and AMS compositions. There is a useful general glossary and an index to steel standard numbers.

A good British 2-volume handbook is *Metals*, edited by J. Dancy, Vol. 1 being concerned with ferrous metallurgy (Morgan Grampian, Design Engineering Series). As well as the technical aspects of iron and steel material, it has a useful product locator section which assists designers and engineers in finding suitable suppliers and a range of different metal working firms for particular types of iron and steel. Stainless steels are dealt with in the next chapter.

A symposium on *Formable HSLA and Dual-Phase Steels* and *Structure and Properties of Dual-Phase Steels* was published by the TMS-AIME and edited by A. T. Davenport and R. A. Kot/J. W. Morris respectively. *Steels for Linepipe and Pipeline Fittings* is the title of a specialized pipeline-steel-property proceedings published by the Metals Society as Book 285. It covers steel property requirements, advances in physical metallurgy, fracture problems, actual grades developed, and welding aspects. The *Guide to the Structure and Properties of Steel* (British Steel Corporation) is an illustrated booklet describing what steel is, its properties, a glossary of terms and bibliography. The *Encyclopaedia of Metallurgy and Materials*, edited by C. R. Tottle (MacDonald and Evans, 1983), is an extremely useful general book. *Cast Iron, Physical and Engineering Properties* by H. T. Angus (Butterworths, 1976) succeeded the 1960 version *Physical and Engineering Properties of Cast Iron* and contains a wealth of information compiled by the author who was the authority on cast irons for many years at the British Cast Iron Research Association. Virtually the authority on cast iron, the book is a good companion on the subject to the *Gray and Ductile Iron Castings Handbook* (Gray and Ductile Iron Founders Society of America). It covers all aspects of the material from its constitution and structure, properties, treatment, components, application and design data. The AFS (American Foundrymen's Society) *Cast Metals Handbook* (1957) covers grey, white malleable and ductile iron and steel castings. The metallurgy, properties and heat treatment are all dealt with and an equivalent ferrous casting alloy specification cross index is included. The same body published a handbook on the *Analysis of Casting Defects* 2nd edn (1964). Numerous types of defects are described and well illustrated. *The Casting Design Handbook* (ASM, 1966) is a more comprehensive compilation of some 18 committees and comprises 2 parts: Part 1 Design Process Relations and Part 2 Properties and Selections of Cast Metals. Iron and Steel Castings are represented in both parts. Finally in this section on ferrous castings sources, the book *Directional Solidification of Steel Castings* by R. Wlodawer (Pergamon Press, 1966) and translated by L. Hewitt and R. Riley has long been highly regarded among foundry managers, foremen and technicians.

A basic textbook for students on physical ferrous metallurgy is *The Physical Metallurgy of Steels* by W. C. Leslie (McGraw-Hill, 1982) dealing with the properties of high porosity iron, carbon steel, HSLA steels, heat-treated steels, very high strength and tool steels, magnetic and electrical steels and stainless steels. For the desulphurization of iron and steel, it is worth consulting the 1980 Iron and Steel Society of AIME monograph *Desulphurisation of Iron and Steel and Sulfide Shape Control* by W. G. Wilson and A McLean. *Advances in the Physical Metallurgy and Applications of Steels*, based on the proceedings of an international conference organized by the Metals Society and held at the University of Liverpool in 1981, is one of the Institute of Metals 'special' publications, namely Book 284 (1982). It outlines recent advances in physical metallurgy particularly in the areas of strength, fracture resistance and corrosion. Well over 30 papers are presented with discussion. A book that highlights important aspects of physical metallurgy and heat treatment in terms of transformations is *Physical Metallurgy of Iron and Steel* by R. Kumar (Asia Publishing House, 1968). As such, it is a good basic guide to the transformations behaviour of the iron-carbon system and the alloying behaviour of iron. There are many other specialist books on physical metallurgy, *Martensitic Transformation*, for example, is fully covered in the book of that title by Z Nishiyama translated into English, edited by Fine, Meshii and Wayman (Academic Press, 1978). The Institute of Metals Book 194 *Non-metallic Inclusions in Steel* by R. Klessling and N. Lange (1978) is one of the best collections of monographs on the subject and began as ISI Special Report 90 London 1964. It contains over 369 references and is the major reference tool in this particular area of steel behaviour.

Other books on properties and related subjects

Many books have been written on the various engineering properties of steels over the years, for example, *The Brittle Fracture of Steel*, by W. D. Briggs (Macdonald and Evans, 1960) which brings together the different approaches to the subject by engineers, metallurgists and physicists. Like many subjects, brittle fracture forms a major section of the modern-day metallurgical literature regularly updated by published articles and conference papers. *The Corrosion and Oxidation of Metals* by U. R. Evans (Edward Arnold) covers a great deal of knowledge on the theory of all types of corrosion mechanism as they affect iron and steel materials. Another standard work on corrosion is the *Corrosion Handbook* by H. H. Uhlig fully described in Chapter 13 on corrosion. The International Conferences on Hot Dip Galvanizing referred to in Chapter 9 on zinc are a useful source.

An excellent handbook on sampling and analysis of iron and steel and the materials and by-products related to their production is the *Handbook of Analytical Control of Iron and Steel Production* by T. S. Harrison (Ellis Horwood, 1979). This comprehensive volume covers sampling, chemical methods, physio-chemical technique and physical methods. Actual methods are described for elements in iron and steel, gases in steel, ferro-alloys, iron ores, sinter, slag, refractories, fuels, CO by-products, waters, effluents and even steelworks' lubricants. There are numerous references given at the end of each chapter. Spark testing is covered by G. Tachorn's *Spark Atlas of Steels, Cast Iron, Pig Iron, Ferro-alloys and Metals* (Pergamon, 1963). Book 262 published in 1979 by the Metals Society entitled *Direct Reduction of Iron-ore: A Bibliographical Survey* lists and describes 54 DR processes, grouped according to industrial significance. This review, well produced and very comprehensive technically, is a translation by GKN of the 1976 Verlag Stahleisen *Direktreduktion von Eisenerz: Eine Bibliographische Studie*. The publication of conferences and symposium proceedings as special publications is a common practice. A useful one is *Ferro-alloys and Other Additives to Liquid Iron and Steel* (ASTM, Special Technical Publications (STP) series, STP 739, 1981).

The *World Survey of Ferro-Alloy Markets and Directory of Producers* published in 1979 by Metal Bulletin and edited by John Parry discusses current ferro-alloy developments, N. J. G. Pounds has written an excellent little book on the steel industry from the geographical aspects and covers the UK, Western Europe, America, the Eastern bloc and the Third World. It contains some interesting maps and useful statistics and is entitled *The Geography of Iron and Steel* (Hutchinson, 1971, 5th edn). In the field of pollution, the *Steel Industry and the Environment* edited by J. Szekely (Marcel Dekker, 1973) is a fairly comprehensive book on the physiological, sociological, legal, technical and engineering aspects of pollution from iron and steel works. As in the case of all the related topics mentioned there are many other books on both the geographical and historical aspects of iron and steel production from general books such as *History of the Iron and Steel Industry* by H. R. Schubert (Routledge & Kegan Paul) and the *History of British Steel* by J. Vaizey (Weidenfeld & Nicolson) to more specific publications on particular companies works or individuals.

The manufacture, structure and applications of iron and steel have changed drastically over the centuries and worth noting are C. S. Smith's *History of Metallography* and the publications of the *Historical Metallurgy Group* of the former Iron and Steel Institute (now Historical Metallurgical Society, Institute of Metals).

Periodicals

Periodicals frequently overlap each other in coverage and papers are increasingly published in several journals of different geographical origin. For this reason, any attempt at structuring has been abandoned and key journals are described alphabetically. It is often better to concentrate as much as possible on a few 'core' journals and supplement these by the various abstract services mentioned later. The number of journals of potential interest in the field of iron and steel metallurgy is enormous (the 1986 edition of *Source Journals in Metallurgy* lists 1,300 scientific, engineering and trade journals which are regularly scanned and indexed by Materials Information) but the following provide a basis for selecting a shortlist of 'core' sources.

Acta Metallurgica (Pergamon Press) includes international review papers and original articles on the material science aspects of metals and alloys. The papers are of a high standard and with good abstracts in German, French and English, well reproduced figures and adequate bibliographical references. *Archiv für das Eisenhuttenwesen* (Stahleisen for the VDEh and Max Planck Inst.) publishes papers in the behaviour of steel, testing and analysis of steel, process metallurgy of steelmaking and fundamentals of working steel. The *BCIRA Journal*, although confidential to member firms and organizations, is an excellent journal on cast-iron research, includes good articles on the various types of cast iron, together with excellent abstracts of the international foundry literature. *The British Foundryman* is the Official Journal of the Institute of British Foundrymen. This publishes papers on the theory and practice of foundry technology, e.g. ferrous castings production, properties and applications. The *Canadian Metallurgical Quarterly* (Pergamon) contains good quality papers by authors on practice or theory of iron and steels. *Concast News* (Concast Service Union AG of Zurich) is a specialist publication on current continuous casting technology.

Foundry (Management and Technology) is published monthly by Penton/IPC. This is the American equivalent of the *Foundry Trade Journal* which is published twice a month and monthly in January, August and December. It is concerned with the casting industry, incorporates the *Iron and Steel Trades Journal* and covers steel castings and ductile irons. *IISS Commentary* is published by the Institute for Iron and Steel Studies which produces very useful techno-economic reports on iron and steelmaking. *Iron Age (Metals Producer)* (twice monthly, Chilton Co., USA) is a chatty trade journal, rather than a technical periodical as such, but it contains useful news items; *Iron Age Metalworking International (IAMI)* (Chilton) is another American trade journal on metal working only of fringe interest to the materials scientist. *Iron and Steel Engineer* (monthly, US Association of Iron and Steel Engineers) covers general developments in the iron and steel industry and specific topics on process and product aspects of iron and steel. It includes news sections. *Ironmaking and Steelmaking* (Institute of Metals) is an international journal of technological advance in iron and steel production. It is published six times/year and is a continuation of what was the main iron and steel journal, the *Journal of the Iron and Steel Institute (JISI)*.

Met. Trans. of AIME (A) (Met. Soc. AIME/Iron and Steel Soc. of AIME) covers physical metallurgy and materials science and the papers are of a very high standard. It is published monthly. *Metals and Materials*, the journal of the Institute of Metals incorporating *Metallurgical and Materials Technology* and *Metals Society World* is published monthly by Mechanical Engineering Publications and includes manufacturing process articles and materials reviews, news items and book reviews. *Metals and Minerals International* is a useful quarterly journal for the iron and steel industry published by the London and Sheffield Co. covering ores and minerals aspects of iron and steel. *Metals Technology* is an Institute of Metals journal which is of a high standard and *Modern Castings* is published monthly by the American Foundrymen's Society. It contains articles and papers in foundry castings including those of iron and steel.

Nippon Kokan Technical Report Overseas (NKK) is a periodical of extremely high quality both in terms of content and presentation. The NKK research reports included are well written in English, the photographs, tables and figures are excellent and the articles include author's designations, good abstracts and literature references. The work published includes the whole gamut of NKK interests, including steel properties and applications. *Nippon Steel Technical Report* (NSCR) is a semi-annual journal which contains reports which are English versions of some appearing in 'Seitetsu Kenkyu' (quarterly) and also English abstracts of the remainder. It is of an extremely high standard in terms of quality of the material on NSC steel products and processes and its overall presentation, abstracts, index etc. *Practical Metallurgy*, a bi-lingual (German/

English) monthly, includes papers on the structural aspects of iron and steel with numerous micrographs, electron photographs etc. It is published in cooperation with the German Society for Metals. *Revue de Métallurgie* is a French-language publication on the metallurgy, properties and applications of iron and steel produced in collaboration with CDS, ATS and IRSID.

Scripta Metallurgica (monthly, Pergamon Press) is the companion journal to *Acta Metallurgica* and is a reputable international materials science journal covering properties of micro alloyed steels, stainless alloy creep behaviour, fatigue cracking of C-steel etc. (an excellent source of ferrous materials science). *SEAISI Quarterly* is the journal of South East Asia Iron and Steel Institute and mainly represents the activities of that body. Articles cover practice in the SEIASI countries, but conference papers can be international in coverage, e.g. 'Energy in iron and steelmaking' by the Chairman of BSC Overseas Services. *Stahl und Eisen* published in German by the VDeH and Wirtschaftsvereinigung Eisen und Stahlindustrie in conjunction with CRM, DFB, ECCA and VFE, is the best of the West German iron and steel journals. It covers the products and process metallurgy of iron and steel with English abstracts and general news features on the DBR iron and steel scene. *Stainless Steel Industry* is a journal for stainless-steel manufacturers, stockholders and fabricators. Published bi-monthly by Modern Metals Publications it contains occasionally good materials papers despite its 'trade' type format. *Steel in the USSR* (Institute of Metals by management with the BLLD) is a monthly selection of translated material from STal and Izvestiya Vysshikl Uchebnykl Zevedenii Chernaya Metallurgiya. It is a regular source of Soviet iron and steel papers.

Steel Times International is the monthly journal of the European Iron and Steel Industry containing a number of general industry (process and plant) articles, but frequently contains papers on materials aspects also e.g. ferro-alloys, continuously cast steel, steel quality, etc. *33 Metal Producing* (McGraw-Hill, New York) is a newsy/magazine-type treatment of the iron and steel industry, very American in style, but useful for general industrial comments and a list of recent US patent titles on steel products and processes. *Trans. of the I & S. Inst. of Japan (ISIJ)* is a publication in English and contains excellent research articles and technical reports aspects of iron and steel: obviously with an emphasis on Japanese practice.

The reader soon becomes accustomed to identifying those periodical sources that give the best current awareness combination. See examples in Table 3.2. In addition to the above core journal descriptions the reader may also require to consult

Table 3.2. Examples of possible journal combinations to give basic current awareness

	Core journals	*Other key journals*	*Fringe journals*
Corrosion Scientist	*Advances in Corrosion Science and Technology* *Anti-Corrosion Methods and Materials* *British Corrosion Journal* *Corrosion Science* *Corrosion*	*Finishing* *Industrial Finishing and Surface Coatings* *Metal Finishing* *Materials Performance*	*Steel Times* *British Journal of NDT* *Coating* *Surface Engineering*
Iron and Steel Process Engineer	*Iron and Steel Engineer* *Iron and Steelmaker* *Ironmaking and Steel making* *Steel Times International* *AIME Proceedings*	*Journal of Metals* *British Steelmaker* *Stahl und Eisen* *Steel in USSR*	*Hutnik* *Seasi Quality* *Bolletino Technico Finisider* *CRM Met Report*
Steel Product Engineer	*Iron and Steel Engineer* *British Welding Journal* *Metal Construction* *Metallography* *Steel Times International* *Metals and Materials Scripta Met*	*Acier Stahl Steel* *Aciers Speciaux* *Concast News* *NKK Tech Rep Overseas* *Nippon Steel Tech Report*	*Architects Journal* *Automotive Engineer* *Chartered Mechanical Engineer* *Building with Steel* IISI Statistics etc. Special product journals (*Tube, Wire, Strip* etc.).

periodicals in other subjects with which iron and steel products and processes are associated e.g.: instrumentation, testing, pollution, and commercial topics. The golden rule is to select, evaluate and control to a minimum the periodical literature that needs to be read cover-to-cover. The user should rely on abstracting and other professional current awareness services to do the bulk of the non-essential scanning. Quality of information is better than quantity and in the case of periodical-based literature, this fact will prevent the user becoming overwhelmed by unnecessary primary literature scanning.

Finally, there has been over the years, a confusing number of cessations, title changes and mergers among many of the journals. Familiar titles such as *Journal of the Iron and Steel Institute*, for example, are no longer current. Most library listings will show all past and present titles and give details of relationships of the different or combined titles.

Abstract journals and conference proceedings

Traditional abstracting tools are an important source of current awareness and, despite more recent mechanized databases which can be 'scanned' online, there are many who maintain that regular browsing or searching through at least some sections of abstract journals is a worthwhile productive means of literature research. The literature of iron and steel materials and related topics is covered in a wide-range of abstracts. The most important of these are:

Metals Abstracts;
Engineering Index;
Chemical Abstracts.

These have already been discussed in Chapter 1. Most of the other abstracting services mentioned there contain information of value to users of iron and steel literature and the reader.

ANBAR (Management) Abstracts cover the important financial and economic aspects of iron and steel materials/products. The Iron and Steel Institute used to produce ABTICS (Abstract and Book Title Index Card Service). This was the main abstract service in the iron and steel field prior to the publication of *Metals Abstracts* and covered some 1,000 journals. The abstracts were produced on 5 × 3 in cards and classified by the UDC. This service has now been replaced by *Steel Alert.*

This is a new current awareness service from Materials

Information which is based on items selected from the Metadex files. Unlike earlier current awareness services provided for the iron and steel industry, however, *Steel Alert* pays much closer attention to 'technocommercial' information.

Conferences

Another source of iron and steel literature is conference proceedings. One of the most comprehensive collections of conference proceedings in the world is held by the BLDSC. The (keyword) *Index of Conference Proceedings Received* contains details of the well-known AIME conference series and conferences by various other metallurgical societies, e.g. Canadian Institute of Mining and Metallurgy Steel Founders Society of America etc. The five well-known AIME proceedings referred to are:

(1) Blast Furnace Conference Proceedings (1941–62) (ironmaking topics).
(2) Electro Furnace Conference Proceedings (1974–to date) (EF steelmaking topics)
(3) Ironmaking Conference Proceedings (1963–to date).
(4) Open Hearth Conference Proceedings (1944–to date) (OH steelmaking topics).
(5) Steelmaking Conference Proceedings (1978–to date) (includes oxygen BOS or (BOF) steelmaking).

Many conference proceedings are published in book form retrospectively or are held as preprints by organizations/individuals who contributed. The Institute of Metals (previously *Iron and Steel Institute) Publications* make up the core reference collection of significant ferrous metals-related conference and committee proceedings and spans a period from 1931 to date, the first being *1st Report of the Corrosion Committee* (ISI SP1). The Institute of Metals also publishes *World Calendar*, a quarterly journal listing all metals-related conferences, events and meetings taking place around the world. The US Special Library Association publishes a *List of Scientific Meetings* and the ASTM (American Society for Testing and Materials) publishes many conference proceedings in its STP (Special Technical Publications) series, for example:

STP 575 Bearing Steels: The Rating of Nonmetallic Inclusions
STP 619 Structure, Constitution. . . . Wrought Ferritic Stainless (DEMO)
STP 794 Through-Thickness Tension Testing (GLODOWSKI Ed)
STP 706 Toughness of Ferritic (LULA)

STP 644 Rail Steels Development, Processing and Use (STONE/KNUPP)
STP 498 Ultra-High-Strength Structural Steel
STP 679 Properties of Austentitic Stainless Steels (BRINKMAN/GARVIN)

Restructuring Steel Plants for the Nineties is the proceedings of an international conference which examined the problems of modifying and updating existing steelplants which was published by the Institute of Metals in 1987.

Standards

Standards have been discussed in general terms at some length in Chapter 14 on Design. The *BSI Yearbook*, apart from listing current and defunct standards, also gives details of British Standards handbooks (e.g. No. 19 Methods for the Sampling and Analysis of Iron, Steel and Other Ferrous Metals); Codes of Practices (e.g. CP 143 Galvanised Corrugated Steel); and European Standards, corresponding ISO standards and Sectional Lists e.g. SL 24 *Iron and Steel* are included. The main BSI standards related to steels are:

BS 4 Structural steel sections
  Part 1 Hot-rolled sections
BS 18 Methods for tensile testing of metals
  Part 2 Steel (general)
  Part 3 Steel, sheet and strip (less than 3. mm and not less than 0.5. mm thick)
  Part 4 Steel tubes
BS 131 Methods for notched bar tests
  Part 2 The Charpy V-notch impact test on metals
BS 1449 Steel plate, sheet and strip
  Part 1 Carbon steel plate, sheet and strip
BS 1639 Methods for bend testing for materials
BS 1837 Methods for the sampling of iron, steel, permanent magnet alloys and ferro-alloys
BS 3894 Method for converting elongation values for steel
  Part 1 Carbon and low alloy steels
BS 4360 Weldable Structural Steels
BS 4848 Hot-rolled structural steel sections
  Part 2 Hollow sections
  Part 4 Equal and unequal angles
BS 5135 Metal-arc welding of carbon and carbon manganese steels

Other standards relate to cast and ductile irons and the associated products, galvanized and coated iron and steel products and engineering products of iron and steel.

Excellent *DIN Handbooks* on iron and steel, welding and steel pipelines are published in English, e.g. Pocket-books 4, 28 and 155 which are quality and dimensional standards for steel and iron. Verlag Stahleisen MBH publish an excellent *Stahl-Eisen Liste*. DIN standard specifications themselves e.g. DIN 17100 Steels for General Structural Purposes supply details of the scope, definition, dimensions and weight, grade classification and designation code with properties, compositions, product process suitability, requirements for delivery, manufacture properties etc. and testing/marketing.

A catalogue of *American National Standards* is published by American Technical Publishers of Hitchin. This lists ANSI, AWWA (cast iron), AWS (American Welding Society) and other designations. Actual AWWA specifications can be consulted in the *Journal of the American Water Works Association (JAWWA)*. As far as the API (American Petroleum Institute) standards are concerned, they are listed in the *API Catalogue*. Here the most relevant are those for tubular goods, e.g. casing, tubing, drill pipes, linepipe, NDT, steel valves, structural piping etc. Other standard issuing organizations will be mentioned later, but as far as national standards are concerned, the various national standardization organizations produce handbooks and lists of standards and other publications e.g. codes etc. produced by them. The *JIS Handbook of Ferrous Materials and Metallurgy* (Japanese Stds Association, 1973) lists national (JIS) standards of Japan as they relate to ferrous metallurgy materials. A brief appendix compares JIS specifications with other international standard specifications.

Another organization issuing important standard specifications on steel topics is the American Society for Testing and Material (ASTM). It publishes a 33-part *Annual Book of Standards*, the main ones concerned with iron and steel being:

Part 1 Steel Piping, Tubing and Fittings (95 standards)
Part 2 Ferrous Castings, Ferroalloys (70 standards)
Part 3 Steel, Strip, Bar, Rod, Wire, Chain and Spring; Wrought Iron; Metallic Coated Products (169)
Part 4 Structural Steel, Rails, Plate Bearings, Forgings, etc. (150)
Part 3.1 Metals, Physical, Mechanical, NDT and Corrosion tests, metallography etc.
Part 3.2 Includes chemical analysis of metals and sampling and analysis of metal bearing ores (127).

The Pocket Book of AISI Standard Steels published by the AISI includes some 14 tables giving chemical compositions of those grades of steel designated as standard by the American Iron and Steel Institute extracted from the AISI Steel Products Manual. SAE, UNS and ASTM designations are also shown where appropriate. The annually published *SAE Handbook* (Society of Automotive Engineers) covers a number of ferrous metals compositions, general data, methods of testing, quality requirements and lists numerous standard SAE designations.

The codes and standards published by the American Society of Mechanical Engineers are listed in the annual ASME publications catalogue. The main reference standards for iron and steel users is the *ASME Boiler and Pressure Vessel Code*, a loose-leaved publication comprising the following sections:

Section II	Part A Ferrous Metals Part C. Welding Rods Electrodes and Filler Materials
Section III	Nuclear Power Plant Components (steel and steel alloys)
Section IV	Boilers (wrought materials and cast iron)
Section V	NDT (e.g. of iron and steel materials)
Section VIII	Pressure Vessels (including steel and cast iron)
Section IX	Welding and Brazing Qualification

Companion volumes include *History of the ASME Boiler Code* and *Interpretations of the ASME Boiler and Pressure Vessel Code*. A number of other relevant standards are published by ASME e.g. B36.10 *Welded and Seamless Wrought Steel Pipe* and B36.19 *Stainless Steel Pipe*; B16.1 *Cast Iron Pipe Flanges and Fittings* and many more.

Among other bodies producing important standard documents is Lloyds Register of Shipping (London) which publishes the *Rules and Regulations for the Classification of Ships*. This includes an updating service and covers grades of steels, castings and forging materials. Likewise, the American Bureau of Shipping, Italian Navy and many other similar organizations worldwide publish ship material standard specifications. A particularly relevant one publishes the *Rules for the Construction and Classification of Steel Ships* namely, *Det Norske Veritas* (Novik, Norway). Chapter X deals with the quality and testing of steel materials.

A further source of standards are government agencies, e.g. the UK Ministry of Defence publishing the DEF series which can be 'picked up' in the *Standards in Defence News* published by the Directorate of Standardisation (Glasgow).

Steel specifications cross-referencing

It is always most essential to read any steel specification correctly as they are not always what they appear. They quote minimum properties as a rule and only give ranges for alloying elements. An understanding of the relationship between specification and actual composition will help to avoid misinterpretations and costly mistakes or conclusions.

Two problem areas frequently encountered in the purchasing, researching and selection are cross-referencing of designations and specifications; and matching equivalent alloys from differing countries. Presented below is a list of reference books and other data sources that are useful resources to consult for such information. *Stahlschlussel, Key to Steel* (12th edn, Verlag Stahlschlussel Wegst GmbH, 1980) contains over 40,000 brands and designations of steels. It includes the material numbers of German and other foreign data, the standards of other countries are related and set opposite to the German number. It also includes ASTM, Military, Federal QQ, AMS-AISI and SAE designations and is tri-lingual (German, French and English). It has become a recognized tool for the companion of the various national and other standard designations for some 15 countries. The designations are arranged in country order and numerically with many cross-references. Stahlschlussel covers structural steels, tool steels, valve steels, high temperature, non-magnetizable, heat-resisting, heat conducting and stainless steels. It also contains an international list of steel suppliers, a table giving details of product shape/condition in respect of German suppliers and lists brand names for steels. As new editions of Stahlschlussel are published users must resist the temptation to dispose of old versions. These are invaluable sources for tracking down obsolete designations and, like many old handbooks or indeed old standard specifications themselves, should not be discarded.

The *Engineering Properties of Steel* (ed. P. D. Harvey) (ASM, 1982) is a handbook which provides data on the composition mechanical and physical properties, fabrication characteristics, machinery data and typical uses of steels ranging from C. steels to maraging grades. The book is useful to engineers and commercial users seeking information on alternative steel grades. The steels are cross-referenced to US and non-US standard specifications. The *ASTM/SAE Unified Numbering System for Metals and Alloys* (4th edn, American Technical Pubs, 1987) correlates nationally (United States) used numbering systems currently administered by societies, trade associations including (ferrous and nonferrous), AA, ACI, AISI and SAE, AMS, ANSI, ASME, ASTM, AWS,

CDA, QQ and WW and MIL. A valuable current source is the *SAE Handbook*, the 1987 edition of which is now in four volumes with Vol. 1 being devoted to materials. The *Worldwide Guide to Equivalent Irons and Steels* (American Society for Metals, 1987) provides chemical compositions, mechanical properties, available product forms and designations for over 20,000 irons and steels from 21 countries. Alloys are sorted by descending carbon content, thereby drawing alloys with equivalent compositions close together in the listing. In addition, the *Handbook of Comparative World Steel Standards* (International Technical Information Institute, Tokyo, 1974) covers steel standards and specifications from Japan, US, UK, West Germany, France and the USSR tabulated in a convenient, comparative style. Another standards book is the *Metallic Materials Specification Handbook*, 3rd edn, by R. B. Ross (E and F. N. Spon, 1979) which provides ingredients and properties of a symbol representing a specification or trade name. It includes 50,000 trade names, specifications or symbols relating to metals. The *Soviet Alloy Handbook* (Battelle Metals and Ceramics Information Center, Columbia, Ohio, Oct. 1980, 358 pages) on the other hand consists almost entirely of tables of data and cross-indexes. A further standard comparator is *Alloy Digest* (Engineering Alloy Digest Inc., 356 North Mountains Ave, Box 823, Upper Montclair, NJ 07043). *Iron and Steel Specifications*, 5th edn (British Steel Corporation), gives a summary of published BSS at 1978 and foreign standards published up to 1974. Steels are grouped according to chemical analysis, not physical properties or end use. *The Jahrbuch Stahl* also published by Verlag Stahleisen gives details of DIN specifications and relevant werkstoffe numbers.

Finally, on standards, in 1964 three volumes were produced by the British Iron and Steel Research Association and published by Pergamon Press entitled *The Mechanical and Physical Properties of the British Standard EN Steels* based on BS 970 1955. This work compiled by Woolman and Mottram, despite having been superseded by later standard specifications, is still invaluable since many people still quote EN grades and many engineers and marketing experts brought up with EN steel specifications rightly or wrongly still use them as a reference point when specifying new materials. A recent publication worth noting is the *Steelmaking Data Sourcebook* from the Japan Society for the Promotion of Science, the 19th Committee of Steelmaking (Gordon and Breach, 1988).

Specialised databases and services

Databases

It is not the purpose of this section to discuss the techniques of online searching or comment on the merits or otherwise of the various commercial information retrieval services offering online databases for searching. Suffice to say that many bibliographic and non-bibliographic databases relevant to iron and steel materials (not to mention the iron and steel industry as such) are available to users. These are available on systems such as Lockheed-Dialog, SDC Orbit, Data-star, Pergamon Orbit Infoline, ESA/IRS, etc. The 'alias' problem is something online users have to beware of, for example, the databases produced by Chemical Abstract Service is called Chemname on the Lockheed Online Service and Chemdex on the SDC Search Service. This 'alias' problem is fairly common and databases offered by each service should be carefully checked as to their source, content and completeness. Sometimes they are identical, sometimes not.

Bibliographic databases include:

Metadex (Metals Abstracts)
Compendex (Engineering Index)
Chemical Abstracts Condensates (ACS)
Ismec (Inst. Mechanical Engineers)
Predicasts PROMPT (Predicasts Technical System)
Weldasearch (Welding Institute)

A typical search output is shown below:

Table 3.3. Example of a search for information on steel plates, rails, section etc. – markets and products on Predicast Prompt Product Code

B16	*Set*	*Items*
SPC=331236 plates	1	128
SPC=331237 sections	2	84
SPC=331238 rails	3	45
CIOR20R3 plate/sections/rails	4	236
SEC 6 market information	5	92490
SEC 3 products/processes	6	70125
C5OR6 markets/products etc.	7	157406
C4AND7	8	120
.SORT8/1–120/CC	9	Sorted by country
T9/8g1–120 journal rep T9/4/1–120 abstracts		Print outs

There were 120 items relating to marketing or product aspects of steel plates, rails and sections

In addition to these bibliographic databases, the following non-bibliographic databases are of relevance to the user interested in iron and steel materials.

Metals Datafile
CRIB
Manlabs
PTS (Predicasts)
Steel Data Bank

Metals Datafile contains numerical data on the physical and mechanical properties, compositions and specifications of ferrous and non-ferrous alloys. The file can be used to select alloys for specific property requirements, to determine property variations due to processing, environmental and/or temporary variations and to local equivalent alloy specifications. Metals Datafile contains data derived from publications, handbooks and data collections and is available online through the System Development Corporation (SDC Search Service) network. The Crib database contains 30,000 records of metallic and non-metallic mineral resources information for 95 countries. Manlabs comprises NPL assessed thermochemical data on over 200 metals, compounds and alloys. Using data on main metals (Fe, Cr, Ni, Co, Ti, C, Nb, Mo and W) over 100 ternary phase diagrams can be calculated with numeric or graphic display. Predicasts PTS (Time Series and Statistical Abstracts) holds intelligence data for planning and forecasting (US and International) and contains production statistics e.g. German steel output, plant developments, investment etc. of steel products in EEC etc. more industry than economic based. See example of output at Table 3.4. Steel Data Bank covers 20,000 annual, quarterly, monthly and weekly time series on the steel industry. It includes production figures and figures on imports, exports, raw steel, pig iron, steel products etc.

Table 3.4. Example of output from 'Non-bibliographic' database

(i) Tables (Time Series)
e.g. *West Germany*
Raw steel production
mil. m tons

1964	37 339
1965	36 821
1966	35 318
etc. (PTS Composites Lockheed-Dialog)	

Special collections and sources

The publications, abstracts and databases of the Institute of Metals have already been mentioned as part of the Institute of Metals and American Society of Metals joint Metals Information activities but, separately administered, is the Joint Library which is a major source on metals in its own right and was formed originally as the Library of the Iron and Steel Institute. In 1938 the Library was amalgamated with the old Institute of Metals Library to form the basis of the present joint library. The Institute of Metals publishes an excellent *List of Serials Held in the Metals Society*. The International Iron and Steel Institute (IISI) also possesses a Library Services department based in Brussels, Belgium. The IISI is more of an economic research organization whose members are actual steel companies. The IISI undertake technological and financial research and collects, evaluates and disseminates world steel statistics.

Its publications range from reports such as *Future Western World Supplies of Iron Ore* (1983); *Vanadium, Niobium, Molybdenum and the Steel Industry*; and *Steel Statistics of Developing Countries* to the actual *Proceedings of the IISI Annual Conference*. Its most famous booklet is probably *World Steel in Figures* which is an annual publication comprising a mixture of tables and graphs on steel, production, trade and consumption. Its journal, *The IISI Bulletin*, is published three times a year. The IISI Committee publish a directory of *World Steel Statistical Sources* to guide the reader to sources of information on the steel industries of various countries. The IISI list of publications gives details of publications and maps available together with prices.

The Institute of Iron and Steel Studies IISS is not to be confused with the IISI, however, and is an educational non-profit-making organization based in New Jersey, USA which offers membership to companies and industrials and provides extremely well-produced worldwide techno-economic reports and maps on iron and steel products and plant. These are published in monthly issues of *The IISS Commentary* which is available to members. The IISI conducts surveys and studies and issues technical information bulletins, produces technical papers and publishes techno-economic books on all aspects of the ferrous metals industry from sponge iron and ferrous scrap to continuous casting and finished steel products. Some of the topics covered are:

iron ores and pellet production;
iron and steel scrap;
raw steel;
ferro-alloys;

steel rails;
steel sections;
steel pipes and tubes;
steel rods etc.;
reports and location maps on companies and plant concerned with iron and steel processing/production are a speciality of the IISS.

Company sources

Valuable sources of information regarding iron and steel are the variety of publications and services emanating from businesses connected with the products and processes related to these materials. Thus books produced by, for example, the British Steel or Bethlehem Steel Corporation, either as simple guides or technical manuals, provide useful information at a variety of levels for 'open' consumption. Since there are literally hundreds of commercial organizations in or connected with the steel industry the trick is to identify those companies which can make a real contribution to the literature/information needs of the user. Publications like *Iron and Steel Works of the World* (Metal Books Publications) or *Steel Traders of the World*, ed. R. Cordero (Metal Bulletin, 1980) are useful here. Furthermore, by examining the IISI Top 20 or so steel companies or checking the main companies covered in the *Steel Times Annual Review* a checklist of main companies can soon be compiled. Once this is done, some productive research in relevant local libraries, steel company libraries themselves or commercial trade literature services like the Technical Indexes Ltd Micro Data Systems will track down the wealth of information available. Booklets produced by BSC and BISF have already been mentioned.

Similarly, Bethlehem Steel produced a 208 page handbook in the 1970s entitled *Modern Steels and Their Properties* with quite a good index, properties table and black and white photographs. As well as these general works, companies produced some excellent textbooks (cf. *Making Shaping and Treating of Steel* already referred to) such as the *Atlas of Continuous Cooling Transformation Diagrams for Engineering Steels* by M. Atkins which presented information gathered by BSC on the continuous cooling transformation behaviour of a wide range of engineering steels to help in the practical aspects of heat treatment and hot processing. Twelve tables linking many national specifications to the relevant diagrams are included with the main 172 diagrams divided into alloy groups. In addition to such books many companies produce bibliographies and again for example British Steel has produced several related to iron and steel materials (not just the industry generally), for example: *Directly Reduced Iron*, *Nonmetallic Inclusions*, *Continuously Cast Steel*, *Oxygen Steelmaking*, *Controlled Rolling of Steels* etc.

House magazines/journals can provide a wealth of data on the plant, processes and products of the company as will the *Annual Report* of the company, or *The Annual Research Report* (if separate) and other research literature. For example, all open research reports published by British Steel are deposited with the main UK Copyright Libraries including the BLL. BS also publishes an annual R. and D. review known as *Steelresearch* which summarizes current research work, lists published papers and discusses possible future developments in iron and steel. Specialist literature on research services related to ferrous raw materials, pilot steelmaking, continuous casting, and metallography of iron and steel products is available from BS and many iron and steel companies maintain a directory of plant and equipment which may or may not be available generally. From the point of view of practical usefulness to practising engineers or materials scientists, the general commercial literature available from companies is also most valuable. This can take the form of product literature and booklets on product applications. Training literature is also useful for students and teachers, but due to demand outstripping supply, coupled with economies in publishing, the amount of literature available in this way is diminishing.

Another important source is films and videos produced by the manufacturers of ferrous materials although it is fair to say that these are concerned mainly with plant and processes rather than materials as such and, again, are mainly for training purposes. Most large iron and steel companies have lists of relevant film/video productions usually available from film libraries or photographic departments. Many companies no longer have the staff or resources to deal with enquiries from the public and users will find that they will achieve far better results if they direct their enquiries for company literature via a local library or the library of an appropriate university, college, research association or institution. A good example is the Sheffield City Libraries (Science Commerce and Technology Branch) which has good contacts with a number of iron- and steelmaking companies and have a large collection of steel company literature. Most county and large municipal libraries have links with local cooperative schemes and the BLDSC, and it cannot be emphasized too often that, in the UK, users should avail themselves of the excellent interlending schemes available throughout the UK library network. Direct approach to companies for anything other than sales literature should be avoided unless absolutely essential and enquirers should be prepared to have to pay for reports and other handbooks previously available at no charge. The days of free information are rapidly fading into the past.

Translating dictionaries

The following dictionaries are listed as being those most relevant to iron and steel. These are regarded as practically essential to anyone with a regular need to translate into English foreign terms in iron and steel and associated subjects:

Iron and Steel Dictionary (Stahleisen Worterbuch) (German–English/English–German), 3rd edn (Verlag Stahl Eisen, 1977).
De Vries and Herrman Technical and Engineering Dictionary (McGraw-Hill, 1972).
Concast Dictionary (English, French, Spanish, Italian, Russian, German, Japanese) (Concast AG Zurich, 1975).
Kettridge Technical Dictionary (metallurgy, metal processing, materials science), (French–English/English–French) (Routledge and Kegan Paul, 1959).

Dictionaries discussed in Chapter 1 are of use for iron and steel terminology and should be examined. The following pocket dictionaries could be described as useful for student and casual users.

Concise Iron and Steel Dictionaries (German, French, Russian, Spanish) (BISF, 1957; 1961). Useful single diagrams and conversion tables.
Vocabulary of Iron and Steel Terms (German–English/English–German and Italian–English/English–Italian) (BSC).
Concise Iron and Steel Dictionary (English–German/German–English) (BSC).
Glossary of Terms used in the Steel Tube Division (Tube Investments Ltd.).

Statistics

Statistics are a most vital source of information without which no user can support his or her work. The statistical publications of the UK Iron and Steel Statistics Bureau, London are particularly important, namely the *UK Iron and Steel Industry: Monthly Statistics* which also includes selected statistics for major overseas countries; *UK Exports of Iron and Steel*; *UK Imports of Iron and Steel*; a quarterly book detailing the exporting trade in 28 products for 14 major steel-producing countries *World Steel Trade* and *World Trade* stainless, high speed and other alloy; and the *Annual Statistics for the UK* by process, product and region. The bureau also publishes the very informative *International Steel Statistics* with special country books and summary tables. Iron and steel statistics publications are also available from a number of other bodies such as the AISI e.g.: *Annual Statistics Reports*, statistics of the Japanese Iron and Steel Federation and the *International Iron and Steel Institute*, as mentioned previously. Other statistical sources are discussed in Chapter 1.

Finally, the Iron and Steel Institute of Japan (ISIJ), Japan Institute of Metals and South East Asia Iron and Steel Institute (SEAISI) are also important sources of information on iron and steel. However, previous general guides to the iron and steel literature are still worth consulting, firstly because they all differ considerably in their approach, demonstrating to the user the diversity of the problems concerned and secondly, because, despite their age, a great deal of the content matter is still useful. In addition, the guides stand up as worthwhile contemporary attempts in their own right. The first compilation was by Gibson and Tapia in 1965 as part of a general *Guide to Metallurgical Information* (SLA Bibliography No. 3). Contributors included librarians of US Steel Corporation, Bethlehem Steel and the American Iron and Steel Institute. The second *How to Find out in Iron and Steel* (Pergamon Press, 1970) was written by an ex-steel company information officer, David White. This excellent little book was the best available guide for many years complementing perfectly the more academic approach of the UNIDO Guide (No. 26) *Information Sources on the Iron and Steel Industry* published some two years later.

CHAPTER FOUR

Stainless steels

SUSAN TUPHOLME

The term stainless steel covers a wide range of corrosion resistant iron-based alloys which contain a minimum of 11 per cent chromium. Other alloying elements such as nickel, molybdenum, niobium and titanium may also be present and they determine the structure of the steel and its subsequent properties.

Stainless steels were first discovered around 1912. Since then an increasing number of different types have been developed. They are used especially where good corrosion resistance together with high strength both at elevated and cryogenic temperatures are required. They are also suited to forming and joining by a variety of techniques. Consequently stainless steels find many applications such as process plant, car exhaust systems and in the oil and nuclear industries.

Key works of reference

Stainless Iron and Steel. 3rd edn, J. H. G. Moneypenny (Chapman and Hall, London, 1951).

Probably the classic reference in the literature on stainless steel, albeit somewhat dated now. Vol. 1, Stainless steels in industry, covers their working and fabrication, resistance to corrosion and behaviour at elevated temperatures. Vol. 2, Microstructure and constitution, particularly covers the effects on structure and properties of alloying elements.

Metals Handbook (American Society for Metals, Metal Park, Ohio).

This is in fact a series of books covering all aspects of metals; properties, inspection, fabrication, heat treatment etc. The handbooks are in the process of being revised and details of the latest volumes concerning stainless steels are given below.

8th edn, Vol. 3, 1967, 'Machining', pp. 375–403. Includes details of machining criteria for all commonly used machining techniques for stainless steels including turning, boring, drilling, milling, sawing, grinding and electrical machining.

9th edn, Vol. 14, 1988, 'Forming and forging', pp. 222–31 and 759–79. Covers all techniques for the bending and forming of stainless steels including blanking and piercing, press-brake and press forming, deep drawing, spinning, stretch forming and the forming of tubes. Gives details of forging techniques for stainless steels particularly with regard to equipment and techniques.

9th edn, Vol. 3, 1980, 'Properties and selection: stainless steels, tool materials and special purpose metals', pp. 1–124. Gives details of the properties of both wrought and cast stainless steels. It covers corrosion resistance and fabrication including welding, machining and cleaning.

9th edn, Vol. 4, 1981, 'Heat treating', pp. 621–49. Describes annealing procedures for austenitic and ferritic stainless steels. The effects of austenitizing and tempering variables on the properties of martensitic stainless steels are presented. Heat treatments for precipitation hardening stainless steels and stainless steel castings are given. Procedures for stress relieving austenitic stainless steels are listed together with the reasons for such treatment.

9th edn, Vol. 5, 1982, 'Surface cleaning, finishing and coating', pp. 549–68. Describes the various available finishes for stainless steels. Procedures and equipment for descaling by abrasives and acid pickling are given as well as grinding, buffing, etching and electroplating.

9th edn, Vol. 6, 1983, 'Welding, brazing and soldering'. Covers all aspects of welding and brazing stainless steels. Details of the different techniques are given together with the special requirements for all the classes of stainless steel.

9th edn, Vol. 7, 1984, 'Powder metallurgy'. Describes the production of stainless steel powders together with their applications.

9th edn, Vol.13, 1987, 'Corrosion'. This covers all aspects of corrosion of stainless steels such as testing, types of corrosion and materials selection and design. It also deals with corrosion in specific industries and environments.

Handbook of Stainless Steels. D. Peckner and I. M. Bernstein (eds) (McGraw-Hill, Maidenhead, 1977).

This large volume describes many aspects of both cast and wrought stainless steels. The areas covered are physical metallurgy, corrosion resistance, physical and mechanical properties at cryogenic, room and elevated temperaures, fabrication and design practice and the application of stainless steels in various industries. Two appendices cover specifications of the world's stainless steel producing countries and of proprietary alloys.

The Metallurgical Evolution of Stainless Steels. F. B. Pickering (ed.) (ASM, Ohio and Metals Society, London, 1979).

This contains a personal selection of 21 key papers given over the last twenty years on the physical metallurgy of stainless steels.

Source Book on Stainless Steels. Compiled by the Periodical Publication Department of the American Society for Metals (ASM, Ohio, 1976).

A comprehensive collection of articles covering stainless steel types and characteristics, design and cost factors, corrosion resistance and protection, forging and heat treatment, forming, welding, machining and grinding, cleaning and finishing.

Source Book on the Ferritic Stainless Steels. R. A. Lula (ed.) (ASM, Ohio, 1985).

A comprehensive collection of the outstanding articles from the periodical and reference literature which gives an up-to-date survey of the subject.

R. A. Lula has edited a number of other volumes dealing with stainless steels such as *Toughness of Ferritic Stainless Steels* (ASTM, STP 706, 1978) and *New Developments in Stainless Steel Technology* (ASM, 1985). Topics in the latter volume include fundamental research, new alloys, selection and application in advanced technological fields. Also covered are new stainless steels for applications requiring high corrosion resistance, improved wear and galling properties, better fracture toughness and high strength. A further volume by Lula is *Stainless Steel* (ASM, 1985) which is a concise treatment of the properties and selection of stainless steels. The book covers physical metallurgy as well as processing and service characteristics. The ASTM, incidentally, lists several publications dealing with stainless steel in its annual catalogue. Another volume worth noting is P. Marshall *Austenitic Stainless Steels: Microstructure and Mechanical Properties* (Elsevier, 1984).

Chromium, Nickel and Other Alloying Elements in US-produced

Stainless and Heat-resisting Steel by J. F. Papp was produced by the US Bureau of Mines in 1988 and *Nickel Alloys and High Alloy Special Stainless Steels* by Ulrich Haubner was published by Sindelfingen Expert Verlag in 1987.

The Super 12% Cr Steels. J. Z. Briggs and T. D. Parker (Climax Molybdenum Co., Ann Arbor, 1965).

An extensive report on the physical metallurgy, mechanical properties, corrosion resistance and processing and fabrication of the 12% Cr–Mo steels, including commercial grades. A comprehensive list of references is given.

Super 12% Cr Steels – An Update. J. Z. Briggs and T. D. Parker (Climax Molybdenum Co., Ann Arbor, 1982).

This brochure is not a replacement for the above, but rather an updating and collection of new data on these materials compiled from commercial sales brochures and national standards and specifications.

'The physical metallurgy of 12% Cr Steels'. K. J. Irvine, B. A. Crowe and F. B. Pickering, *Journal of the Iron and Steel Institute*. *195* (1960), 386–405.

High Strength 12% Cr Steels. K. J. Irivine and F. B. Pickering, ISI Special report no. 86 (London, 1964).

These two articles contain classic work on this family of steels.

'Physical Metallurgy of Stainless Steel Development'. F. B. Pickering, *International Metal Reviews* no. 211. *21* (1976) 227–68.

This review covers the physical metallurgy of four main types of stainless steels (martensitic, controlled transformation, ferritic and austenitic) and considers the effects of composition on the structure and properties. Quantitative relationships are summarized and an outline of welding and certain corrosion effects given. Alternative methods of increasing the strength of ferritic and austenitic steels are outlined. 240 references are reviewed.

An Introduction to Steel Selection: Part 2, Stainless Steels. D. Elliott and S. M. Tupholme, Engineering Design Guide no. 43 (OUP, 1981).

This guide describes the properties of a number of grades of wrought stainless steels which are fairly readily available commercially. Corrosion resistance, physical and mechanical properties are covered together with manufacturing requirements such as formability, joining, machinability and costs.

Structure, Constitution and General Characteristics of Wrought Ferritic Stainless Steels (American Society for the Testing of Materials STP 619, ASTM, New York, 1977).
General discussion and listing of basic properties.

A Review of Worldwide Developments in Stainless Steels. R. M. Davison, T. R. Laurin, J. D. Redmond, H. Watanabe and M. Semchyshen (Climax Molybdenum Co., Ann Arbor, 1983).
The corrosion resistance of new grades of ferritic, austenitic and duplex stainless steels is discussed. These new grades have been produced as a consequence of the availability of new refining techniques.

An Introduction to Stainless Steel. J. G. Parr and A Hanson (American Society for Metals, Ohio, 1965).
This book discusses the heat treatment, mechanical properties, fabrication and corrosion resistance of wrought and cast stainless steels.

Stainless and Heat Resisting Steels. L. Colombier and J. Hochmann (Edward Arnold, London, 1967).
This book describes the influence of various alloying elements on structure and discusses the general properties and corrosion resistance of various types of stainless steel together with corrosion resistance and mechanical properties at elevated temperatures.

Mechanical and Physical Properties of the Austenitic Chromium-nickel Stainless Steels at Ambient Temperatures. Mechanical and Physical Properties of the Austenitic Chromium-nickel Stainless Steels at Elevated Temperatures (International Nickel Ltd., 1966).
Both these publications give extensive details on the mechanical and physical properties of many standard grades both in wrought and cast form. Some data is also given for cryogenic applications.

Corrosion Resistance of the Austenitic Chromium-nickel Stainless Steels in Chemical Environments (International Nickel Co., London, 1966).
The various forms of corrosion are discussed and the susceptibility of various standard grades to corrosion in a wide range of environments is listed.

Intergranular Corrosion of Ferritic Stainless Steels. A. P. Bond and E. A. Lizlovs (Climax Molybdenum Co., Ann Arbor, 1969).
A classic paper on the susceptibility of various commercial and experimental steels to intergranular corrosion and the effect of Mo, Ni, Ti and Nb additions.

H_2S Corrosion in Oil and Gas Production. R. N. Tuttle and R. D. Kane (eds) (National Association of Corrosion Engineers, Houston, 1984).

This is a compilation of classic papers and includes much engineering data on stainless steels as well as non-ferrous metals. Guidelines are given for the selection and testing of materials.

Stainless Steel and the Chemical Industry (Committee of Stainless Steel Producers, American Iron and Steel Institute, Climax Molybdenum Co., Ann Arbor, 1966).

This book describes how to get the most from stainless steel equipment by proper design, selection of material, fabrication and maintenance.

High Performance Stainless Steels for Seawater Service. J. D. Redmond and K. H. Miska (Climax Molybdenum Co., Ann Arbor, 1983).

This describes a relatively new class of high performance austenitic and ferritic stainless steels that provide high levels of resistance to pitting and crevice corrosion.

Publications of the ASM such as:

Forming of Stainless Steels (1968);
Joining of Stainless Steels (1967);
Machining of Stainless Steels (1968);
Selection of Stainless Steels (1968)

give detailed practical advice to the fabricator and user of stainless steels.

Welding Metallurgy of Stainless and Heat Resisting Steels. R. J. Castro and J. J. de Cadenet (Cambridge University Press, 1975) discusses the applications of the different welding processes, the metallurgical effects arising during welding and the subsequent corrosion resistance. The application of weldability tests is also covered.

'Developments in Stainless Steel Production'. J. C. C. Leach. *Steel Times 210* (10) (1982), 577–82 summarizes the developments in stainless steel production particularly with respect to Argon Oxygen Decarburization (AOD) and continuous casting. It also lists principal AOD and continuous casting plants throughout the world.

Making, Shaping and Treating of Steel. H. E. McGannon (ed.) (United States Steel, 1971) gives information on the manufacture, heat treatment, fabrication, mechanical properties and corrosion resistance of stainless steels.

Metallic Materials Specification Handbook. 2nd edn, R. B. Ross (E. and F. N. Spon Ltd., London, 1972), gives specifications and trade names for many metals including stainless steels. It discusses their general characteristics, heat treatment, welding, machinability, corrosion resistance and uses.

Metals Reference Book. 6th edn, by C. J. Smithells (Butterworths, London, 1983) includes detailed mechanical and physical properties, corrosion resistance and welding of stainless steels.

Periodicals

There are constant developments in the metallurgy of stainless steels; new grades are produced for specific applications, new data is available on corrosion resistance and there are new methods of steelmaking. All these developments and others are covered by a variety of periodicals and it is essential to consult these to keep abreast of current data and state of the art technology.

Stainless Steel Industry, published bi-monthly by Modern Metals Publications Ltd., Dorking, Surrey, is one of the few journals devoted exclusively to stainless steels. Its reports range from technical papers on new grades, practical advice on fabrication and welding, and applications in particular industries to reports on exhibitions and conferences. An important item is the product profile, which lists suppliers of a particular stainless steel product form such as wire or bar. There is also a buyers guide which gives a list of manufacturers for various product forms and finished components. *Stainless*, published quarterly by British Steel Stainless, covers various applications of stainless steels and includes information on fabrication, joining and maintenance. *Stainless Steel and High Performance Alloys*, published bi-monthly by South African Stainless Steel Development Association, covers the technology and development of stainless steel, design and fabrication, marketing of products and the economics of using stainless steel. *Aciers speciaux*, published quarterly in Paris, is the main French journal on technical aspects of stainless steel. *Journal of Materials for Energy Systems*, published quarterly by the American Society for Metals, covers the structure, properties and use of materials including stainless steels in coal conversion systems, power plant and the oil and gas industries.

The following journals are concerned with fundamental aspects of corrosion and often include articles on stress corrosion cracking, intergranular corrosion and pitting of stainless steels. *Corrosion Science*, published monthly under the auspices of the Institution of Corrosion Science and Technology by Pergamon Press, Oxford; *British Corrosion Journal*, published quarterly by the Institute of Metals, London; *Corrosion*, published monthly by the National Association of Corrosion Engineers, Houston; and *Materials Performance*, published monthly by NACE, Houston. This periodical presents papers on the performance of materials including stainless steels in a variety of corrosive environments as well as giving reports on related conferences.

All aspects of stainless steelmaking are dealt with in the following two journals.

Steel Times, published monthly by Fuel and Metallurgical Journals, Redhill, now incorporates *Iron and Steel International* and covers the European iron and steel industry. There are regular features on stainless steelmaking and an annual review from the British Independent Steel Producers Association. Tube and wire manufacture are also covered.

Metal Progress, published monthly by American Society for Metals, reports on new types of materials, processing, fabrication and testing. Also, *The Materials and Processing Databook* is published annually. This gives data sheets on metals including stainless steels, heat treatment, welding and joining, and testing and inspection.

Fundamental research into stainless steels and their properties also takes place in other countries and their literature should also be consulted. Some of the important sources are listed below:

Steel in the USSR (monthly, Metals Society, London).
Stahl und Eisen (fortnightly).
Archiv für das Eisenhüttenwessen (monthly, Verein Deutscher Eisenhüttesleute and Max-Planck Institute in Düsseldorf).
Transactions of the Iron and Steel Institute of Japan (monthly, ISI Japan, Nippon Tekko, Kyokai).
Scandinavian Journal of Metallurgy (bi-monthly, Jernkontoret, Stockholm).

The addresses of the publishers of periodicals listed in this section and many other metallurgical journals may be found in *Source Journals in Metallurgy*, 3rd edn (Materials Information, a joint service of the Institute of Metals and American Society for Metals, 1986).

Abstracting services

The following services index stainless steel literature:

Metals Abstracts (Metals Information) gives abstracts of articles published worldwide. These are listed under subject headings and are cross-referenced.

Current Technology Index (Library Association Publishing, London in annual volumes from monthly compilations).
Engineering Index (H. W. Wilson).
Applied Science and Technology Index (H. W. Wilson Co., New York).

Databases and datafiles

Metadex

The Materials Information Metadex database was founded in 1966, is based on Metals Abstracts and is discussed in Chapter 1. Relevant keywords include austenitic stainless steel, ferritic stainless steel, martensitic stainless steel and various terms for corrosion.

Metals Datafile

Metals Datafile is an online source for numerical data on the mechanical and physical properties, composition and specification of ferrous and other alloys. It provides a rapid means of identifying new grades of stainless steels, locating similar alloy specifications in different countries and selecting alloys with specific properties. It is available through Metals Information.

Steel Supplements

These are published monthly by Materials Information and the information is not part of the main Metadex file. The supplements give details of all the techno-commercial advances in areas such as product and process developments, applications and competitive materials.

Stainless Steels Digest

Published monthly by Metals Information, these contain abstracts on developments in austenitic, ferritic and martensitic stainless steels.

Standard specifications and codes of practice

Typical British Standards concerned with stainless steel are indicated below:

Materials

BS 1449 Steel plate, sheet and strip
Part 2: 1983. Stainless and heat resisting plate, sheet and strip.

BS 1554: 1981 (1986) Specification for stainless steel and heat resisting steel round wire.
BS 2056: 1983 Specification for stainless steel wire for mechanical springs.
BS 5312: 1976 Stainless steel catering containers and lids.
BS 5577: 1984 Specification for table cutlery.

Other British Standards cover the use of stainless steel for fasteners, medical and domestic equipment.

ASTM standards

American Society for Testing of Materials standards are used in the USA and are sometimes also specified in other countries. The standards for stainless steel are wide-ranging and cover all the different types as well as numerous product forms. Some of the more frequently used are given below.

A 240 Heat resisting chromium/chromium-nickel stainless steel plate/sheet/strip for fusion welded unfired pressure vessels.
A 297 Steel castings of iron-chromium/iron-chromium-nickel alloy for heat resistant/general applications.
A 312 Seamless and welded austenitic stainless steel pipe.
A 479 Stainless/heat resisting steel bars.
A 511 Stainless steel seamless tubes for mechanical applications.

Other specifications, in addition to material properties for different product forms and applications, also cover testing of stainless steel particularly corrosion tests and analysis techniques.

Pressure Vessel Codes

Stainless steels are frequently used in the construction of pressure vessels. The most widely used design code for these is the ASME Boiler and Pressure Vessel Code. Stainless steel requirements are given in Section II Part A Ferrous Materials. Some examples of the more important specifications for stainless steels are given below:

SA 193 Alloy steel and stainless steel bolting materials for high temperature service.
SA 240 Heat resisting chromium and chromium-nickel stainless steel plate, sheet and strip for pressure vessels.
SA 312 Seamless and welded austenitic stainless steel pipe.
SA 351 Austenitic steel castings for high temperature service.
SA 403 Wrought austenitic stainless steel pipe fittings.
SA 479 Stainless and heat resisting bars and shapes for use in boilers and other pressure vessels.

SA 484 General requirements for stainless and heat resisting wrought steel products, except wire.

Other specifications cover the requirements for the other principal types of stainless steel as well as precipitation/age-hardening steels and for stainless steel cladding.

Other standards

The other major stainless steel producing countries have their own series of standards and their equivalent to BS designation is as follows:

West Germany	Werkstoff number
France	Afnor grade
Italy	UNI grade
USA	AISI grade
Sweden	SIS grade
Japan	JIS grade

Conferences

The most recent conferences held concerning stainless steels were:

Stainless Steels (University of York, 1988) sponsored by the Institute of Metals (1988) covering physical metallurgical development of stainless steels, passivation, welding and weldability, stainless steels as structural components in thermal nuclear reactors and stainless steels in combustion systems.

New Developments in Stainless Steel Technology (Detroit, 1988) sponsored by American Society for Metals. Placed emphasis on the new types of stainless steel now required with improved corrosion resistance, better fracture toughness, higher strength and weldability to meet the most stringent performance standards.

Stainless Steel '77 (London), sponsored by Climax Molybdenum and AMAX Nickel. Proceedings edited by Robert Q. Barr, published by Climax Molybdenum, Ann Arbor. Discussed the physical metallurgy of stainless steels, the effects of processing on the properties of stainless steels and new applications.

Alloys for the Eighties (Ann Arbor, Michigan, 1980) sponsored by and proceedings published by Climax Molybdenum. Covered four main areas; alloys for transportation, alloys for energy conversion, alloys for fuel production and distribution and alloys for process industries. Papers on stainless steels predominated in the fourth area.

Duplex Stainless Steels (St Louis, Missouri, 1982). Proceedings edited by R. A. Lula, published by American Society for Metals, Ohio. Covers all aspects of the duplex (austenitic/ferritic) stainless steels including corrosion resistance, mechanical properties, welding and applications particularly in the oil industry.

Journées des aciers speciaux (St Etienne, France, 1980) sponsored by Cercle d'Études des Métaux. Presented papers on the physical metallurgy of stainless steels, corrosion in various environments including chloride media and mechanical properties and applications.

Ferritic Steels for High Temperature Applications (Warren, Pa, 1981) proceedings published by American Society for Metals. Was concerned largely with the metallurgy of ferritic stainless steels of the 12 per cent Cr Mo V. type, particularly with regard to their use in nuclear applications.

Stainless Steel Castings (Bal Harbor, Florida, 1980). Proceedings edited by V. G. Behal and A. S. Melilli, ASTM STP 756. Describes all the developments to that date.

Developments in stainless steelmaking are also discussed in *Secondary Steelmaking* (London, Metals Society, 1977) and *Electric Furnace Conference* (St Louis, Missouri, 1976).

An annual corrosion conference is held in the UK and this deals with many aspects of the corrosion resistance of stainless steels and other materials. This is sponsored by the Institution of Corrosion Science and Technology, a joint venture with NACE. Many aspects of the stress corrosion cracking of stainless steels were covered in *Fundamentals of Stress Corrosion Cracking* (Ohio State University, 1967). The chapter on corrosion should be referred to by those interested in this aspect.

Advisory bodies, trade and research associations

British Steel has a Stainless Steel Advisory Centre which is equipped to answer telephone questions 24 hours a day. Advice is available on all aspects of steel standards and specifications (both British and foreign), material selection, fabrication and manipulation. There is also advice for buyers on price, delivery and product forms available.

Stainless Steel Advisory Centre
British Steel Stainless
Shepcote Lane
PO Box 161
Sheffield S4 7TL.
Telephone: 0742 440060/441224

The British Independent Steel Producers Association also has an enquiry service for trade and technical problems.

BISPA
5 Cromwell Road
London SW7 2HX
Telephone: 01 581 0231/5; 01 581 4224/8

The principal commercial research laboratory in the UK involved with research into stainless steel is British Steel Swinden Laboratories in Rotherham. In addition to carrying out work on behalf of British Steel the laboratory is available for contract research on all aspects of stainless steel fabrication, corrosion, mechanical properties and other metallurgical studies. In conjunction with the Stainless Steel Advisory Centre (through whom initial contact should be made) free advice is available on metallurgical problems. Details of the contract research facilities can be obtained from:

British Steel Technical
Swinden Laboratories
Moorgate
Rotherham
S. Yorks S60 3AR
Telephone: 0709 820166

Information on stainless steel castings may be obtained from:

Steel Castings Research and Trade Association (SCRATA)
5 East Bank Road
Sheffield S2 3PT
Telephone: 0742 728647

Problems on the welding of stainless steels can be referred to:

Welding Institute
Abington Hall
Abington
Cambridge CB1 6AL
Telephone: 0223 891162

They also produce a research bulletin which is circulated to members.

Commercial aspects of stainless steels

Around 50 per cent of stainless steel in the UK is sold by stockholders and enquiries for supplies ex-stock can be directed to:

National Association of Steel Stockholders (NASS)
Gateway House
High Street
Birmingham B4 7SY
Telephone: 021 632 5821

NASS is really a number of committees which work in various fields relating to steel stockholding. These committees are classified into various product groups, one of which is the Stainless Steel Product Group. This group committee liaises between producers, such as British Steel and BISPA, and stockholders to exchange information and observe trends. They also have a representative on the relevant British Standards committees.

The Stainless Steel Fabricators Association (SSFA) is particularly concerned with developing the use of stainless steels. It has around 150 members who are concerned with all aspects of stainless steel processing, production of all types of equipment from nuts and bolts to heavy fabrications, and stockholding. The SSFA Council maintains regular contact with British Steel and BISPA. The Technical Committee can advise members on problems encountered in stainless steel fabrication and represent their interests on British Standards committees. The Technical Committee and Council also arrange meetings, which are open to non-members, on all aspects of the fabrication of stainless steels. Summaries of papers given at these meetings are usually printed in *Stainless Steel Industry* (see above under Periodicals, p. 92). An annual directory is published which gives a list of members and details of their products. The directory also contains a classified list of components with a list of their suppliers and gives details of steel specifications and standards. SSFA can be contacted at:

Stainless Steel Fabricators Association of Great Britain
14 Knoll Road
Dorking
Surrey RH4 3EW
Telephone: 0306 884079

Metallurgical research to improve cutting performance and

corrosion resistance of stainless steel with respect to cutlery is carried out by:

Cutlery and Allied Trades Research Association (CATRA)
Henry Street
Sheffield
Telephone: 0742 769736

Metallurgical research into springs is carried out by:

Spring Research and Manufacturers Association
Henry Street
Sheffield
Telephone: 0742 769736

The World Metals Index at Sheffield Central Library contains over 70,000 grades of ferrous and non-ferrous materials and can provide a grade identification service for British, American and Continental materials including stainless steels. The Index can be contacted at:

World Metals Index
Sheffield Central Library
Surrey Street
Sheffield
Telephone: 0742 734744

Manufacturers handbooks

All the major producers of stainless steels issue a range of publications which give technical data on the various grades of stainless steel and assist with materials selection for particular applications.

For example British Steel Stainless have brochures and data sheets on their products such as plate, sheet and strip, on corrosion resistance, fabrication and cleaning. Brochures are also available for particular applications such as building, cladding and offshore. These may be obtained from the Stainless Steel Advisory Centre (see above).

Some of the more detailed brochures issued by some other major producers are described below. Other brochures on new grades etc. may be obtained direct from all major producers.

Technical Data, 2nd edn (TI, Chesterfield, Stainless Steel Division, 1975). This book provides information about the stainless steels used by Tube Investments, Chesterfield, Stainless Steel Division, for the manufacture of seamless and drawn welded tubes and tubular products. The manipulation, welding and machining of stainless steel tubes is also described. Characteristics such as properties, corrosion resistance, weldability and machinability to aid choice of grade is covered. There is also a section on the physical metallurgy of stainless steels.

Stainless Steel Products for Oil and Gas Production (Sandvik Ltd, Halesowen, West Midlands). Sandvik has produced a book which reviews the applications of stainless steels for oil and gas production both downhole and topside, together with case histories and performance records. Welding techniques and consumables for stainless steels are also discussed. Sandvik also produces brochures on stainless steels for the process industry and in nuclear engineering.

Working Instructions: Welding Brazing, Soldering and Riveting (Uddeholm Ltd., Crown Works, Rubery, Birmingham). Gives a survey of the welding characteristics of stainless steels. It discusses the usefulness and application of various welding methods on these materials. Uddeholm has also produced brochures on stainless steels for process plant and nuclear plant and also on stainless steel powder products.

A Handbook for the Welding of Stainless Steels is produced by Avesta, a member of the Axel Johnson Group, Aldwych House, Aldwych, London. They also produce brochures on stainless steels for forgings and for welding wire.

Product information

Information on international production and trade in stainless steel can be obtained from *World Stainless Steel Statistics* (Inco Europe Ltd. and the World Bureau of Metal Statistics). This volume covers international trade from 1973 to 1982 and gives a detailed breakdown according to product form. It is useful for those engaged in marketing stainless steel on a national or international basis.

A recent publication is the *Stainless Steel Databook* (Metal Bulletin, 1988). A publication covering similar ground is *World Trade – Stainless, High Speed and Other Alloy Steels* (Iron and Steel Statistics Bureau, Croydon), a quarterly publication dealing with the export trade of the leading steel producing countries.

Iron and Steel Works of the World, 9th edn (Metal Bulletin Plc, 16 Lower Marsh, London, 1987) is a worldwide buyers guide to all product forms of stainless and other steels. It also lists the steel-making and rolling equipment at every steel plant in the world, so it can be seen where the most modern refining techniques are in use; and *Stainless Steel Directory 1987–8* (Modern Metals publications, Dorking, Surrey) has a section listing suppliers of wrought and cast products (such as plate, bar, wire and sand castings). The manufactured products section contains over 800 headings of items made wholly or partially from stainless steel, from pressure vessels to tea pots, and gives their suppliers. Other sections cover processing services, alloy and scrap suppliers, trade names and locations of stockholders.

Most stockholders publish lists of their product range often with some technical data. Two examples are given below. Other lists can be obtained direct from the stockholders.

Stainless Steel Products Manual (RGB Stainless Ltd., Smethwick, West Midlands) is an extensive manual with sections on pipe, flanges, sheet and plate, with details on specifications, dimensions and weight. Information on properties of stainless steels, corrosion resistance and welding procedures is also included; and *Guide to Stainless* (GKN Steelstock Ltd., Kidderminster) lists the company's stock range together with technical data, corrosion data, a fabrication guide and summaries of British Standards.

Stainless Steel, an International Survey and Directory, M. Nurse and D. Hargreaves, (eds). (Metal Bulletin plc, 1985), looks at stainless steel production throughout the world and lists producers. It also outlines usage of stainless steels and has a product-by-product directory.

CHAPTER FIVE

Aluminium

D. KEEVIL

Introduction

Aluminium is the most abundant metallic element. It is estimated that it forms about 8 per cent of the Earth's crust and is exceeded in quantity only by oxygen and silicon. Despite this natural abundance the industry, as we understand it today, is barely one hundred years old. Inherent difficulties in isolating the metal from the naturally formed alumina-bearing ores and rocks inhibited all but the most tentative early exploration. In 1886 two young men, in different continents and working quite independently, discovered the basic electrolytic reduction method for aluminium. They were Charles Martin Hall of the USA and Paul Heroult of France. The process discovered by these two workers is still the most important means of aluminium production.

Once a reliable method of large-scale production had been established, attractive properties of the metal and its alloys ensured a dynamic future of exploitation and development. Aluminium alloys have a high strength-to-weight ratio, they resist corrosion to a high degree and are conductive, reflective and nontoxic. Alloys can be worked by every form of metalworking into almost every required shape. Objects made from aluminium can be easily recycled at the end of their useful life.

The rapid development of the aluminium industry has brought in its wake a concomitant development in recorded information. A requirement for information can quite easily range from a simple verification of an alloy composition to a detailed study of some specific topic.

The sources are varied, and it is the intention of this chapter to provide a route which will, it is hoped, allow the most relevant source to be identified and therefore used. Where there is a wide choice, emphasis is placed on UK sources.

Abstracting services

It is fairly obvious that information needs can range from the straightforward verification of a simple technical fact through to comprehensive surveys of the literature. It is, however, reasonable to suppose that most enquirers will, at some time or other, need a source which attempts to survey most of the information for a given subject area. An appropriate example of such a source is the typical abstract service. As is generally appreciated, the object of such a service is to provide a means for examining the primary literature (books, articles, papers from conference proceedings etc.) by means of summaries or abstracts of the original sources with full document citations. The abstracts are arranged in a convenient order, together with subject and author indexes. Publication mode is normally in printed format but in many instances there is also online access facilities.

Although the large, general abstracting service such as *Engineering Index*, *Chemical Abstracts*, or *Metals Abstracts* should by no means be ignored, the aluminium industry has been particularly fortunate in having several excellent sources produced for its own purposes over the years.

First in the field were the abstract journals produced by individual companies as a response to their own corporate needs. Although excellent publications, their use is limited by their scarcity in publicly accessible collections. Perhaps the best example of the genre is the *Abstract Bulletin of Aluminium Laboratories Limited* which was produced between 1929 and 1959.

The early literature of aluminium is covered quite adequately by a number of publications which allow easier public access than the *Abstract Bulletin*. *The ASM Review of Metal Literature* (1944–67), The Institute of Metals' *Metallurgical Abstracts* (1909–67 in two series, 1934 being the first year of the second series), and the British Non-Ferrous Metals Research Association's *BNF Bulletin* (1921–83) are the most important publications.

In 1962 the Centre International de Développement de L'Aluminium (CIDA) was set up with the general aim of assisting in the development of new uses, and the expansion of existing uses of aluminium. It was perceived that one way to effect these aims would be to rationalize the rather disjointed system of company abstract services typified by the *Abstract Bulletin of Aluminium Laboratories Limited*.

The rationalization produced *Aluminium Abstracts* published twice a month from 1963 until 1970, when CIDA was closed down. In addition to standard articles and papers, a considerable number of short unsigned articles and marketing information was included. Unfortunately the publication suffered from a rather crude subject index which makes comprehensive and accurate searching a difficult task.

In 1968, the Aluminum Association (USA) in conjunction with the American Society for Metals launched the *Aluminum Technical Information Service Journal* (ATIS). The launch date more or less coincided with the period of time when the CIDA was being considered for closure. After a period of discussion the two abstract services were merged in 1970 to form the *World Aluminum Abstracts* (WAA). To this present day WAA remains the premier source for information on aluminium. As with the ATIS *Abstract Journal*, the physical preparation, editorial work and printing of the abstracts is done by the American Society for Metals. Policy and quality control is effected by two working panels representing American and European interests respectively. The WAA is published as a monthly abstract journal featuring computer-generated monthly and annual subject indexes. There is a separately produced annual bound volume. The American Society for Metals also provides magnetic tapes of the database for in-house computer searching and in line with many other abstract services, the database is also available for interactive online searching with the following information systems suppliers:

DIALOG Information Services, California (accessed in the UK through DIALOG, PO Box 8, Abingdon, Oxford).

Space Documentation Service, European Space Research Association, Frascati, Italy (accessed in the UK through DIALTECH, Department of Trade and Industry).

It is a common failing of many abstract services that the documents they abstract can prove difficult to obtain in their full version, much to the frustration of the enquirer! Although the WAA cannot claim to be perfect in this respect, great efforts are made by the policy control panels to ensure that as many originals as possible are available from one or the other of the photocopying services linked with the Abstracts. Currently these services are the Aluminium Federation, the Institute of Metals and the American Society of Metals.

The source coverage of the WAA is wide. Abstracts are prepared by appropriately qualified people from more than 1,200 current scientific and technical journals. Also covered are patents, government and agency reports, conference proceedings, theses, books, pamphlets etc. Each abstract is provided with a complete bibliographic citation and a unique serial number.

It was intended from the onset that WAA be as complete a guide as possible to the information produced by the industry. It is useful to look at the categories covered:

Section number	*Title*
1	General
2	Ores, alumina production, extraction
3	Melting, casting, foundry
4	Metalworking, fabrication, finishing
5	Physical and mechanical metallurgy
6	Engineering properties and tests
7	Quality control and tests
8	End-uses

Journals

A vital and dynamic press is an essential part of the development of any industry. Enquirers and researchers need sources that provide them with news of current events, technical and commercial progress. The aluminium industry has been singularly well-blessed in the quality of the journals that have served its interests over the last half-century although recent economic recession has meant that a number of important titles have disappeared.

The premier journal in the field is without doubt *Aluminium*[1] (Aluminium Verlag, Düsseldorf, West Germany, 1918–). It publishes technical and economic papers of the very highest quality, some in English. The contributions are drawn from worldwide sources. Typically each issue includes excellent short pieces on the latest machinery and equipment, new products, detailed statistics, good book reviews, and conference notices. Until recently, there were English cover-to-cover translations available at an additional charge, but this service was considered to be uneconomic and has ceased. It has a very good annual index.

In terms of seniority, the French journal *Revue de l'Aluminium* is next in line (1929–83). Although very much linked to the Pechiney aluminium company, it maintained an independent editorial voice and featured key papers in addition to a well-balanced news and commercial section. *Alluminio* (1932–84) was associated with a variety of publishers in its native Italy. Always beautifully produced, with first class artwork and a strong bias towards articles that featured new uses of aluminium, it ceased publication in 1984.

The United States of America hosts two journals which deal closely with the industry. *Modern Metals*[2] is a monthly publication with a heavy commercial flavour aimed at the 'sharp end' of the industry, i.e. the distributive trade, stockholders and actual users of aluminium. Thus its articles feature market surveys and evaluations, commercial trends, new uses etc. It is a very parochial publication and scarcely ever looks beyond the US boundaries for its material. *Light Metal Age*[3] is published six times a year, a fact which makes its news sections rather redundant; but it is one of the best international sources for descriptions of new plant layouts and industrial development.

At one time the United Kingdom was the home of several important journals especially in the period following the Second World War when the domestic industry was expanding. Titles such as *Light Metals* and *Metals Industry* were excellent independent channels of communication. Their demise left only the light metals section of *Metal Bulletin*[4] and *Metal Bulletin Monthly* to cater for aluminium. Although a major source of commercial and statistical information, the *Metal Bulletins* have never claimed to cover technical information very extensively and it was left to a new venture *Aluminium Industry*[5] (1982–) to redress the balance. This latter publication has made a good start and its authority grows with each issue (currently ten per annum).

Among the remaining developed nations the USSR and Japan produce important titles, but their use is restricted by language barriers.

Handbooks

An enquirer often needs to be able to locate quickly various items of standard specific information such as numerical data, mechanical and physical properties. Published handbooks aim to provide such information in a convenient format but collections of data are obviously expensive to produce and keep updated and economics rule that much of the information needed on aluminium is in fact provided in sources of a more general nature.

Smithell's Metals Reference Book[6] edited by E. A. Brandes, 6th edn (1982) is a major publication. Specific information on aluminium is tapped by use of subject index or by direct use of individual chapters. The American Society for Metals produces its comprehensive *Metals Handbook*, a constantly revised and frequently published series of volumes, described in Chapter 1. The most important volume for aluminium information is *Vol. 2: Properties and Selection: Non-ferrous Alloys and Pure Metals*.

The ASM were also involved in the production of one of the best collections of data devoted entirely to aluminium. This was the 3 volume set *Aluminum*[7] edited by Kent R. Van Horn (ASM, 1967), prepared by the staff of the Aluminum Company of America and the publication provides an encyclopaedic guide to aluminium. The three volumes cover respectively: properties, physical metallurgy and phase diagrams, design and application and fabrication and finishing. *Aluminum* was for many years a key source of data verification, but its value decreased with time and the Aluminum Association in the USA undertook the sponsorship of a revised edition. Consequently *Aluminum: Properties and Physical Metallurgy*[8] edited by J. E. Hatch was published by ASM in 1984 as the first part of the new edition. The Aluminum Association's contacts with its member companies has ensured a valuable collection of data being made available to a wider audience.

Probably the best single-volume collection of essential data is the *Aluminum Taschenbuch*[9] now in its 14th edition, 1984. The literal translation of the title 'Aluminium Pocketbook' is something of a misnomer as its 1,094 pages are packed with technical data. It is published by Aluminium Verlag in Düsseldorf. Currently under the editorship of W. Hufnagel, it provides a distillation of the expertise of the staff of the Aluminium Zentrale, also in Düsseldorf, and the German aluminium industry.

Standards

Standards, both domestic and foreign, are the lifeblood of the manufacturing engineer. The present-day level of sophistication in the various national standards organizations (particularly the British Standards Institution) in the way they propagate and disseminate their information makes superfluous any attempt to provide even a thumbnail sketch of the material available.

Assistance in the interpretation and understanding of the standards is, however, another matter and help of this nature can be found by recourse to the organizations described elsewhere.

Standards and specifications can cause great problems in the identification of different grades of materials and alloys. A large body of useful literature has grown to meet this demand and is described below.

Alloys, grades and materials – data sources

There is an ever present need to be able to identify accurately the range of alloy grades and proprietary names in everyday use.

Woldman's Engineering Alloys[10] edited by Robert C. Gibbons (American Society of Metals, 1979) is a most useful general source and is discussed in Chapter 1. Another general collection of data that is extremely useful is the *Metallic Materials Specification Handbook*[11] by Robert B. Ross and now in its 3rd edition (1979) – again discussed in the introductory chapter. In the field of alloy nomenclature, the aluminium industry can lay claim to be an innovator in the rationalization of systems on an international basis. In 1954 the Aluminum Association of the USA launched a logical numerical system to describe the alloys offered by the US industry. Each registered alloy was described by a four-digit number, which was a meaningful shorthand description of the main alloying constituents. The alloy designation was married to further alpha-numerical suffixes describing the particular condition or heat-treatment of the alloy. So successful was the system that the major producing countries decided to adopt it as a basis, firstly for an international system, and secondly for their own national systems. The international designation system is now well-established and is a great boon for the international specifier. Its use means that an alloy (e.g. 6061) is produced and described as such in the USA, UK, France, Germany, Japan, Australia etc., with obvious advantages in cross-frontier understanding between companies and individuals. The system is still monitored and controlled by the Aluminum Association and regular lists are published under the title *Registration Record of International Alloy Designations and Chemical Composition Limits for Wrought Aluminum and Aluminum Alloys*.

All the major aluminium producing countries are hosts to representative organizations which produce alloy guides for their own countries. These guides are among the most useful and readily accessible sources and with the English-speaking user in mind the three most important are listed below.

Aluminum Standards and Data (The Aluminum Association, 1984).
The Properties of Aluminium and its Alloys (The Aluminium Federation, 8th edn, 1983).
Key to Aluminium Alloys Designations, Composition, Trade Names of Aluminium Materials. English edn by W. Hufnagel (Aluminium Zentrale, July 1982).

The most ambitious undertaking in this field is *The Handbook of International Alloy Compositions and Designations Vol. 3, Aluminium*.[12] This monumental work provides an alpha-numeric listing of over 9,500 alloys and 3,600 trade names. There is also a sequence of alloys listed alpha-numerically by country of origin (34 countries) and several useful appendixes containing addresses of manufacturers and standard organizations. The collection of data was originally the work of one man – W. F. Kehler – although the

Aluminum Association added to, and completed the files before jointly publishing the tables with Battelle's Metals and Ceramics Information Centre as part of that organization's International Alloys Handbook Series. The publication has also been processed for the American Society for Metals on *MDF/1 Metals Data File* which is a commercially available numeric online database. The Handbook was first published in 1980 and it was intended that frequent revisions would keep the publication fully up-to-date. This has not taken place in the printed version, but some updating has taken place on the online file.

Conferences

Volumes of proceedings of conferences are increasing in value as primary sources of information in a specific subject field. Conferences are usually organized around themes of current professional interest and, outside the patent literature, they often offer the earliest publication of research and development effort.

For a first-use information source the most valuable conference proceedings are those that have achieved serial status. Their status is self-evident and they are most valuable statements of current state-of-the art. The most important conference series in the aluminium industry is without doubt *Light Metals* which has been held annually in the USA since 1962. The sessions are sponsored by the Light Metals Committee of the Metallurgical Society of AIME (American Institute of Mining, Metallurgical and Petroleum Engineers Inc.). Collectively, the annual papers present information on the extractive metallurgy of aluminium, alumina and bauxite and aluminium casting. The papers are well produced and edited and publication normally follows in the same calendar year of the conference. Typically, the 1985 volume contained 110 papers presented at a total of 23 graded sessions and the authors represented 20 different countries.

Aluminium Technology '86 are the proceedings of the international conference sponsored by the Institute of Metals and Associazione Italiana di Metallurgia, held in London in 1986, edited by T. Sheppard and published by the Institute. Another prestigious series is the *International Light Metals Congress* held about every seven years in Austria as a tribute to Karl Bayer who invented the alumina process bearing his name in 1888. The range of the conferences covers the whole spectrum of aluminium metallurgy and applications: the 7th Congress was held in 1981 and the next is scheduled for 1988 to coincide with Bayer's centenary.

The major aluminium industry associations are very much involved in the organization of useful conferences on behalf of their members. They are usually called to meet a demand for information on some application of aluminium and seek to bring together previously scattered information and experience or, alternatively, to communicate the activities of a particularly lively sector of the industry. Two associations are particularly active and as their publications are those most readily available to a European audience they should receive most attention. The Aluminium Zentrale have organized and published several important conference proceedings in recent years. In response to an upsurge of interest in the 1970s, they held two symposiums on the use of aluminium in automobiles. *Aluminium and Automobil: International Conference, Düsseldorf 1976* and *Aluminium and Automobil, Düsseldorf, 1980*. The Aluminium Zentrale also published in 1983 the proceeding of their conference on *Aluminium Boats*.

The Aluminum Association has been involved in many extremely important conference ventures. Perhaps the most important is the series reflecting advances in extrusion technology. The third conference in this series was the most ambitious to date: *ET84 Extrusion Productivity through Automation: Proceedings of the Third International Aluminium Extrusion Technology Seminar*, 2 vols (1984) presented 124 papers on various aspects of extrusion. The Aluminum Association has adopted a similar stance with finishing of aluminium and has published the results of three conferences in this area. The last was held in 1982 and presents papers on anodizing, organic coatings, coil coating, painting etc. Another annual series of interest are the papers presented at the Association's *Aluminum Industry Energy Conservation Workshops*. The eighth volume was published in 1984 and deals mainly with advances in furnaces and melting technology. The Association is also notable for recognizing areas of impending importance for the industry and for arranging for special conferences to deal with possible problems arising from new technical advances or legislative changes. A recent example of such a meeting is the *Shape Control Workshop Papers, 1984* dealing with new or emerging technology in rolling mill quality control procedures.

Not all conferences are confined to the scientific and technical advances in the industry. The needs of the economist and businessman are met by Metal Bulletin Publications Ltd. in their series of congresses held approximately every two years. Papers are presented by a wide range of speakers from the business and financial community and the series is being increasingly used as a platform for the launching of innovative economic strategies. Metals Bulletin's *Third International Aluminium Congress* was held in Germany in 1984.

Standard texts

Despite the existence of the sources indicated above it is necessary for an enquirer to have recourse to convenient standard textbooks or compilations. The aluminium literature is quite rich and varied, with some aspects better covered than others. Examined below are some well-defined areas of the industry which experience has shown to be of the most interest.

Economics

In 1978 aluminium was accepted as a trading commodity on the London Metal Exchange. This action created a demand for economic information on the industry from a wide cross-section of the financial community in search of investment data. There has been a proliferation of literature aimed at satisfying this demand, but unfortunately much of it has been of an ephemeral nature. Currently there are signs that some companies are now beginning to produce documents that are carefully constructed and offering accurate information and analysis.

One criterion that the discerning enquirer should look for is some evidence of longevity and continuous production. There are two major sources that meet this standard. *The Economics of Aluminium*, 3rd edn (1988),[13] compiled and published by Roskill Information Services of London, is an excellent compilation of statistical and economic data. The industry is examined country by country and company by company. There are invaluable tables, and a special feature is the excellent collection of end-use statistics which are normally rather difficult to come by. *The World Aluminium Industry* (1981) is an impressive collection of data published by Australian Mineral Economics Pty Ltd. and made available in Europe through Aluminium – Verlag GmbH. Volume 1 covers the supply side of the industry in great depth – bauxite, alumina, primary and secondary aluminium production, world capacity and planned expansions. Volume 2 covers past, present and future supply/demand balances. Volume 3 is essentially a collection of tabular data, updating Volumes 1 and 2.

Metallurgy

There are several good books available dealing with the metallurgy of aluminium. For those who require an introductory level there are two excellent modern publications in the English language. I. J. Polmear has written *Light Alloys: Metallurgy of the Light Metals*[14] published by Edward Arnold as part of a series of student texts in metallurgy and materials science. More recent is Frank King's *Aluminium and Its Alloys* (Ellis Horwood, 1987) and *Production of Aluminium and Alumina*, edited by A. R. Burkin (Society of Chemical Industry, Wiley, 1987). In English translation there is *Aluminium Viewed from Within: An Introduction into the Metallurgy of Aluminium Fabrication*[15] by D. Altenpohl. This is a well-illustrated book which seeks to provide the practical metal user with an insight into the theoretical metallurgy. A rather more substantial statement can be found in *The Metallurgy of Aluminium* by Dr Marc Van Lancker. Although almost twenty years old it is regarded by many as the standard work. *Aluminium Alloys: Structure and Properties* by L. F. Mondolfo is a monumental source presenting comprehensive data on the metallurgy and properties of aluminium and its alloys. Some two-thirds of the book are devoted to the elaboration of equilibrium diagrams and alloy systems and there are extensive lists of references.

Surface finishing and anodizing

Aluminium naturally forms an oxide coating on exposure to the atmosphere and this natural coating gives aluminium many of its attractive anti-corrosion properties. The anodizing process artificially enhances the natural oxide coating and additionally much use is made of painting, electro-plating, powder coating, chemical conversion coatings etc. There is a great requirement for information on the surface finishing of aluminium and consequently the sub-literature reflects that demand. The key text is the *Surface Treatment and Finishing of Aluminium and Its alloys*[18] in two volumes by S. Wernick and R. Pinner. This eminently practical reference is crammed full of information of great value to the practical metal finisher. All the main processes are described: mechanical finishing, electrolytic and chemical processes, anodizing (both decorative and protective) colour anodizing, tests and testing of coatings, electrodeposition etc. Proprietary processes are fully described and documented and a most valuable feature is the listing of various standard formulations for the different processes. It is a unique work and the appearance of the fifth edition, which is under active preparation, is eagerly awaited.

Of more modest size and presentation are two books which are nevertheless of great value. *The Technology of Anodizing Aluminium*[19] by Arthur W. Brace and Peter G. Sheasby and *Anodic Oxidation of Aluminium and Its Alloys*[20] by V. F. Henley. Both sources are highly practical and useful descriptions of the anodizing process, with Henley's book perhaps being the better introductory text.

Metalworking processes

The various practices concerning the working and shaping of aluminium are unfortunately not well represented in book form. In former times the best books on such processes as rolling, extrusion, casting, welding etc. were provided by the large aluminium companies, especially in the heyday of the development phase of the industry in the 1950s and 1960s. During this period companies such as Kaiser, Reynolds, Alcan, and Alcoa produced excellent guides far above the normal level associated with company product literature. Realignment of company resources and priorities towards the end of the 1970s were probably instrumental in the decline of company publishing. Most of the up-to-date information is now contained in the periodical literature and the various handbooks listed above. However, an excellent general overview of the metal working processes can be found in a compilation published by the American Society for Metals *Source Book on Selection and Fabrication of Aluminium Alloys*.[21] This excellent publication draws its materials from a number of previously published sources.

Building and structural engineering

As a field of application aluminium in building and structural engineering accounts for the largest annual tonnage use of aluminium. The situation for books is somewhat similar to that indicated for the metalworking processes, i.e. good material originally published by the larger companies, but, eventually discontinued with increasing economic pressure and replaced by a plethora of catalogues and manufacturers' technical literature. There are still some good general sources, however, especially, in so far as stress/strain data, good building practice and formulae are concerned.

The Aluminum Association has published the *Construction Manual Series*, which taken collectively, forms an excellent reference. Alcan Canada Products produces an almost lone survivor from the traditional series of fine company texts in its *Strength of Aluminum* now in its 5th edition. Although now rather old, profitable study can be made in *Aluminium in Building*[22] by E. I. Brimelow. This book still forms an excellent introduction. The most recent textbook is *Aluminium Alloys Structures* by F. M. Mazzoloni, resulting from ten years' practical research arising from the author's chairmanship of the European Convention for Constructional Steelwork Committee on Aluminium Alloy Structures. Readers interested in the use of aluminium in the construction industry should also refer to Chapter 17 on construction.

Aluminium in human health

The use of aluminium alloys for food packaging reaches back into the last century. Since those very early days there has existed a lobby of opinion opposed to the use of aluminium with foods although the industry has always responded with documentation supporting the continued use of aluminium as a perfectly safe product. In the last fifteen years or so the health position has become much more complex, with additional problems other than food contact being added to the equation e.g. the role of aluminium in dialysis patients, aluminium smelter air pollution etc. Of all the areas examined this is the one that least lends itself to a 'do-it-yourself' approach, but there are some sources which can be quoted as reasonably straightforward and authoritative.

Aluminium: Its Application in the Chemical and Food Industries[24] by P. Juniere and M. Sigwalt is available in English translation and is a well-organized, well-indexed source. The Aluminum Association is responsible for *Guidelines for the Use of Aluminum with Food and Chemicals* and this booklet on compatibility data is frequently revised (5th edn, 1984) and is the most readily available source. The Aluminum Association is also responsible for funding surveys of published literature on the human health question. Their findings are available and have been published in two main series of documents. *Aluminum in the Environment and Human Health*[25] by J. R. J. Sorenson *et al.* (1974) and *Literature Review (for the Period 1973–1980) Health Effects of Aluminum Compounds in Mammals*[26] by Geraldine Krueger and co-workers. It is intended that the latter series be kept revised and up-to-date, probably on an annual basis.

The Associations

Despite the existence of the numerous sources mentioned above and the presumed ability of the enquirer to extract the information he or she needs from those sources, there are many occasions when more personal contact is needed. One answer is to take a problem to an appropriate trade or industry organization. Almost all of the aluminium-producing countries have established such organizations and although their roles may differ in some details, and taking into consideration the different financial provisions available, the enquirer will find abundant help from these bodies. It is not practical to provide a description of all the associations, but the following selection can be considered those most accessible in the UK.

The Aluminium Federation (ALFED)

Alfed is the major UK trade association for the domestic industry. It was formed in 1962 by the amalgamation of the Aluminium Development Association and the Aluminium Industry Council. Its roles include the representation of the industry to government both at home and abroad. Additionally, the Alfed collects and publishes statistics and provides an information service through its executive staff and the library. The Alfed is in fact a federation of specialist, product-orientated associations, and from the headquarters in Birmingham there operates the Aluminium Extruders Association, the Aluminium Coatings Association, the Aluminium Powder and Paste Association, the Aluminium Rolled Products Association, the Aluminium Primary Producers Association and the British Anodising Association. The affairs of the Federation and its member associations are effected through a network of council and committees. The Alfed publishes a wide variety of handbooks, pamphlets, leaflets and wallcharts, many of which are free of charge. A list of publications is available. The various executives servicing the different associations are also available to answer questions of a technical nature. This is an invaluable resource for the solving of all manner of practical problems that can arise. Contact by telephone, letter or telex is welcomed.

The Library was originally founded in 1940 with the setting up of the Wrought Light Alloy Development Association in Birmingham. After the Second World War the organization changed its name to the Aluminium Development Association and moved to London, where the Library underwent a period of rapid expansion, the objective being the creation of the most comprehensive collection of information possible on aluminium and its alloys. The policy has always been to make the collection available to all those who have a genuine interest in the metal and its uses. The Library is probably the largest collection in the world that is freely accessible to all, whether they be member companies, their customers, government department, educational institutions or the general public.

Some data on the stock may be of interest. There are over 40,000 separate documents including over 2000 books, 5000 translations of foreign papers, standards and specifications, 15,000 examples of trade literature, technical reports from various organizations and some 200 current periodicals. Special indexes maintained by the Library include the Products and Services Index a unique and detailed guide to companies and their activities and the Alloys and Trade Names Index which is an index designed to supplement the information available in published handbooks and therefore concentrates on non-standard grades. There are about 7000 entries in this index.

All the services normally associated with a library are available. Visitors are welcome but should make a prior appointment through Alfed.

Aluminium Federation
Broadway House
Calthorpe Road
Birmingham
B15 1TN
Telephone: 021 456 1103
Telex 333349 ALFED G

The Aluminum Association (AA)

This was founded in 1933 and represents the US aluminium industry as a whole. It is a prime source for information on aluminium and the aluminium industry and its membership comprises virtually all US domestic primary producers and the majors among the other producing sectors. Like the Alfed in the UK, the Association provides a forum for its members to exchange experiences and talk to government. Its objectives include the provision of information, the development of voluntary standards and technical data and the collecting and publishing of statistical information.

The AA maintains a small library, but its primary responsibility is to the staff of the Association and enquirers should contact the Association at the address below. There have been many references to the major publishing ventures of the Association throughout this chapter, but there are many, many more documents available on request. A list of publications is available. Provision policy varies from time to time, but currently an effort is being made to recover some of the cost of the publishing programme by making a charge.

The Aluminum Association
900, 19th Street NW
Washington DC
20006 USA
Telephone: (202) 862 5100

Aluminium Zentrale (AZ)

The Zentrale was established in Berlin before the Second World War and, after the war, it re-established offices in Düsseldorf, West Germany. The aims of AZ are broadly similar to the Alfed and the AA, but there is perhaps more emphasis on promotional work and technical information. A wide range of useful documents is published. A well-appointed library is maintained, but as with the Aluminum Association its primary purpose is service to its own members. Financially separate from the Zentrale, but housed at the same address is Aluminium-Verlag, which can be described as the publishing arm of the AZ. Aluminium-Verlag is the only publisher in the world specializing in the aluminium literature and as can be seen from this paper they publish many specialist texts, brochures, monographs and journals in both German and English. An excellent list of publications is available on request from:

Aluminium Zentrale Ev
PO Box 1207
Konigsallee 30
D-4000 Düsseldorf 1
FDR
Telephone: (1211) 320821

The Japan Light Metal Association (JLMA)

The Japanese have developed a highly proficient and technically advanced aluminium industry. While there is an impressive body of literature available to the domestic user, language difficulties inhibit the use of much of that literature in the West, although important journals such as the *Journal of Japan Institute of Light Metals* and *Sumitomo Light Metals Technical Review* are available with English abstracts preceding the main papers. Numerical data are nearly always presented in English and there is much sponsorship of translated journals or documents. The JLMA publishes a useful *Japan Aluminium News* as an English language contact journal and as an association its work can closely be compared with that of Aluminium Zentrale.

The JLMA publishes many technical reports and surveys. The Association co-operates with the editorial staff of the *World Aluminum Abstracts* and is responsible for supplying most of the references to Japanese publications that appear in that abstract service. Its address is:

Japan Light Metal Association
Nihonbashi Asahiseimei Building
113 Nihonbashi 2-chome
Chuo-Ku
Tokyo 103
Japan
Telephone: (03) 273 3041

References

1 *Aluminium* Aluminium-Verlag GmbH, Postfach 1207, D4000 Düsseldorf 1, FDR.
2 *Modern Metals* Modern Metals Publishing Co, 211 E. Chicago Ave, Chicago, Illinois, 60611 USA.
3 *Light Metal Age* 693 Mission Street, San Francisco, California, 94105 USA.
4 *Metal Bulletin* Metal Bulletin Journals Ltd, 16 Lower March, London SE1 7RJ.
5 *Aluminium Industry* Light Metals Publications, 60 Arkwright Road, Sanderstead, Surrey, CR2 0LL.
6 Brandes, E. A. (ed.) (1982) *Smithell's Metals Reference Book* 6th edn (London, Butterworths).
7 Van Horn, K. R. (ed.) (1967) *Aluminum* (Ohio, American Society for Metals).
8 Hatch, J. E. (ed.) (1984) *Aluminum: Properties and Physical Metallurgy* (Ohio, American Society for Metals).
9 Hufnagel, W. (ed.) (1984) *Aluminium-Taschenbuch* 14th edn (Düsseldorf, Aluminium-Verlag).
10 Gibbons, R. C. (ed.) (1979) *Woldmans Engineering Alloys* 6th edn (Ohio, American Society for Metals).
11 Ross, R. B. (1980) *Metallic Materials Specification Handbook* 3rd edn (London, Spon).
12 Kehler, W. F. (1980) *Handbook of International Alloy Compositions and Designations* Vol. 3 *Aluminum* (Washington, Aluminium Association and Ohio, Battelle Columbia Laboratories).
13 Roskill Information Services (1988) *The Economics of Aluminium* 3rd edn (London).
14 Polmear, I. J. (1981) *Light Alloys: Metallurgy of the Light Metals* (London, Edward Arnold).
15 Altenpohl, D. (1982) *Aluminium Viewed From Within: An Introduction into the Metallurgy of Aluminium Fabrication.*
16 Van Lancker, M. (1967) *The Metallurgy of Aluminium* (London, Chapman & Hall).
17 Mondolfo, L. F. (1976) *Aluminium Alloys: Structure and Properties* (London, Butterworths).
18 Wernick, S., Pinner, R. and Sheasby, P. G. (1987) *The Surface Treatment and Finishing of Aluminium and Its Alloys* 5th edn (Finishing Publis.).
19 Brace, A. W. and Sheasby, P. G. (1979) *The Technology of Anodizing Aluminium* (Stonehouse, Technicopy).
20 Henley, V. F. (1982) *Anodic Oxidation of Aluminium and Its Alloys* (Oxford, Pergamon).
21 American Society for Metals (1978) *Source Book on Selection and Fabrication of Aluminum Alloys.*
22 Brimelow, E. J. (1957) *Aluminium in Building* (London, MacDonald).
23 Mazzoloni, F. M. (1985) *Aluminium Alloy Structures* (London, Pitman).
24 Juniere, P. and Sigwalt, M. (1964) *Aluminium: Its Applications in the Chemical and Food Industries* (London, Crosby Lockwood).
25 Sorenson, J. R. J. (1974) 'Aluminum in the environment and human health' Washington Aluminum Association and in *Environmental Health Perspectives* **8** 3–95.
26 Krueger, G. L., Morris, T. K. and Widner, E. M. (1982) *Literature Review (for the period 1973–1980) Health Effects of Aluminum Compounds in Mammals* (Ohio, University of Cincinnati).

CHAPTER SIX

Copper and copper alloys

V. A. CALLCUT

Introduction

Copper and its alloys have been at the service of technology since its beginning about 6,000 years ago. From early man using the metal to make crude tools and weapons to today's technologists in a computer-assisted society, copper and its alloys are playing a vital role in our development. Although there have been good results from archaeological work there is a shortage of original records relating to the earliest metallurgical techniques. During the last two centuries this position has been rectified and the modern technology of the non-ferrous metals has been well documented.

Copper and its alloys have kept pace with the demands of users. Copper sheets were available when required as a worm- and biofouling-resistant sheathing for the ships that, during the Napoleonic wars, were required to spend months at sea on blockade duty and still be speedy enough to win battles. Stronger copper was available when needed for the boilers and tubing for the early steam engines. Copper wire of good conductivity was available when Faraday needed it and the strength, corrosion resistance and ease shaping and of machining of brass has been appreciated by scientists and engineers since accurate components and instruments were first required.

This review concentrates on the present situation and includes an extensive list of further possible contacts.

Historical

The word 'copper' is derived from the Latin 'cyprium aes', meaning Cyprian metal or metal of Cyprus, from where the

Romans obtained their main supplies. It is not clear whether the metal was in fact named after the ancient Greek word for the island or vice-versa. The fact that the goddess of love, Venus, was said to have come from Cyprus explains why she adopted the copper 'ankh' ♀ sign as her own.

While originally copper may have been found and used from its native form it has been shown that the technique of smelting it from its ores was developed very soon afterwards. The benefits of additions to copper were also developed during the infancy of metallurgy, sometimes the lessons of experience being hard. It was, for example, found that both arsenic- and phosphorous-containing additions would harden the copper but the early school of metallurgists who chose to use arsenic soon died out. Tin was the first main alloying addition and the extra strength and corrosion resistance of the bronzes continue to be valued.

The most useful description of medieval metallurgy is *De Re Metallica* originally published in 1556 but now widely available in the form of the Hoover translation of 1912. The descriptions of many of the processes are excellent but it is the opening words that sum up the technologists enthusiasm:

> Many persons hold the opinion that the metal industries are fortuitous and that the occupation is one of sordid toil, and altogether a kind of business requiring not so much skill as labour. But as for myself, when I reflect upon its special points one by one, it appears to be far otherwise.

The wealth of literature published since that time confirms his point. Most of the standard works on copper include references to its early history and many of the papers detailing the industrial archaeology of copper and its alloys can be traced through the enthusiasm of the Historical Metallurgical Society.

It is unfortunate that one of the wonders of the world, the bronze Colossus of Rhodes, was eventually used only to demonstrate the ease with which copper can be recycled but many other artefacts have survived to prove the strength, corrosion resistance, durability and versatility of the metal. Some are well catalogued in museums and many others are still in their original allocated place. A random selection could include:

- Bronze-age artefacts such as spear and arrow heads, shields, buckets and brooches;
- Egyptian bronze cats;
- musical instruments such as the Danish lur (a horn) from about 700 B.C onwards;
- Greek and Roman coinage;

- Roman armour;
- ninth-century bronze doors in the Basilica in Milan and the cathedral of Aix-la-Chapelle;
- various English standard weights, such as King Edgar's bushel in Winchester;
- a bronze bell cast in 1380;
- the three pairs of doors of the Baptistry in Florence (1336, 1424 and 1453);
- the massive 'Mons Meg' cannon, now at Edinburgh Castle (c. 1500) and others recovered from years of immersion in the sea by Swedish, Spanish and British ships (such as the Mary Rose wreck of 1545);
- brass monuments in church floors – about 10,000 still exist, some from the thirteenth century;
- clocks, watches, navigational and other scientific instruments dating from the fifteenth century onwards;
- plumbing such as the ¾ in bronze stopcock first installed in Hampton Court in 1539;
- many magnificent and intricate statues cast by the lost-wax process (precision cast) such as those in Italy dating from the fifteenth-century and the sixteenth-century collection in Innsbruck commemorating the Emperor Maximillian.

Structure of the industry

Initially the industry was vertically integrated to the extent that those who mined and smelted copper also made artefacts from it and were able to liaise directly with customers regarding fitness for purpose. This was particularly true of the United Kingdom which was the world's largest producer of copper from the Middle Ages until the 1860s. Now copper is mined extensively on every inhabited continent and is a vital commodity in world trade. Other than for local use, it is not now economic to manufacture components at the source of the copper but there is an excellent understanding between producers and manufacturers regarding the assured quality and quantity of their supplies. Papers and books from statesmen of the industry like Chester Beatty and Sir Ronald Prain detail much of the development of the structure of the copper mines and refineries and the way in which they improved local economies.

At times as recent as the Korean war crisis, supplies of copper were restricted and therefore expensive. Usage then became price-sensitive and there remains a hangover from these times despite the fact that supplies and reserves now exceed foreseeable demand

and that the price of copper in real terms has never been lower. There is a large infrastructure of organizations of producers, financiers, merchants, manufacturers and statisticians who keep commerce supplied with sufficient information to enable progress to be monitored. Much of this information is, of course, valuable and is either kept commercially confidential or available only at realistic costs.

To establish current values, copper is traded as a commodity on the London (LME) and New York (Comex) metal exchanges and the majority of supplies to manufacturers are to contracts based on these prices. Quality standards are well established internationally and producers also maintain a reputation for consistency of the quality of their individual brands of copper registered on each exchange. Statistics relating to prices and tonnages traded are published regularly and available in the national daily press as well as the specialist journals such as *Metal Bulletin*.

The largest tonnages of copper come from such areas as South America, North America and Canada, Zambia, Zaire and other Southern African countries, Australia and Pacific Ocean territories and in Europe, Finland, Poland and Russia have very significant supplies. Many of these producers are vertically integrated to the extent that they have refining, melting, casting and rolling capacity to make wire rod since it is from this form that the majority of primary copper is drawn into the high-conductivity wire which is the biggest end-use. Other than to this extent, the majority of fabricators are not owned by producers but by other large engineering groups.

The absence of significant multinational vertical integration has meant that, in order to continue the general promotion of markets for copper and its alloys, it has been essential to set up and maintain Copper Development Associations in each major industrial country and the names and addresses of each of these are listed in the references section. They have the responsibility of estimating the needs of local markets for information regarding types of coppers and copper alloys available and to promote their properties relating to the suitability of materials for the very wide variety of possible end-uses. They make available basic literature and visual aids describing the materials and other publications covering usage in specific engineering applications. Also in operation are information services answering industrial queries relating to usage. They also organize events such as conferences which enable recent developments to be publicized and discussed.

International Copper Information Bulletin

Published three times each year by Copper Development Association, this is the main source of reference dedicated to current information on copper and its alloys. News is given in the form of brief abstracts of published material under headings which include all aspects such as:

- recent publications and reports from Copper Centres;
- International Copper Research Association (INCRA);
- production, consumption, marketing and statistics;
- national and international standards and related publications;
- recent technical papers and articles;
- new books and published conference proceedings;
- new copper and copper alloy materials, products and processes;
- forthcoming conferences and symposia.

The bulletin was first issued in 1976, superseding previous publications. Large sections such as that covering recent technical papers and articles are subdivided to cover areas such as primary production, foundry practice, fabrication, mechanical and physical properties, available forms, end-use categories, joining, finishing and other considerations. Since there is not yet an index and the material is not yet available in machine-readable format, the bulletin remains very much a publication of current reference. For systematic subject searches covering the many years over which some problems and phenomena have been investigated and re-investigated it is necessary to use the major databases described elsewhere such as Metadex operated by Metals Information (derived from Metals Abstracts), DIALOG or DIALTECH. Copper Development Association, Inc. of New York publishes *Copper Abstracts* which cover the field of American publications well.

Other than publications of the Copper Centres themselves and INCRA research reports, the papers referred to in International Copper Information Bulletin, or in extracts therefrom published elsewhere, are not available from the Copper Centres; they do not have the resources to run library services. They can be obtained directly from the publishers or through local libraries.

Statistics and economics

Although some surveys are carried out for the UK government by the Department of Trade and Industry and similar organizations such as the US Department of the Interior Bureau of Mines do the same elsewhere in the world, the collection and interpretation of statistics is naturally one of the most commercial aspects of modern information technology. Several organizations exist to fulfil this need.

Metal Bulletin, published twice a week, contains useful current information on the mining, extraction, production and usage of metals and alloys together with up-to-date prices for a range of metals and alloys in a variety of forms. The companion *Metal Bulletin Monthly* covers items in greater depth. Also published are annual surveys and lists of producers and approved copper brands.

The World Bureau of Metal Statistics has recently published *World Copper Statistics since 1950* which is a comprehensive statistical compendium of information up to 1976 together with a supplement to bring the data up to date. It draws together in one volume the unique collection of data assembled by WBMS over the last 25 years and includes a useful explanatory commentary. Monthly surveys are covered in *World Metal Statistics*. The same organization also publishes an annual review of the *World Flow of Unwrought Copper*.

Metal Statistics 1975–1985 (updated annually), published by Metallgesellschaft Aktiengesellschaft contains 448 pages showing detailed figures for the production and consumption of ores and metals from 1975 to 1985, with briefer details of production worldwide from 1900. Graphs of metal prices from 1975 to 1985 are also included.

The International Wrought Copper Council has published *World Trade in Copper and Copper-Alloy Semi-Manufactures 1988*, an annual review of trade broken down by shape for both copper and copper-alloy semi-manufactured products in thirty-three tables.

Metals Analysis and Outlook is a half-yearly review published by Metals and Minerals Research Services which contains a series of graphs relating major non-ferrous metal consumption to industrial production, metal prices to the industrial business cycle and investment to consumption. A longer term view is taken in *Metals Analysis Five Year Outlook*.

Copper Studies is published monthly by Copper Studies Inc. and Commodities Research Unit and contains well-researched, in-depth articles on subjects of commercial significance to the industry's commercial interests in exploration, extraction, refining, casting, alloying and fabrication.

Non-Ferrous Alert is one of three abstracts bulletins published by Metals Information, providing coverage of the technocommercial aspects of the industries.

The addresses of the above organizations are given in the references together with the names and addresses of many of the specialist trade organizations. The latter generally publish directories containing details of their members and their production capabilities but do not release production statistics.

Standards and specifications

There is a gradual approach towards the harmonization of materials standards internationally but in the mean time the situation is confusing. For nearly 30 years the International Standards Organisation (ISO) has been working to gain agreement on standards for copper and copper alloys and a great deal has been achieved. Originally the publications were published as recommendations but they are now termed standards and national member bodies are obliged to consider their contents for adoption during the preparation of standard specifications suited to local requirements. The ISO designation system for copper and its alloys effectively abbreviates the composition (i.e. CuZn30 is the 70/30 brass) and has been adopted by many countries although at present there may be slight variations in compositions and properties between different member countries to allow for the fact that those agreed and approved by the ISO member countries are an inevitable compromise. While approximate equivalents do exist, complete correspondence cannot be assumed.

British Standard specifications are revised at intervals and re-issued as required. Those relevant can be ascertained by references to the BSI annual catalogue or to descriptive publications issued by the Copper Development Association.

The German DIN specifications are also frequently quoted internationally. They are issued individually and are available in English translation but the most convenient collection of standards relating to the wrought coppers and copper alloys is contained in their Taschenbuch No. 26.

The American ASTM standards are the least harmonized with ISO and remain an independent but very important set of material and test method standards. The books of ASTM standards are re-issued in updated form each year, that for copper and its alloys being volume 02.01.

Work is starting on the preparation of European (CEN) standards using the ISO documents as initial drafts. When approved under a weighted voting system by a majority of members, it is mandatory that each country will adopt the documents without modification as their national standard specifications. The aim is to abolish technical barriers to trade by 1992 and when this is achieved it will certainly simplify existing equivalents problems. It is, however, extremely unlikely that the present proliferation of specifications prepared to meet their own requirements by individual manufacturers and users will be superseded. The very wide variety of properties of the coppers and copper alloys means that they are used for components in nearly every item of plant, equipment and consumer-durables made. In order to achieve and maintain maximum fitness for purpose and economy it will be essential that there will continue to be special requirements needed and specified. Manufacturers will continue to market materials under their own identifications which may be related to national specifications but claimed to be more suited to certain end uses.

Basic information on many specifications is contained in a number of CDA publications, some of them based on the comprehensive CIDEC data sheets published in the early 1970s. The Metals Information Metals Datafile is a useful machine-readable numerical data bank containing information on the compositions and physical properties of ferrous and non-ferrous alloys.

More detail on company specifications is available in their sales literature. Identification of many of the designations used is contained in books such as those by Ross and by Woldman. A very useful service for identifying specifications is run by the World Metal Index. The Ministry of Defence (Procurement Executive) has its own series of standards such as the DTD and Def Stan series with those specific to naval requirements being in the NES series (replacing DGS specifications) administered from Bath.

Terminology

While the basic shapes in which metals are made are fairly common, the terms applying to them differ between industries and often differ geographically as well. Terms applying to the copper industry have been standardized in the five parts of ISO 197 to meet the requirements of the European Customs Co-operation Council and these have now been largely adopted in a new British Standard specification soon to be published. Language difficulties can be overcome by the use of some of the available dictionaries mentioned in the references but these should be used with care for the same reasons applicable to the English terms. It should also be noted that there will be some differences between, for example, French and French-Canadian usage.

The Institute of Metals services include the MITS Translation Service which publishes more than 1,000 translations each year and also publishes the *Directory of Metallurgical Consultants and Translators*.

Technology

This term is taken to cover the largest section of information requirements dealing with ores, extraction, refining, melting, casting, hot and cold fabrication, heat treatment, joining, mechanical and physical properties, finishing, application-related considerations and environmental implications. Many books have been published on these subjects and only a small selection of these can be covererd in the references. They normally contain an excellent basic summary of their subject but can rarely hope to be as up-to-date as the best of the papers presented at conferences and seminars and published in current journals. No attempt can be made here to review important papers but many of the books and CDA publications are available to cover this subject.

The techniques by which copper is extracted, refined, melted, alloyed and fabricated are being continuously improved. Little can be done to change the basic chemical reactions but the way in which they are achieved is being varied by significant investment in modern plant and equipment in the interests of more economic operation and of improvements to, and better reproducibility of, quality requirements. With the exception of that by West, new books covering the metallurgy of copper are rare. Much of the basic texts of the older ones is still valid but the technology has moved on.

The current situation regarding available materials and recommendations regarding usage can be obtained by reference to the publications and videos of CDA and other copper centres. General reviews on specific subjects are published periodically by journals such as the Institute of Metals *Metals and Materials* and developments are also covered by other publications, of which a small selection includes: *Foundry Trade Journal*, *Metallurgia*, *Welding and Metal Fabrication*, *Wire Industry*, *Tube International*, *Sheet Metal Industries*, *Product Finishing*, *British Corrosion Journal*, *Corrosion Science*, *Corrosion Engineering*, *OEM Design and Materials*, *Design Engineering*, *New Scientist* and a very large selection of other journals published worldwide. References to articles of significance in these publications are usually covered in *International Copper Information Bulletin* within a few months of their appearance.

Much of the important research work on copper and its alloys aimed directly at new and developing markets is that sponsored by International Copper Research Association in research laboratories worldwide. This is reported in INCRA publications and also at seminars and conferences. The BNF Metals Technology Centre (originally the British Non-Ferrous Metals Research Association) carries out many short- and long-term investigations of interest to the industry which may be sponsored by subscribers, INCRA and other organizations. Many of the results are reported shortly after they have been obtained but, like all contract research organizations, the results of some work are confidential to the sponsors for a period of time. Specific applications requirements are also studied within the industry or at, for instance, the Admiralty Research Establishments such as the Admiralty Marine Technology Establishment at Holton Heath. Research conducted at and reported by the universities ranges from the purely academic through to work of direct application such as that on machinability being undertaken at Birmingham.

Analysis

Of the two books mentioned in the references, that by Elwell and Scholes is the more recent. The techniques described are accurate but have been overtaken to a large extent by the instrumental methods now actually in use in modern production control and reference laboratories. This applies not only to the determination of alloying constituents but especially to the determination of impurities in copper where lower limits for some elements can now be expected to be below one part per million. The new LME contract for better purity grade 'A' copper has been made viable by the publication of an amendment to BS 6017 for raw coppers backed up by the development and standardization of the new analytical methods covered in the eight parts of BS DD95.

Alloying relationships

Most of the common binary copper phase diagrams and some parts of the new common ternaries have been published by CDA but copies are now difficult to obtain. Reference should be made to the book by Hansen, the more recent publication by Massalski or to the examples included in Smithells.

The effects of impurities in copper have been studied by many people since the classic summary by Archbutt and Prytherch and are covered in many of the standard books. Most of the conclusions are valid at the levels mentioned but it should again be borne in mind that, with improvements in production techniques made at the refineries in order to meet the demands of their customers, the levels of impurities found now are much lower. Attempts should not be made to extrapolate backwards since at very low levels, inter-element effects are significant. Reference should therefore be made to recent papers published concerning the effect of impurities on the annealability of pure copper. It should be noted that the less-pure grades of copper such as fire-refined tough pitch copper are now very little made. Most of the copper produced is to the grade 'A' standard and meets requirements that it be easily processed down to fine wire. Supplies of oxygen-free high-conductivity copper are now available from several countries and its use is becoming more common (and well described) both in the pure form and as high-purity alloys for special purposes in the semi-conductor electronics industry.

Of the alloys of copper the brasses have recently received much attention in the literature. With the advent of a much wider selection of engineering materials to be covered in educational courses it has been noted that some of the more traditional ones have been receiving less attention than their excellent combinations of possible properties and end uses justify. Besides strengths that can match that of carbon steels, there are the advantages of formability, corrosion resistance, machinability, ductility, conductivity, spark-resistance, wear resistance and low magnetic permeability which can be combined as required in the large variety of compositions and forms of brass available. Special publications are available from CDA as well as two educational videos and there have been several useful summary articles in the technical journals.

The copper-nickel alloys have also been publicized recently because of the increasing importance of their resistance to marine biofouling and corrosion. This is important not only for merchant shipping and naval requirements for seawater piping but also for the more recent developments of the use of 90/10 copper-nickel to sheath the hulls of ships and the legs of offshore structures.

Use of the aluminium bronzes has been increasing over the years with the demand for materials for pumps, valves and bearings that are strong and resistant to seawater corrosion and there has been a consequent increase in the amount of literature available.

Other copper alloys such as the bronzes, gunmetals, nickel-silvers, copper-beryllium and copper-chromium alloys are also covered in the standard books, CDA publications and articles in the technical journals.

Recent new alloy developments which have been reported as achieving industrial significance include those of a dezincification-resistant hot-workable brass, copper-nickel-tin 'spinodal' alloys of great strength, special high-strength high-conductivity alloys for the electronics industry and 'shape memory' alloys invaluable as thermal actuators.

The essential functions of copper as a biocide, fungicide and vital trace element in the development of crops, animals and humans have also been highlighted.

Corrosion

The corrosion resistance of copper and its alloys is one of the many important reasons for their selection for use under arduous conditions. These can range from the use of copper for years of exposure to the weather as one of the best roofing materials to the reliable service given by copper alloys pumping seawater carrying a suspension of abrasive solids. For the most economical usage of materials without risk of expensive failures it is necessary to quantify corrosion resistance as accurately as possible, given that exposure conditions can be correctly predicted. The many variables that have to be considered are described in the references quoted. Some, Rabald for example and the Processing Wall Chart, give useful recommendations to be interpreted with experience. Extra help can be obtained if required from corrosion consultants who have access to many papers and case histories.

Conclusion

The modern copper industry is well supplied with sources of all types of information on its statistics, economics, standards and technology. This brief review has shown ways in which access can be gained and provides a large number of significant sources of information.

Addresses

For others such as embassies and consulates, see also *Shorter Aslib Directory of Inforamtion Sources in the United Kingdom*, edited by E. M. Codlin, 590 pp (ASLIB, London, 1986).

Admiralty Research Establishment, Holton Heath, Poole, Dorset BH16 6JU.
ASM International, 9460 Kinsman Road, Metals Park, Ohio 44073, USA.
American Society for Testing and Materials (ASTM), 1916 Race St., Philadelphia, Pa. 19103, USA (publications available in UK from American Technical Publishers).
American Technical Publishers, 68a Wilbury Way, Hitchin, Herts SG4 0TP.
Asociacion Mexicana del Cobre AV, AB Sonora 166–1er Piso, 06100 Mexico DF, Tel: 528 8651.
Association of Bronze and Brassfounders, c/o Heathcote and Coleman, 136 Hagley Road, Edgbaston, Birmingham B16 9PN.
Association of Consulting Scientists, Owles Hall, Buntingford, Herts, 0763 72665.
BNF Metals Technology Centre (was British Non-Ferrous Metals Research Association, now includes Metal Users Consultancy Service and British Association for Brazing and Soldering), Grove Laboratories, Denchworth Road, Wantage, Oxon, 02357 2992.
British Library, Science Reference and Information Services, 25 Southampton Buildings, WC2. Aldwych Reading Room, 9 Kean Street, WC2. Tel: (for both) 01 636 1544.
British Non-Ferrous Metals Federation (BNFMF), 10 Greenfield Crescent, Edgbaston, Birmingham B15 3AU. Tel: 021 456 3322.
British Plumbing Fittings Manufacturers Association, 10 Greenfield Crescent, Edgbaston, Birmingham B15 3AU.
British Standards Institution, Library Sales, Linford Wood, Milton Keynes MK14 6LE. Tel: 0908 220022.
British Standards Institution (Secretary of Technical Committee NFM/34 Copper and Copper Alloys), 3 York Street, Manchester M2 2AT. Tel: 061 832 3731.
Building Research Establishment, Garston, Watford WD2 7JR. Tel: 0923 676612.
Building Services Research and Information Association, Old Bracknell Lane West, Bracknell, Berks RG12 4AH. Tel: 0344 426511.
Canadian Copper and Brass Development Association, 55 York Street, Toronto, Ontario M5J 1R7, Canada. Tel: 363 8826.
Centre d'Information Cuivre, Laitons, Alliages, 58 rue de Lisbonne, 75008 Paris, France. Tel: 225 25 67.
Centre d'Information des Métaux Non-Ferreux, Informatiecentrum van de Non-Ferrometalen, Rue Montoyer 47, 1040 Bruxelles-Brussel, Belgium.
Centro Brasiliero de Information do Cobre, Caixa Postal 30, 128, 01000 Sao Paulo, Brazil. Tel: 011 258 8041.
Centro Espanol de Information del Cobre, Jose Abascal, 47–3° dcha, Madrid 3, Spain. Tel 91 441 15 76.
Commodities Research Unit Ltd. (CRU), 31 Mount Pleasant, London WC1X 0AD. Tel: 01 278 0414.
Conseil Intergovernemental des Pays Exportateurs de Cuivre (CIPEC), 39 rue de la Bienfaisance, 75008 Paris, France.
Copper Development Association, Orchard House, Mutton Lane, Potters Bar, Herts EN6 3AP. Tel: 0707 50711.
Copper Development Association Inc., Greenwich Office Park 2, Box 1840, Greenwich, Conn 06836 1840. Tel: (203) 625 8210.
Copper Smelters and Refiners Association, 10 Greenfield Crescent, Edgbaston, Birmingham B15 3AU. Tel: 021 456 3322
Copper Studies Inc., 33 West 54th St., New York, NY 10019, USA (also at Commodities Research Unit, London address).
Copper Technical Data Centre, Box 5484 M.C. Townsville, Queensland 4810, Australia. Tel: (008) 075060.
Corrosion and Protection Centre, Industrial Services, UMIST, Sackvill St., Manchester M60 1QD.

Department of Industry, Statistics and Market Intelligence Library, Export House, 50 Ludgate Hill, London EC4M 7HU. Tel: 01 248 5757.
Deutsches Kupfer-Institut EV, Knesebeckstr. 96, D1000 Berlin 12, W. Germany.
DIALOG Information Services, California – UK access through DIALOG, PO Box 8, Abingdon, Oxford.
DIALTECH – access through the Department of Trade and Industry to the European Space Research Association Space Documentation Service (which covers materials).
DIN Deutsches Normenausschuss, 5–8 Kamakestr., D5000 Köln 1, W. Germany.
ERA Technology Ltd., Cleeve Road, Leatherhead, Surrey KT22 7SA.
ESDU International Ltd. (Engineering Sciences Data Unit), PO Box 166, Chalfont St Giles, Bucks HP8 4JG.
Harwell Corrosion Service, Building 393, Materials Development Division, AERE Harwell, Didcot, Oxon OX11 0RA.
Historical Metallurgical Society, 1 Carlton House Terrace, London SW1Y 5508, Tel: 01 839 4071.
Indian Copper Development Centre, 27–8 Camac Street, Calcutta 700016, India. Tel: 44 5724/25.
Institute of Metals, 1 Carlton House Terrace, London SW1Y 5508. Tel: 01 839 4071.
Institute of Metal Finishing, Exeter House, 48 Holloway Road, Birmingham B1 1NQ. Tel: 021 622 7387.
Institution of Mining and Metallurgy, 44 Portland Place, London W1N 4BR. Tel: 01 580 3802.
Instituto Argentino del Cobre y Sus Aleaciones, Avenida del Libertador 1784, 1425 Buenos Aires, Argentina. Telex: 17621 'For CADIC'.
International Copper Research Association (INCRA), 708 Third Avenue, New York, NY 10017 and Brosnan House, Darkes Lane, Potters Bar, Herts EN6 1BW.
International Wrought Copper Council, 6 Bathurst St., Sussex Square, London. Tel: 01 723 7465.
Istituto Italiano del Rame (IIR), via Zuimbini 25, 201430 Milano, Italy. Tel: (02) 4233713.
Japan Copper Development Association, Konwa Building, 12–22 1-chome, Tsukiji, Chuo-ku, Tokyo 104, Japan. Tel: Tokyo 6631 6633.
Metals and Minerals Research Services Ltd. (also Metals and Minerals Publications), 222–225 Strand, London WC2R 1BA.
Metals Information – see Institute of Metals and ASM International.
Metal Finishing Association, 27 Frederick Street, Birmingham B1 3HJ.
Metal Roofing Contractors Association, Braby Carter Ltd., 33 Chase Road, London NW10. Tel: 01 965 1161.
Micronutrient Bureau, MB House, Wiggington, Tring, Herts HP23 6ED.
National Corrosion Service, National Physical Laboratory, Teddington, Middx TW11 0LW. Tel: 01 977 3222.
Scandinavian Copper Development Association, Outokumpo Oy, 28101 Pori 10, Finland.
Science Reference Library – see British Library.
South African Copper Development Association (Pty) Ltd., PO Box 14785, Wadeville, S. Africa 1422.
Technical Report Centre (for MoD Specifications), Station Square House, Station Square, St Mary Cray, Orpington, Kent BR5 3RE. Tel: 0689 32111.
UMS Usines Metallurgiques Suisses SA, Kollerweg 32, CH-3006 Berne, Switzerland. Tel: Berne 443251.
Warren Spring Laboratory (for information on mineral and scrap processing), Stevenage, Herts SG1 2BX. Tel: 0438 3388.

World Bureau of Metal Statistics (WBMS), 41 Doughty St., London WC1N 2LF.
World Metal Index, Sheffield City Libraries, Surrey St. Sheffield S1 1XZ. Tel: 0742 734742.

Bibliography

Directories, suppliers, statistics and economics

Binary Alloy Phase Diagrams, edited by T. B. Massalski, 2200 pp. (ASM International, Ohio, 1987).
Constitution of Binary Alloys, edited by M. Hansen, 1301 pp. (McGraw-Hill, New York, 1958).
Directory of American Research and Technology, 730 pp. (R. R. Bowker Ltd., Sevenoaks, 1986).
Directory of Metallurgical Consultants and Translators, 150 pp. Metals Information (ASM, Ohio and Institute of Metals, London, 1986).
Finishing Handbook and Directory, edited by J. E. Bean. Product Finishing (Sawell Publications, London, 1987) (updated annually).
Foundryman's Handbook, 436 pp. (Pergamon, Oxford, 1986).
Foundry Yearbook and Castings Buyers' Directory 1987 (Fuel and Metallurgical Journals Ltd., Redhill, 1987).
Kempe's Engineers Yearbook, edited by C. Sharpe. 2,400 pp. (Morgan-Grampian, London, 1988) (updated annually).
Kompass Buyers Guide – Metal Industries, 974 pp. (Kompass Publishers, East Grinstead, 1987) (Products, services, company information).
Machinery Buyers Guide, 1200 pp. (Findlay Publications, Horton Kirby) (updated annually).
Metal Bulletin's Prices and Data Book, edited by R. Packard. 364 pp. (Metal Bulletin Books Ltd., Worcester Park, 1986).
Metallstatistik 1975–1985, edited by W. Bauer. 448 pp. (Metallgesellschaft Aktiengesellschaft, Frankfurt, 1986) (updated annually).
Metal Traders of the World, ed. D. Gilbertson and R. Packard. 735 pp. (Metal Bulletin Books Ltd., Worcester Park, 1983).
Mining Annual Review. 516 pp. (Mining Journal, London) (revised annually)
Sheet Metal Industries Yearbook, 296 pp. (Fuel and Metallurgical Journals Ltd., Redhill, 1987).
Smithells Metals Reference Book, edited by E. A. Brandes. 2,400 pp. (Butterworths, London, 1983).
Wire Industry Yearbook, edited by P. A. Clayton. 384 pp. (Magnum Publications Ltd., Oxted, 1986).
World Copper Databook, edited by R. Packard. 410 pp. (Metal Bulletin Books Ltd., Worcester Park, 1986) (Producers, brands, traders).
World Non-Ferrous Metal Production and Prices, 1700–1976. 432 pp. (Cass, London, 1979).

Historical and surveys

Agricola, G, *De Re Metallica*, translated from 1556 edition by H. C. and L. H. Hoover, 638 pp. (Dover Publications, New York, 1950).
Prain, R, *Copper – the Anatomy of an Industry*, 298 pp. (Mining Journal Books, London, 1975).
Webster Smith, B, *Sixty Centuries of Copper*, 96 pp (CDA Publication No. 69, Hutchinson, London, 1965).

Standards

Annual Book of ASTM Standards, Vol. 02.01, Copper and Copper Alloys, 1271 pp. (ASTM, Philadelphia, and American Technical Publishers). Published annually.
BSI Catalogue (annually) (British Standards Institution, Linford Wood, Milton Keynes MK14 6LE).
Nichteisenmetalle, DIN Taschenbuch No. 26, 358 pp. (Beuth, Berlin, 1984) Compilation of DIN standards for wrought non-ferrous metals.
Engineering Alloys, edited by N. E. Woldman and R. C. Gibbons, 1427 pp. (Van Nostrand Reinhold, New York, 1973).
International Standards Organisation (ISO) Catalogue – from BSI or other national standards organizations.
Metallic Materials, edited by R. B. Ross, 935 pp. (Chapman and Hall, London, 1968).
Worldwide Guide to Equivalent Non-ferrous Metals and Alloys, 626 pp., edited by P. M. Unterweiser (ASM, Ohio, 1980).

Terminology

Freiwillig, R, *Dictionary of Physical Metallurgy* (English, German, French, Russian and Spanish), 192 pp. (Elsevier, Amsterdam, 1987).
Merriman, A. D. and Bowden, J. S., *A Dictionary of Metallurgy*, 321 pp. (Metal Treatment and Drop Forging, Feb 1952 – March 1957).
A Vocabulary of Copper and Its Alloys (English and French-Canadian) (Canadian Copper and Brass Development Association and Canadian Government Publications Centre, 1982).

Analysis

Elwell, W. T. and Scholes, I. R., *Analysis of Copper and its Alloys*, 183 pp. (Pergamon Press, Oxford, 1967).
Oozinel, C. M. and Andraso, G. E., *Modern Methods of Analysis of Copper and its Alloys*, 234 pp. (C. M. Oozinel, Brussels, 1960).

Copper

Archbutt, S. L. and Pryrtherch, W. E., *The Effect of Impurities in Copper*, 134 pp. (BNFMRA Monograph No. 4, 1937).
Atkinson, R. L., *Copper and Copper Mining* (Shire, Princes Risborough, 1987).
Biswas, A. K. and Davenport, W. G., *Extractive Metallurgy of Copper*, 438 pp. (Pergamon, Oxford, 1980).
Copper for Busbars, 51 pp. (CDA Publication No. 22, 1984).
Ehrlich, R. P. (ed.), *Copper Metallurgy* (proc. Extractive Metallurgy Symposium), 371 pp. (Metallurgical Society (USA), 1970).
The Electrorefining and Winning of Copper. Proceedings of the symposium sponsored by TMS Copper, Nickel, Cobalt, Precious Metals, and Electrolytic Processes Committees, (Denver, Colorado, 24–6 February 1987) edited by J. E. Hoffman *et al.* (Warrendale, PA., Metallurgical Society, 1987).
Finlay, W. L., *Silver-bearing Copper*. 356 pp. (Corinthian Editions, New York, 1968).
High Conductivity Coppers – Properties and Applications, 33 pp. (CDA Technical Note TN 29, 1981).
High Conductivity Coppers – Technical Data, 123 pp. (CDA Technical Note TN 27, 1981).

Hoffman, H. O. and Hayward, C. R., *Metallurgy of Copper*, 409 pp. (McGraw-Hill, New York, 1924).
Newton, J. and Wilson, C. L., *Metallurgy of Copper*, 518 pp. (Wiley, New York, 1942).

Copper and copper alloys

Aluminium Bronze Alloys for Industry, 16 pp. (CDA Publication No.83, 1986).
Aluminium Bronze Alloys – Technical Data, 28 pp. (CDA Publication No. 82, 1981).
Aluminium Bronze Corrosion Resistance Guide, 28 pp. (CDA Publication No. 80, 1981).
The Brasses – Properties and Applications, 24 pp. (CDA Technical Note TN 24, 1987).
The Brasses – Technical Data, 112 pp. (CDA TN 26, 1980).
Butts, A, Copper – The Science and Technology of the Metal, Its Alloys and Compounds, 936 pp. (Reinhold, New York, 1954).
Copper and Copper Alloys – Composition and Properties, 28 pp. (CDA TN 10, 1986).
Copper – Nickel Alloys, Properties and Applications, 29 pp. (CDA Technical Note TN 30, 1982).
Design in Brass, 16 pp. (CDA Publication No. 84, 1986).
Le May, I and McDonald Schetky, L. (eds), *Copper in Iron and Steel*, 423 pp. (Wiley, New York, 1982).
Source Book on Copper and Copper Alloys, 414 pp. (American Society for Metals, Ohio, 1979).
West, E. G., *Copper and Copper Alloys*, 241 pp. (Ellis Horwood, Chichester, 1982).
Wilkins, R. A. and Bunn, E. S., Copper and Copper Based Alloys, 353 pp (McGraw-Hill, New York, 1943).

Castings

Copper Base Alloy Foundry Practices, 223 pp. (American Foundrymen's Society, Chicago, 1952).
Hudson, F. and D. A., *Gunmetal Castings*, 420 pp. (Macdonald, London, 1967).
Macken, P. J., *Copper Alloy Casting Design*, 68 pp. (CDA Publication No. 76, Copper Development Association, Potters Bar, 1970).
Meigh, H, *Designing Aluminium-Bronze Castings*, 8 pp. (Engineering Technical File No 116, The Design Council, London, 1983).

Fabrication

ASM Handbooks, 9th editions: *No. 2 Properties and Selection – Non-Ferrous Alloys*, 855 pp.; *No. 6 Welding, Brazing and Soldering*, 1152 pp.; *No. 9 Metallography and Microstructures*, 795 pp.; *No. 10 Materials Characteristics*, 761 pp., *No. 13 Corrosion* (ASM International, Ohio).
Fuller, J. R., *Art of Coppersmithing*, 324 pp. (David Williams, New York, 1894).
Greaves, R. H. and Wrighton, H, *Practical Microscopical Metallography*, 256 pp. (Chapman and Hall, 1933).
Hughes, R. and Rowe, M, *The Colouring, Bronzing and Patination of Metals*, 372 pp. (Crafts Council, London, 1982).
Jevons, J. D., *The Metallurgy of Deep Drawing and Pressing*, 739 pp. (Chapman and Hall, 1949).

Pearson, C. E. and Parkins, R. N., *The Extrusion of Metals*, 336 pp (Chapman and Hall, London, 1960).
Recent Progress in Metal Working, 135 pp. (Institution of Metallurgists, Iliffe Books, London, 1963).

Corrosion

Butler, G. and Ison, H. C. K., *Corrosion and its Prevention in Waters*, 281 pp. (Leonard-Hill, London, 1961).
Corrosion Prevention Directory, 136 pp. (Department of Industry, HMSO, London, 1975).
Evans, U. R., *The Corrosion and Oxidation of Metals: Scientific Principles and Practical Applications*, 1092 pp. (Arnold, London, 1960).
Laque, F. L. and Copson, H. R., *Corrosion Resistance of Metals and Alloys*, 712 pp. (Rheinhold, New York, 1963).
Leidheiser, H., *The Corrosion of Copper, Tin and Their Alloys*, 411 pp. (Wiley, New York, 1971).
Rabald, E., *Corrosion Guide*, 900 pp. (Elsevier, Amsterdam, 1968).

CHAPTER SEVEN

Lead

M.J. CONWAY

Introduction

Lead is one of the oldest known and used metals. One of the earliest known lead artifacts, dating back to around 5000 BC, is a spinning bobbin found in Merw, USSR, while other finds of decorative and artistic works from the same period were discovered in Germany. A thousand years later, the Egyptians and Sumerians were using lead oxide in make up. Lead was also being used as a glaze on pottery and to win silver, as well as for projectiles in slings. Other uses of the same period were for charms, models, writing tablets and sinkers. There is also evidence that the Egyptians used lead pipes. Lead coils were minted in both China and Numidia in Asia Minor, around 3000 BC. The Phoenicians mined lead at Rio Tinto in Spain around 2300 BC, and also traded lead with England. The Hanging Gardens of Babylon (500 BC) had floors of sheet lead soldered together to retain the moisture needed for the plants. By 100 BC the Romans were using lead pipes to suply the million-strong population of Rome with water. Other uses, such as lead wire, and lead barrel hoops were also common. The Romans mined lead in Sardinia, Spain, Gaul, England and Germany. In AD 1200, huge quantities of lead were required for the stained glass windows in the Gothic cathedrals and buildings then under construction. By AD 1400 a more modern use was found for the metal, as lead shot for the newly conceived musket, and later, as a type metal by Gutenberg. In 1675, lead glass was discovered in England, and this formed the basis for lead crystal which is still popular today. After 1910, most of the world's lead mining was carried on outside Europe, and after the First World War, world production rose to some 1.5 million tonnes per year. Most of the expansion was due to the use of lead-acid batteries in the new automobile industry, and also as lead sheathing on electric cables. Most of the modern uses are of some antiquity, though in this century new and unique uses have been found for lead and its compounds, namely for radiation shielding and as a fuel additive. The importance of lead lies in its range of useful properties. It is corrosion-resistant, has a high density and low melting point, is easy to form, and protects against radiation. It also readily forms alloys with other metals, and its compounds find many and varied uses as pigments, as stabilizers in plastics, as driers or catalysts for unsaturated oils used in coatings and inks, and as biocides.

General information

In common with most metals, lead is a prime commodity and there is therefore a great deal of information on it in a wide variety of sources. The majority of the sources quoted below are books, periodicals, reports and conference proceedings.

Information on standards is obtainable from many national centres. Those most often encountered are from the British Standards Institution in the UK, from two sources in the USA, the American National Standards Institute (ANSI), or the American Society for Testing and Materials (ASTM), and in West Germany the Deutsches Institut fur Normung e.V. (DIN). Details of other countries can be found in *Use of Engineering Literature* edited by A. J. Anthonu (Butterworths, 1986). Patents are best obtained from the appropriate national Patent Office, which will have records of the most recent, relevant applications. The most important sources are the British Patent Office, the US Patent Office and the German Patent Office. Patents may also be searched online, a major database being Derwent's World Patents Index, available through Dialog and Oribit. Databases such as CA Search and BNF Metals also cover patents in depth. Chapter 1 has an introduction to patent literature.

Books, reports, proceedings and periodicals

All inclusive works

The best and most comprehensive single-volume work on the chemistry and metallurgy of lead and its alloys is *Lead and Lead Alloys: Properties and Technology* by W. Hofmann (English translation of 2nd rev. German edition), Springer Verlag, 1970. The book is divided into two main sections, the first dealing with physical metallurgy and metallography of binary and ternary alloys, with reviews of published and unpublished work on the mechanical properties and corrosion resistance of the alloys. The second section deals with melting, casting, batteries, printing metals, forming and metallic coatings. Extraction and production are also dealt with briefly. A bibliography of 1308 references is included. Another primary source worthy of note is Gmelin's *Handbuch der Anorganischen Chemie* (Springer Verlag, 1974). A more comprehensive but also more cumbersome work than Hofmann, it is multi-volume and written in German with German–English tables of contents and English marginal headings and subheadings. Journal, report and patent literature to 1970 is critically reviewed. As the breakdown of the volumes is somewhat complex, the major contents of each of the volumes as it relates to lead, is given in Appendix A. *Lead in Modern Industry* (Lead Industries Association, New York, 1952) though somewhat dated now, does provide an excellent introduction to all aspects of lead technology. A survey of production, applications and properties of lead and lead alloys is provided in *The Technology of Heavy Non-Ferrous Metals and Alloys* (J. T. Cairns and P. T. Gilbert, George Newnes, 1967). For a historical overview of the development of the technology of metals, *Metallurgy in Antiquity* (R. J. Forbes, R. J. Brill, 1950) is worthy of note, and also Gmelin, Vol. A1.

Information sources arranged by major area of interest or uses

Analysis

'Lead' by T. W. Gilbert, pp. 69–175, in *Treatise on Analytical Chemistry of the Elements*, Vol. 6 edited by I. M. Kolthoff, P. J. Elving and E. B. Sandell (Wiley Interscience, 1964) gives full details of the wide range of conventional analytical techniques available for lead. *Chemical Anaylsis in Extractive Metallurgy* (Charles Griffin & Co., 1971) and *Chemical Phase Analysis* (Charles Griffin & Co., 1974) both by R. S. Young, also covers this subject, as does Gmelin, Vol. B1.

Periodicals which deal with this topic are many and varied; only those that are most often encountered are cited here. *Analytica Chimica Acta*, *Analytical Chemistry*, *Atomic Spectroscopy*, *Electrochimica Acta*, *Environmental Technology Letters*, *Journal of the*

Association of Official Analytical Chemists, *Journal of Crystal Growth*, *Journal of the Electrochemical Society*, *Journal of Materials Science*, *Metall*, and *Physica Status Solidii Parts A and B*.

Batteries

The best English language source available on the subject is *Storage Batteries: A General Treatise on the Physics and Chemistry of Secondary Batteries and Their Engineering Applications*, 4th edn by G. W. Vinal (Wiley, 1955). Recent books on the subject are *Bleiakkumulatoren*, 9th rev. edn by W. Garten (Varta AG, 1968) and *Storage Batteries; Including Operation, Charging, Maintenance and Repair*, 2nd edn by G. Smith (Pitman, 1971). Current research on lead-acid batteries is dealt with in many conferences throughout the world, including the triennial Lead Conferences, held since 1962, for which edited proceedings are available, published by the Lead Development Association (LDA). They often have additional contributions and an edited account of follow-up discussions. LDA also publishes several booklets, many the result of specific seminars, on applications and specific uses of batteries as follows. *Lead Power News*, which highlights new and interesting uses of lead batteries in many industrial fields. *Expanded Lead Grids for Batteries* discusses the role of these grids in cost reduction and in the manufacture of maintenance-free batteries. *Wrought Lead Strip for Batteries* is an illustrated report reviewing the technology of making automotive battery grids from lead strip. *The Battery Truck Book* is a booklet reviewing advantages and ways of using battery electric trucks. *Advanced Lead Alloys for Maintenance-Free Batteries* is a report on a technical seminal which discussed new low-antimony and non-antimony lead alloys available for maintenance-free batteries. *Marketing Maintenance-free Batteries in the United States* is a report of a 1981 seminar, as is *Battery Electric Boats for Inland Waterways*. *The Quiet Connection* discusses new and interesting applications of lead batteries in power boats for use on inland waterways, and *Battery Electric Vehicles for Aircraft Ground Support* highlights new airport equipment including loaders, tugs etc., and specialized battery electric equipment for handling aircraft, cargo and luggage. *Electrolytic Method for Recovery of Lead from Scrap Batteries* by A. Y. Lee *et al.* was published by the US Department of the Interior Bureau of Mines (1984).

Periodicals include *Batteries Today*, *Battery Man*, *Electric Vehicle Developments*, *Electrical Review*, *Electrochimica Acta*, *Journal of the Electrochemical Society*, and *Journal of Power Sources*.

Building

Lead Development Association's *Lead Sheet in Building* is the prime source for information on the specification and design of lead sheet for roofing and cladding, flashings, weatherings and the techniques of leadwelding. Also available from the same source are *Lead Sheet Flashings*, which covers the design and specification for traditional flashing applications, plus an illustrated guide to new flashing details associated with the rehabilitation of old houses. *Leadwork* is an annual bulletin which updates recent technical developments in the application of lead sheet, and *Control of Lead at Work* advises on the interpretation of the Control of Lead at Work Regulations 1980.

References to lead in building in journals are not frequent but may occur in any of the wide range of building industry periodicals.

Cables

The book *Lead and Lead Alloys for Cable Sheathing* by S. A. Hiscock (Ernest Benn, 1961) gives a historical survey of lead sheathing for cables, and provides excellent descriptions of the various alloys and their properties, and the design and operation of extrusion presses. The *Lead Conferences*, referred to in Batteries above, also contain some papers on this subject. The LDA also produced a report, *Lead Cable Workshop* (1978), which discusses day-to-day problems that arise during manufacture, installation and use of lead-sheathed cables. Topics covered include lead quality, production equipment, sheath quality control, lead alloys and service experience. Periodicals relevant to cables would be those dealing with electrical transmission and electrical problems in general such as *Archiv fuer Elektrotechnik*, *Electric Machines and Electromechanics*, *Electric Power Systems Research*, *Electrical Engineer*, *Electrical Review*, the *IEEE Transactions* series, the *International Journal of Electrical Engineering Education*, *Power Line*, and *Siemens Forschungs und Entwicklungsberichte*.

Chemicals

Gmelin's-Handbuch Vols C1–C4 give the most complete survey of information on lead chemicals. The best English survey of the properties and industrial applications of lead chemicals is *Lead Chemicals* by D. Greninger, V. Kollonitsch, and C. H. Kline (International Lead Zinc Research Organization (ILZRO), 1975). Organic compounds of lead are dealt with in detail in *The Organic*

Compounds of Lead by H. Shapiro and F. W. Frey (Interscience, 1968), and two ILZRO publications, *Investigation in the Field of Organolead Chemistry* by L. C. Willemsens and G. J. M. van der Kerk (1965), and *Organolead Chemistry: A Concise Review with Special Reference to the Literature Covering the Period January 1953 to July 1963* by L. C. Willemsens (1964). A literature survey on organometallic lead compounds, covering the perod 1837 to 1958, has been produced by M. Dub in Vol. 11 *Organometallic Compounds of Germanium, Tin and Lead* (Springer-Verlag, 1961). There are two other multi-volume series which include good sections on lead chemicals. *Nouveau traite de chimie minérale. Tome VIII. Germanium, étain, plomb*, edited by P. Pascal (Messon et Cie, 1963). has a bibliography of 3,905 references, giving a complete survey of the literature to 1957. *Comprehensive Inorganic Chemistry* Vol. 2 by R. Nyholm *et al.* (eds) Pergamon Press, 1973) contains 'Lead' by E. W. Abel, Ch. 18, pp. 469–803.

Most of the periodicals mentioned in the other sections contain information on lead chemicals.

Coatings

There are a large number of comprehensive books on electroplating which contain sections on electroplated lead and lead alloy coatings. A good basic introduction is provided in H. K. Weisner, 'Lead', in *Modern Electroplating*, 3rd edn by F. Lowenheim (ed.) (Interscience, 1974), Ch. 11, pp. 266–86. Similarly, a chapter by A. H. DuRose, 'Blei' Ch. 17.09, pp. 377–91, in *Handbuch der Galvanotechnik Band 2 Verfahren für der galvanische und stromlose Metallabscheidung* (Carl Hanser Verlag, 1966) also provides a good introduction to the subject. A Brenner has critically surveyed the literature on alloy plating in *Electrodeposition of Alloys: Principles and Practice*, Vols 1 and 2 (Academic Press, 1963). Volume 1 covers lead–copper and lead–silver alloys, while Volume 2 deals with lead–tin, lead–nickel and lead–bismuth alloys. W. H. Safranek gives a useful summary of the mechanical properties of lead electrodeposits in *The Properties of Electrodeposited Metals and Alloys* (American Elsevier, 1974), pp. 202–12. Dip coating, lead lining and other specialized coating techniques are described in *Lead and Lead Alloys* by W. Hofmann (Springer Verlag, 1970). Electrochemical aspects of lead electrodeposition are found in Gmelin, Vol. B2, and in 'Lead' by T. F. Sharpe, in the *Encyclopedia of Electrochemistry of the Elements*, Vol. 1 (Marcel Dekker, 1973), Ch. 5, pp. 236–347. Periodicals such as *Journal of Coatings Technology*, *Metal Finishing*, *Plating and Surface Finishing*, and *Transactions of the Institute of Metal Finishers* have occasional articles on lead coatings, mainly electroplating.

Corrosion

Here also there are a large number of reference works. E. Rabald, in *Corrosion Guide* 2nd rev. edn (Elsevier, 1968), subdivides the book by the metal concerned. F. L. LaQue and H. R. Copson (eds), in *Corrosion Resistance of Metals and Alloys*, 2nd edn (Reinhold, 1963), summarize the corrosion resistance of lead to various chemicals in Ch. 12, pp. 285–303. The subject is also dealt with in detail by Hofmann, and a book by the Lead Industries Association, *Lead for Corrosion Resistant Applications* (1974), is the first monograph on the subject, giving the resistance of lead to corrosion by a wide range of chemicals.

There are no particular periodicals concerning themselves with corrosion as applied to lead, but those mentioned in Coatings above are likely to have occasional articles, as are any other journals dealing with electroplating or hot coating. A comprehensive list of these can be found in the chapter on Zinc in this book.

Electrochemistry

Gmelin Vol. B2 deals in detail with this, while T. F. Sharpe in 'Lead', *Encyclopedia of Electrochemistry of the Elements*, Vol. 1, Ch. 5, pp. 236–347 (Marcel Dekker, 1973), concerns himself more with the theoretical aspects.

Periodicals dealing with electrochemistry overlap greatly with the Batteries literature. *Electrochimica Acta*, *Journal of Applied Physics* and the *Journal of the Electrochemistry Society* are most useful.

Environment

This is an area of current topicality and interest, and there is a considerable body of work on the subject. It would be impossible to cover the whole field adequately, so listed here are some of the most relevant books. The subclinical effects of lead are discussed in *Sub-clinical Lead Poisoning* by H. A. Waldron and D. Stofen (Academic Press, 1974). In 'Lead', in *Metals in the Environment* by H. A. Waldron (ed.) (Academic Press, 1980), Ch. 6, pp. 155–97, Waldron discusses the chemical behaviour of lead in the environment. *The Proceedings of the Heavy Metals in the Environment Conferences* held in 1979, 1981, and 1985, published by CEP Consultants Ltd., Edinburgh, are another valuable source. *Lead versus Health: Sources and Effects of Low Level Lead Exposure* by M. Rutter and R. Russell Jones (John Wiley, 1983) gives a comprehensive and up-to-date account of the subject, and *Low Level Lead Exposure: The Clinical Implications of Current Research* by H. L. Needleman (ed.) (Raven Press, New York, 1980) stems from a symposium designed to communicate new data on the effects of lead in the environment to clinicians and public health officials. A National Academy of Sciences publication, *Airborne Lead in Perspective* (NRC, Washington DC, 1972) examines this aspect while a Department of the Environment booklet *Lead in the Environment and Its Significance to Man* (HMSO, 1974) gives a good overall view. *Lead in the Environment* by W. R. Boggess and B. G. Wixson (Castle House Publications, Tunbridge Wells, 1979) gives a complete survey of lead in the environment together with control strategies. Likewise *Lead Pollution: Causes and Control* by R. M. Harrison and D. P. H. Laxen (Chapman and Hall, 1981) examines the sources and environmental pathways of lead. *Lead in the Marine Environment* by M. Branica and Z Konrad (eds) (Pergamon Press, 1980) is the proceedings of a 1977 conference held in Yugoslavia and discusses the problems of lead pollution in the marine environment.

The two volume work by J. O. Nriagu, *The Biogeochemistry of Lead in the Environment* provides an overall view of the chemical behaviour of lead in the environment. Volume 1 deals with the ecological cycles, Volume 2 with biological effects. Two interim reports from the Colorado State University, *Impact on Man of Environmental Contamination Caused by Lead, 1 July 1971–30 June 1972*, and *Environmental Contamination Caused by Lead, 1 January 1974–31 Dec. 1974*, both edited by H. W. Edwards, give details of the University's research programme on the environmental effects of lead. The LDA also produces some publications on the subject. *Lead and the Environment: Guidelines for Monitoring Emissions from Leadworks* (1977) proposes an emissions monitoring programme with standardized equipment and techniques suitable with certain adaptations for all lead works, and enabling interplant comparisons to be made and the value of improved control techniques demonstrated. *Control of Lead at Work: Report on Seminar Proceedings March/May 1981* reviews papers presented and discussions following them at six seminars on the (then) new Control of Lead at Work Regulations 1980. It gives detailed guidance to all users of lead affected by the Regulations.

Numerous conferences have been held on lead in the environment, under the auspices of a variety of bodies. Recent ones include: *Lead in the Environment* (Institute of Petroleum, London,

27 Jan. 1972). Edited proceedings in association with the Chemical Society and British Occupational Hygiene Society; *International Symposium on Environmental Health Aspects of Lead* (US-EPA and the EEC Amsterdam 2–6 Oct. 1972, proceedings published May 1973); papers from the *Conference on Low Level Lead Toxicity* (National Inst. Environmental Health and US-EPA, Raleigh NC, Oct. 1973), and published in *Environmental Health Perspectives* May 1974, Issue 7, pp. 1–252; and the EEC, US-EPA and WHO symposium on *Recent Advances in the Assessment of the Health Effects of Environmental Pollution* (Paris, 24–8 June 1974).

Periodicals dealing with the environment are many and varied, but the major ones of importance are: *Archives of Environmental Health*, *Bulletin of Environmental Contamination and Toxicology*, *Environmental Health Perspectives*, *Environmental Pollution Parts A and B*, *Environmental Research*, *Environmental Science and Technology*, *Environmental Technology Letters*, *Journal of the Water Pollution Control Federation*, and *Science in the Total Environment*.

Joining

The Soldering Handbook by B. M. Allen (Iliffe Books, 1969) and *Solders and Soldering Materials, Design, Production and Analysis for Reliable Bonding* by K. H. Manko (McGraw-Hill, 1964) give good surveys of solders and soldering, the latter reflecting American practice. Properties and applications of tin-based lead solders are discussed in *Tin and Its Alloys* by E. S. Hedges (ed.) (Edward Arnold, 1960). Periodicals include *Brazing and Soldering*, *Plating and Surface Finishing*, and *Tin and Its Uses*.

Metal working

W Hofmann's *Lead and Lead Alloys* (Springer Verlag, 1970) deals with the subject in detail, while J. T. Cairns and P. T. Gilvert in *The Technology of Heavy Non-ferrous Metals and Alloys* (George Newnes, 1967) cover only some aspects. Extrusion for cable sheathing is covered comprehensively by Hiscock (see Cables section above). The use of lead in the construction of chemical plant is dealt with in 'Lead and lead alloys' by H. C. Wesson in I. L. Hepner's *Materials of Construction for Chemical Plant* (Leonard Hill, 1962), Ch. 3. Its use in other industrial environments is covered in Hofmann, and two LIA books *Lead in Modern Industry* and *Lead for Corrosion Resistant Applications*. Lead Development Association produces Leadwork, *Lead for Industrial Sound Insulation* (1975), and *Lead for Radiation Shielding* (1972), as well as specific guides as detailed above under Building. Periodicals would be the same as for Building and Joining above.

Phase diagrams and thermodynamics

Good, annotated phase diagrams can be found in Hofmann and in the series *Constitution of Binary Alloys* 2nd edn, M. Hansen (McGraw-Hill, 1958); 1st supplement 1965, R. P. Elliott; 2nd supplement 1969, F. A. Shunk. They are also contained in two AMS books, *Metals Handbook Vol. 8 Metallography, Structures and Phase Diagrams* (1973), and *Selected Values of the Thermodynamic Properties of Binary Alloys* by Hultgren *et al.* 1973). Gmelin Vol. A1 and Vols C1 to C4 also have critical reviews of thermodynamic data. Microstructural and metallographic information is contained in the AMS *Metals Handbook*, and phase diagrams for ceramic systems have been compiled by the American Ceramic Society in *Phase Diagrams for Ceramists* (1964). Periodicals include *Analytical Chemistry*, *Applied Physics Letters*, *Journal of Applied Physics*, *Journal of the Electrochemical Society*, *Journal of Materials Science*, *Metallurgical Transactions A*, *Physica Status Solidii A*, *Transactions of the Institute of Mining and Metallurgy*, and others which also deal with physics, chemistry or other properties of materials.

Pigments

The primary source is the *Pigment Handbook* by T. C. Patton (ed.), 3 vols (John Wiley, 1973), though this reflects American usage and terminology. Individual pigments are considered in detail, including a historical introduction, methods of manufacture and applications. An introduction to the subject of pigments can be found in *Pigments* by J. S. Remington and W. Francis (Leonard Hill, 1954), and Part 6 of the *OCCA Paint Technology Manual* series gives brief details of lead pigments (Chapman and Hall, 1966). Periodicals include *Anti-Corrosion*, *Journal of the Oil Colour Chemists Association*, *Korrosion*, *Pigment and Resin Technology*.

Printing metals

Books on lead-based printing metals have been produced by the main alloy producers, such as Capper Pass and Son Ltd: *Printing Metals, Their Production, Nature and Use*, 5th edn (July 1955); Fry's Metal Foundries Ltd., *Printing Metals* rev. edn, 1959; and Paul Bergsoe and Son A/B, *Printing Metals*, 2nd edn (Glostrup, Denmark, 1958).

Statistics

Lead mining, production and end-use statistics are collated by many organizations. The most reliable sources are monthly bulletins published by the International Lead Zinc Study Group (ILZSG), which produces *Lead and Zinc Statistics*, and the World Bureau of Metal Statistics, producing *World Metal Statistics*. The American Bureau of Metal Statistics produces the *Yearbook of the ABMS*, which gives lists of major smelting and refining plants, and includes mine production figures for individual US mining companies. Metallgesellschaft AG also produces a yearbook called *Metal Statistics*, which contains a historical graph showing price changes of lead since 1850.

Periodicals for this section cover a wide field, and include most mining and engineering journals, such as the *Canadian Mining Journal*, *Engineering and Mining Journal*, *The Journal of Metals*, *The Mining Magazine*, and the *US Bureau of Mines Mineral Commodity Profiles and Reports*.

Miscellaneous

Lead Isotopes by B. R. Doe (Springer-Verlag, 1970) is a comprehensive work on the occurrence and characterization of lead isotopes. *The Reactor Handbook Vol. 1 Materials* (US-Atomic Energy Commission, Interscience, 1969) provides radiochemical data for lead and covers nuclear radiation shielding. *The Use of Lead as a Shielding Material*, Ch. 59 by G. L. Stuckenbroeker, C. F. Bonilla and R. W. Peterson. *USAEC Reactor Materials Handbook*, 3rd edn (Nuclear Engineering Design, 1979), is a very good survey of lead as a shielding material.

There are a large number of journals which deal with lead and related subjects, in addition to those mentioned above. A full list of all the journals monitored for *Leadscan* is available from the Lead Development Association.

Organizations, research associations etc.

The Lead Development Association (LDA) is engaged in promoting the use of lead and its compounds and derived products. To this end it provides a technical service and an information service, run from 42 Weymouth Street, London. It has close links with similar organizations throughout the world, and is actively engaged in organizing conferences and meetings for and between producers and present and potential end-users of the metal. As part of the information service an abstracts journal, *Leadscan* (formerly *Lead Abstracts*) is published quarterly.

Leadscan stemmed from the original *ZDA Abstracts* which began in 1943. *Lead and Zinc Abstracts* were first published as separate journals in 1962. The major areas covered by the journal are batteries (mainly lead-acid), battery-powered vehicles, coatings with the metal or its compounds (including paint), lead sheet in buildings, joining and metal working, and corrosion and cathodic protection. There are many other smaller uses of lead, such as in pigments, as stabilizers in rubber and plastics, and as cable sheathing. Regional information centres are listed below.

Information centres

Asia

Zinc and Lead Asian Service, 95 Collins Street, Melbourne, Victoria 3000, Australia.

Australia and New Zealand

Australian Lead Development Association, 95 Collins Street, Melbourne, Victoria 3000, Australia.

Brazil

Instituto Brasiliero de Informacao do Chumbo, Niquel e Zinco, Avenida Nove de Julho, 4015, 01407 Sao Paulo.

France

Centre d'Information du Plomb, 79 Av. Denfert-Rochereau, 75014, Paris.

India

Indian Lead Zinc Information Centre, B–6/7, Shopping Centre, Safdarjung Enclave, New Delhi 110029.

Japan

Japan Lead Zinc Development Association, New Hibiya Building, 3–6 Uchisaiwaicho, 1-chome, Chiyoda-Ku, Tokyo.

Mexico

Instituto Mexicano del Zinc, Plomo y Coproductos, AC Av. Sonora No. 166–1er Piso, Colonia Hippodromo Delegacion Cuauhtemoc, 06100 Mexico DF.

Scandinavia

Scandinavian Lead Zinc Association, Sturegatan 22, S–114 36 Stockholm, Sweden.

United Kingdom

Lead Development Association, 42 Weymouth Street, London W1N 3LQ.

Zambia

ZCCM Ltd., PO Box 48, 74 Independence Avenue, Lusaka.

USA

Lead Industries Association Inc., 292 Madison Avenue, New York, NY 10017.

ILZRO (International Lead Zinc Research Organization)

Lead Industries Association Inc., 292 Madison Avenue, New York, NY 10017.

Other areas

Contact Lead Development Association, 42 Weymouth Street, London W1N 3LQ, UK or the nearest Regional Centre.

CHAPTER EIGHT

Tin and tinplate

ROGER L. DAVIES

Introduction

Tin

Tin is one of the oldest metals known to man; bronze artifacts (bronze being an alloy of copper and tin), dating from about 3500 BC, have been found at the ancient site of Ur of the Chaldees in present day Iraq. From Roman times, most of the world's tin ore was mined in Cornwall but from the early part of the nineteenth century it has also been mined by the Malaysian States. Their monopoly was broken in the 1870s when new ore deposits were found in Australia.

The principal naturally occurring ore of tin is cassiterite or tin oxide (SnO_2). Tin deposits are confined to comparatively small areas within South-east Asia, South America, Africa, Australia, China, USSR and Western Europe, It is found in vein and placer deposits. Vein (or lode) deposits are mined by conventional hard-rock underground mining techniques. Placer (or residual) deposits are won by dredging off- and onshore, gravel pumping, and open-cast mining methods. After mining, the run-of-mine ore is concentrated, and smelted at the mine or exported to other countries, for further processing.

The crude impure tin is cast into slabs or ingots for further refining. Fire or electrolytic refining can be used, and the final 50 kg ingots will have a purity of 99.75 – 99.85 per cent tin after fire refining and at least 99.99 per cent tin after electrolytic refining.

Another source of tin is so-called secondary tin. This is recovered or recycled from three main sources: detinning of tinplate and other tin coated products; recycling of high tin alloys (pewter, solder, white bearing metal etc.) and the reuse of low tin alloys.

Tin is a valuable metal being generally more costly than other major base metals such as lead, zinc and copper. The metal, in its usual silvery white form, is relatively soft and ductile. A second unstable allotrope exists. Tin has the chemical element symbol Sn, derived from the Latin word stannum. It has an atomic weight (or mass) of 118.69, and an atomic number of 50. Tin's melting or fusion point is 231.88°C and it is in Group IV of the periodic table of the elements. Tin forms two series of inorganic chemical compounds – tin (II) or stannous (Sn^{2+}) compounds and tin (IV) or stannic (Sn^{4+}) compounds. It also forms a number of complex stannate compounds. Divalent stannite compounds are also possible. Organotin compounds can also be synthesized, in which at least one tin-carbon bond exists, the tin usually being present in the +4 oxidation state. Tin is able to form binary and tertiary metal alloys with a variety of metals including lead, antimony, copper, zinc, nickel and iron.

Consumption of tin

UK tin consumption in 1984 was 5838 tonnes and market sector five year averages were: tinplate and tinning (including tin-lead or terne plating, tin-zinc and tin nickel plating) 44 per cent; alloys 33 per cent; wrought tin 1 per cent; solder 9 per cent and chemicals 13 per cent.

Most of the countries who are major consumers of tin have little or no available tin ore deposits. Conversely, the major tin producers are relatively small consumers. Tin is therefore an important international trading commodity and until late in 1985 tin trading was subject to a Commodity Agreement – the 6th International Tin Agreement. The Governing body for the Agreement was the International Tin Council (ITC) which was established in 1956. There were six Agreements in the period to 1985, the objective of each Agreement being to achieve the long-term balance between world production and consumption of tin, to alleviate serious difficulties arising from an actual or anticipated surplus or shortage and to prevent excessive fluctuations in the price of, and export earnings from, tin metal.

Although both China and the USSR are substantial producers and consumers of tin, they do not publish official tin trading statistics. The USA General Services Administration (GSA) maintains a strategic stockpile of the metal and makes disposals on to the open market as required.

Prior to the collapse of the 6th Agreement tin was traded in two key markets, where the official daily cash and Forward prices were fixed. They were the London Metal Exchange (LME) and the Kuala Lumpur 'Straits' Tin Market. Prices were also quoted in New York – the Metals Week average daily composite price and American Metal Market price. The USA GSA in Washington also quoted a cash price.

Trading on all markets ceased in late October 1985, following the ITC's suspension of tin buffer stock maintenance and price support operations, when the Council ran out of funds. An audit showed gross liabilities of nearly £900m. Tin had peaked at £10,265 in February 1985 and was £8140/tonne three months forward when the market closed.

Tinplate

Tinplate can be defined as unalloyed low carbon (0.04–0.15 per cent C) cold rolled or reduced mild steel (sometimes called blackplate in the uncoated state), coated very thinly with commercially pure tin. The tin coating can be applied electrolytically on an electro-tinning production line (ETL) or by hot dipping in molten tin. Tinplate combines the strength and formability of steel with the corrosion resistance, solderability and good appearance of tin metal. It should be noted that the term 'tinplate' is often used as a general term (particularly in statistical publications) to include tinplate, a tinplate substitute – called electrolytic chromium/chromium oxide coated steel (ECCS) – formally known as tin-free steel (TFS), and blackplate, although this usage is now not strictly correct. It is more usual to refer to tinplate, ECCS and blackplate individually as such and collectively as tinmill products. This terminology is used through this chapter.

Electrolytic (ET) tinplate is available commercially in sheet or coil or single reduced or rolled (SR) or double reduced (DR), in a gauge or thickness range of 0.16–0.49 mm. The tin coating can be of equal weight on both surfaces – (E designation) or with different tin coating weights on each surface – Differential Tinplate (D designation). Electrolytic tinplate can be supplied in matt, stone, or bright reflective finishes. Hot dipped (HD) tinplate is available in sheet form only, in similar gauges to that obtainable for electrolytic tinplate. HD tinplate is now somewhat difficult to obtain worldwide, having been largely superseded by electrolytic tinplate.

Tinmill products

These are sold or traded primarily by area, rather than by weight. The unit used at present is the Systeme Internationale Tinplate

Area (SITA) of 100 sq. m of tinplate (or 200m^2 of tinned surface). The amount of tin electrolytically applied to each surface or side of the steel base, is defined in grammes per square metre. Therefore the designation E2.8/2.8 corresponds to 2.8g/m^2 of tin per surface. The Basis or Base Box is an obsolete trading unit which refers to an area of 31,360 sq. in (20.2325m^{-2}) or historically 112 sheets – 20 × 14 in. The tin coating weight is defined in pounds per basis box (lb/bb). Therefore 1 lb/bb is equal to a total weight of 22.4 g of tin on both surfaces of a square metre of tinplate (11.2g/m^2 per surface). A SITA is equivalent to 4.943 Basis Boxes.

Packaging is the most important use of tinplate (and tinmill products generally), accounting for about 90 per cent of UK consumption.

Information sources

Books and related publications

The books mentioned below constitute the 'core' publications which a librarian or information officer, would need to acquire, to put together a good general library section on tin – its occurrence, mining, refining and its many uses, including tinplate. They are mostly recent and readily available in print. For those wishing to make a more detailed study, particularly of the historical development of tin and tinplate, some publications, including many which are out of print, are listed in the reference list.

The history of UK tin mining in the last 100 years or so is adequately covered in the series of publications by both Trouson and Morrison (1980, 1984); the publications of the Cornish Chamber of Mines, Truro (1978); Tylecote (1986); and Atkinson (1985). Penhallurick (1986) has comprehensively reviewed early tin mining and trade in tin in the ancient world and described in some detail the occurrence of tin in Africa, Asia and Europe. A chapter is devoted to tin mining in Southwest England, Cornwall, Devon and the Isles of Scilly. The book contains a useful index and a comprehensive bibliography.

The early history of the utilization of tin in the UK has always been closely linked to that of tinplate. Hedges (1964) considered the patterns of early trade dating from the Bronze age; the international control of the tin trade; tin in coinage, pewter and the decorative arts and music. An appreciation of historical tin prices and production statistics for the period 1700–1976, can be obtained from Schmidtz's (1979) world's non-ferrous metal production and prices survey.

Minchington (1957) has fully reviewed the history of the British tinplate industry from 1600 to 1954 and the publications by Brooke (1932, 1944 and 1949) faithfully list the chronology of the Tinplate Works in the UK (mainly in South Wales), from 1600 to the late 1940s. Consult also Spence Thomas (1913), Jones, J. H. (1924), Jones, E. R. (1959) and Wilkins (1903),

Robertson's (1982) book on tin production and marketing reviews production methods, costs and trends, tin consumption trends, smelting and the marketing system, the ITC Agreements and the future supply and reserves potential. Wright (1982) has comprehensively reviewed the extractive metallurgy of tin.

A detailed history of the international tin agreements made in the inter-war years and up to 1983 is given in the authoritative book by Fox (1974). The full story of the genesis of the current 6th ITA, together with its objectives, definitions and the 61 articles and supporting annexes, is set out in a recent United Nations (1983) publication.

Barry and Thwaite's (1983) excellent treatise on tin and its many alloys and compounds discusses the physical and chemical properties of tin; its occurrence and production, various tin alloy systems; tin and tin alloy applications, solder and bearing alloys, melting casting and fabrication of tin alloys, tin and tin alloy coatings (including terne or terneplate) and tinplate, tin chemicals and lesser known applications. Blunden, Cusack and Hill (1985) have brought together in one volume information on the industrial and commercial uses of both the inorganic and organic compounds of tin. Coverage includes industrial manufacture, toxicology, anti-fouling systems, agrochemicals, wood preservatives, PVC stabilizers, catalysts, glass, pharmaceuticals, ceramics and many other uses. Manufacturers and trade names of biocidal formulations are also mentioned. Evans and Karpel (1985) and Zuckermann (1977) cover similar ground. The World Health Organisation (1980) has produced a volume which contains the collective views of a group of world experts on the environmental health aspects of this useful group of chemicals. The report is based on original publications and reviews, and includes a bibliography of more than 300 references.

Price's (1983) work on tin and tin-alloy electroplating considers the development of the various processes for the electro-deposition of binary alloys of tin with copper, lead, zinc, nickel, cobalt and cadmium, from the early 1920s to the present. A wide bibliography of published sources and a useful index is included. The chapter on tin electroplating in Lowenheim's (1974) general study on electroplating is also worthy of study as it reviews the numerous plating bath configurations available for efficient tin plating.

Hoare, Hedges and Barry's (1965) classic work on the technology of tinplate and Hoare and Hedges' (1945) earlier study are some of the few books devoted exclusively to tinplate. Both are now sadly out of print, but are still essential reading for an appreciation of the early days of hot dipped tinplate production, and the advent of the electrolytic tinning line in the mid 1930s. They outline the history and development of the tinplate manufacturing sequence; the structure of the tin coating; the corrosion of tinplate; tinplate testing techniques and the commerce of tinplate. The only modern equivalent is Morgan's (1985) excellent study on tinmill products and canmaking technology, which is an up-to-date publication of manageable proportions and of reasonable cost. It covers both production techniques and technology, and the conversion of the material into a diverse and technically complex range of containers. It is a reliable reference work for students and for anyone involved or interested in the manufacture and use of tinmill products. Coverage includes: the manufacture of tinplate from ironmaking to electroplating; tinplate properties and testing; modern canmaking systems; can and container corrosion; and waste tin recovery and recycling. Future developments are also reviewed. It is well illustrated and cites numerous references to sources of further information. A chapter on the manufacture of tinmill products in the new 10th edition of *The Making, Shaping and Treating of Steel* (AISE, 1985) presents a good general description of the processing of steel for tinmill products, the essential details of various types of modern electrolytic tinplating lines, and the properties, specifications and utilization of the resultant products. The companion works by Roberts (1978, 1983) on the technology of the hot and cold rolling of steel both provide an excellent source for an appreciation of the rolling and finishing techniques used to produce the steel base material for tinmill products.

There are a number of publications which mention the use of tinplate for cans and containers, including aerosol containers. Boustead and Hancock (1981) have made a comprehensive and unique summary of the energy and material requirements of the beverage container and packaging industry, with a detailed study of the amounts of energy used in iron, steel and tinmill products manufacture, and the subsequent fabrication of metal containers. Any of the books by Hanlon (1984), Paine (1983) or the Open University (1979) will adequately describe the role of tinmill products as modern packaging raw materials. The handbooks devoted to aerosol container technology by Sanders (1979) and Johnson (1985) include chapters on the development and technology of tinplate and ECCS aerosol cans. The use of tinplate in packaging is dealt with at length in a later chapter.

Reference books and handbooks

Smithell's Metal Reference Book (1976, Butterworth) presents much useful information on tin's crystallographic structure, metallography, thermochemical data and the general physical and chemical constants. Also listed are the composition, melting point ranges and typical uses of soft solders.

The *American Society of Metals* (ASM, Metals Park, Ohio), *Metals Handbook Series* makes numerous references to tin, its alloys and uses including tinplate. The ASM also produce a *Worldwide Guide to Equivalent Nonferrous Metals and Alloys* (1980). This publication lists specifications and designations of equivalent nonferrous metals and alloys produced or used in the main industrialized nations throughout the world. Another ASM publication is the *ASM Metals Reference Book* (2nd edn, 1983) which provides information on the physical, chemical and mechanical properties, and compositions standard of ferrous and non-ferrous metals and alloys, including tin.

The Metallic Materials Specification Handbook (3rd edn, 1980, Spon) devotes a full chapter (Ch. 50) to tin and tin alloys, which includes a listing of the main commercial tin alloy composition, and brand names.

The CRC Handbook of Chemistry and Physics, 61st edn (CRC Press, Boca Raton, Fl, USA) or 'Rubber Handbook' contains a wealth of chemical and physical property reference data on tin. Also briefly outlined are details of production methods, properties in general and uses of the metal.

The Gmelin Handbook of Inorganic Chemistry, 8th edn (Springer Verlag, Berlin) consists of twenty-one separate volumes, many of which refer to the literature on tin, its alloys, and tin inorganic and organotin compounds.

The Metal Bulletin Handbook (Metal Bulletin Books, Worcester Park, Surrey) is published yearly in two volumes. Volume 1 gives historical details of trends in tin metal and ore prices on the London and Kuala Lumpur markets and also the USA GSA disposals. Prices for secondary and scrap tin are also listed. Volume 2 presents statistics for world tin production and consumption of prime, secondary and scrap metal, stocks, and import/export data for all the major producers and users. Similar production and consumption information is provided by the current annual *Metallstatistick (1972–1984)* (Metallgesellschaft A.G., Frankfurt).

The third edition of *Non-ferrous Metal Works of the World* (1983, Metal Bulletin Books) is indispensable to anyone interested in non-ferrous production facilities worldwide. It is the only single-source guide to both producers and semi-fabricators of both primary and secondary metals embracing 2,000 companies in 95 countries. Also listed are manufacturers of refined tin, tin alloys and tin powder.

The Sheet Metal Industries Handbook (annual, SMI, Redhill, Surrey) lists useful information on various tin alloy compositions; tin solder constituents and physical properties; terminology used in sheet metal working and metal box making; and conversion factor tables for sheet metals.

The *Metal Finishing Guidebook and Directory* (annual, Metals and Plastics Publications, Hackensack NJ) provides information on the practicalities of electroplating tin from acid stannous, fluoroborate and stannate plating baths and the electroplating of tin-lead, tin-nickel and other tin alloys. The *Finishing Handbook and Directory* (annual, Sawell Publications, London) covers similar subject matter.

Journals

There is only one journal exclusively devoted to tin and tinplate – *Tin International* (incorporating *Canning and Packing* and *Tin Printer and Box Maker*). This monthly, published by Metals and Minerals Publications, London, covers all aspects of tin mining; smelting refining; recycling; tin alloys, tin chemicals, and tinplate and can and container making. Price and production statistics for both commodities are included.

Journals which regularly make reference to tin are listed below:

Bulletin of the Institution of Mining and Metallurgy (monthly, IMM, London).
Industrial Minerals (monthly, Metal Bulletin Books, Worcester Park, Surrey).
Metallurgical Transactions – A – Physical Metallurgy and Materials Science (monthly, Met. Soc. AIME/American Society for Metals, Warrendale, PA, USA).
Metallurgical Transactions – B Process Metallurgy (quarterly, Met. Soc. AIME/ASM).
Metals and Minerals International (quarterly, London and Sheffield Publishing Co., London).
Mining Journal (weekly, Mining Journal Ltd., London).
Mining Magazine (monthly, Mining Journal Ltd., London).
Transactions of the Institution of Mining and Metallurgy (monthly, IMM, London).

Journals which regularly refer to the manufacture, the commerce and/or use of tinmill products are:

Aerosol Age (monthly, Industry Publications, Fairfield NJ, USA).
Aerosol Report (monthly, Hüthig Verlag GmbH, Heidelberg, West Germany).
Bleche Rohre Profile (monthly, Meisenback KG, Bamberg, West Germany).
Emballage Digest (monthly, SEPE, 92100 Boulogne, France).
Good Packaging (monthly, San Jose, CA, USA).

Industrial Conserve (monthly, Stazione Sperimentale per e'Industria delle Conserve, Parma, Italy).
Iron and Steel Engineer (monthly, Assoc. of Iron and Steel Engineers, Pittsburgh PA, USA).
Iron and Steel International (bi-monthly, Business Press, Sutton, Surrey).
Iron and Steelmaker (monthly, Iron and Steel Society, Warrendale PA, USA).
Kawasaki Steel Giho (quarterly, Kawasaki Steel Corp., Tokyo).
Kawasaki Steel Technical Report (quarterly, Kawasaki Steel Corp., Tokyo).
Metal Finishing (monthly, Metals and Plastics Publications, Hackensack NJ, USA).
Modern Metals (monthly, Chicago Ill., USA).
NKK News (monthly, Nippon Kokan, Tokyo).
Nippon Kokan Technical Report Overseas (bi-annual, Nippon Kokan, Tokyo).
Nippon Steel News (monthly, Nippon Steel Corp., Tokyo).
Nippon Steel Technical Report (bi-annual, Nippon Steel Corp., Tokyo).
Packaging (monthly, Turret-Wheatland, Rickmansworth).
Packaging (monthly, Cahners Publishing, Denver Co, USA).
Packaging News (monthly, Maclean Hunter, London).
Plating and Surface Finishing (monthly, American Electroplaters Soc., Orlando Fl, USA).
Sheet Metal Industries (monthly, Fuel and Met. Journals Ltd., Redhill, Surrey).
Stahl u. Eisen (twice monthly, Verlag Stahleisen, Dusseldorf, West Germany).
Steel Times (monthly, Steel and Metallurgical Jnls., Redhill, Surrey).
Steel Times International (monthly, Steel and Metallurgical Jnls., Redhill, Surrey).
Tetsu-to-Hagane (monthly, Iron and Steel Inst. of Japan, Tokyo).
Transactions of the Iron and Steel Institute of Japan (monthly, ISI Japan, Tokyo).
Verpackungs-Rundschau (monthly, P Keppler Verlag GmbH, Heusenstamm, West Germany).

Journals which occasionally mention both tin and tinmill products are:

British Corrosion Journal (bi-monthly, Institute of Metals, London).
International Metals Review (bi-monthly, Institute of Metals, London).
Materials Reclamation Weekly (Maclaren Publishers, Croydon).
Materials Science and Technology (monthly, Institute of Metals, London).
Metal Bulletin (twice weekly, Metal Bulletin Jnls., Worcester Park, Surrey).
Metal Bulletin Monthly (Metal Bulletin Jnls., Worcester Park, Surrey).
Metals and Materials (monthly, Institute of Metals, London).

Newspapers

The following newspapers or periodicals feature tin and its uses from time to time:

American Metal Market (weekly, AMM New York).
Financial Times (daily, FT Ltd., London).
Metals Week (weekly, McGraw-Hill, New York).
New Straits Times (daily, Kuala Lumpur, Malaysia).

Abstracting and indexing publications

The following are good sources of information on all aspects of tin and tinmill products:

Alloys Index (monthly with annual cumulations, Metals Information, London).
Corrosion Abstracts (bi-monthly, National Assoc. of Corrosion Engineers, Houston, USA)
Engineering Index (monthly, Engineering Info. Inc., New York).
Food Science and Technology Abstracts (monthly, Internat. Food Information Service, Shinfield, Reading).
Institution of Mining and Metallurgy Abstracts (monthly, IMM, London).
Metal Finishing Abstracts (bi-monthly, Finishing Publications, Teddington).
Metal Finishing Plant and Processes (bi-monthly, Finishing Pub., Teddington).
Metals Abstracts and Metals Abstracts Index (both monthly, with annual cumulations, Materials Information, London).
Mineralogical Abstracts (quarterly, Mineralogical Soc. of GB/Min. Soc. USA, London).
Non-Ferrous Alert (monthly with six-monthly indexes, Materials Information, London).
Pira Abstracts (monthly, Packing Industries Research Association, Leatherhead).
World Surface Coatings Abstracts (monthly, Pergamon Press, Oxford).

Patents

Patents are rarely accorded their rightful place as a primary information source, possibly because of the legal language, in which they are published. Nevertheless for tin, tin alloys, and tin based chemicals, and for tinmill products, they comprise a useful source of new product information and give commercial companies an indication of what products and processes their competitors are developing.

The well-known official publications published by patent offices around the world are: the *Official Journal (Patents)* (weekly, London) for British applications; the *European Patent Bulletin* (weekly); the *Official Journal of the European Patent Office* (monthly); the *Patents Coooperation Treaty (PCT) Gazette*; (fortnightly Geneva) and the Official Gazette (weekly, Washington DC) for the US patents. Further information on Patents publications and searching aids can be obtained from Fisher's chapter in *Information Sources in Engineering* (1985, Butterworth) or *Guide to Official Industry Property* by Rimmer (1985, Science Reference Library).

There are a few *commercial* organizations which publish abstracts of patent specifications and/or patent searching aids. They are Derwent Publications Ltd. (London) and INPADOC – the International Patents Documentation Centre (Vienna). The Derwent journal the *World Patents Index* (weekly); sections E: Chemdoc and M: Metallurgy, subsets M1 (Metal Finishing) and M2 (Metals) are most useful scanning tools.

There are many conventional abstracting journals that sometimes include patent specifications. Often these journals are the only source of an English-language version of a foreign-language specification. Some examples are *Chemical Abstracts*, *Metals Abstracts* and *Metal Finishing Abstracts*.

There are now over forty databases, available for online searching, which are devoted to or include patents. A selection of those of interest to readers of this chapter are: *The UK Patent Office Database* (Infoline); *INPADOC and INPANEW* (Infoline); *Derwent WPI* (SDC-Orbit and DIALOG); *Patsearch* (US Patents and trade marks on Infoline); *Claims* (US Patents and trade marks on DIALOG); *JAPIO* (Japanese patent abstracts on SDC-Orbit); and *Metadex* (Dialog, ESA-IRS and SDC-Orbit).

Further database sources can be elucidated from *Patent Databases* (1983 Online Info. Centre, Aslib , London; and Kyayyat and Ince's *Brit-Line Directory of British Databases* (1986, EDI/Alan Armstrong, Reading) and online in the *Cuadra Directory of Online Databases* (Data-Star) and the *Database of Databases* (Dialog).

Standards

Most Standards organizations in the tin-producing and consuming countries publish standards pertaining to the quality and methods of analysis of ingot tin, tin alloys and tin and tin alloy coated materials. Countries using and manufacturing tinmill products also quote production standards and test methods for their range of materials. Reference should always be made, for UK standards, to the current *British Standards Institution (BSI) Catalogue (or Yearbook)*, which lists all standards in force and with details of amendments or withdrawals. BSI's monthly publication *BSI News* further updates the Yearbook.

Another valuable source of standards information, in English, is the voluntary standards of the *American Society for Testing and Materials (ASTM)*. They are recognized worldwide and many have acquired the status of international standards. The ASTM Yearly subject and alphanumeric volume should always be consulted for the current ASTM situation.

Some ASTMs double as American National Standards and are therefore designated as such. The American National Standards Institute (ANSI), of New York, is the US national standards body.

Other yearbooks or catalogues which can be consulted if one is interested in the International, European or EEC scene are the *International Standards Organisation (ISO) Catalogue*, published in Geneva, and the appropriate National Catalogue or Yearbook for each individual country under investigation. The European Coal and Steel Community have compiled European Standards for Tinmill Products which, in practice, are very similar to the various European National Standards.

The relevant UK standards are as follows:

BS 219:1977 (1984)	Specification for soft solders
BS 1468:1967	Tin anodes and tin salts for electroplating
BS 1872:1984	Specification for electroplated coatings of tin
BS 2920:1973	Cold-reduced tinplate and cold-reduced blackplate
BS 3252:1960	Ingot tin
BS 3338:1961 (1984)	Parts 1 to 21
BS 3788:1984	Specification for tin coated finish on culinary utensils
BS 5140:1974 (1981)	Pewter
BS 6137:1982	Specification for electroplated coatings of tin/lead alloys
BS 6534:1984	Method for quantitative determination of lead in tin coatings
BS AU90:1965	Soft Solders for Automobile Use (withdrawn 1985)
BS Aerospace Series 2B 21:1933	White Metal (88/8/4) ingots for bearings

A selected list of relevant ISOs, Euronorms and ASTMs follows:

Euronorms

EURONORM 145:1978	Tinplate and blackplate – qualities, dimensions and tolerances
EURONORM 146:1980	Tinplate and blackplate in coil form for subsequent cutting into sheets – qualities, dimensions and tolerances
EURONORM 158:1983	Double reduced electrolytic tinplate in cut lengths

International Standards

ISO R1111/1:1983	Single cold reduced tinplate and cold-reduced blackplate Pt.1 – sheet
ISO R1111/2:1983	Pt.2 – Coil for subsequent cutting into sheet form
ISO 2093:1973	Metallic coatings – electroplated coatings of tin
ISO 4977/1:1984	Double cold reduced electrolytic tinplate. Pt. 1 – sheet
ISO 5950:1979	Continuous electrolytic tin-coated cold reduced carbon steel sheet of commercial and drawing qualities

ANSI/ASTM Standards

ASTM B32–83	Specification for Solder Metal
ANSI/ASTM B339–72 (1984)	Classification of Pig Tin
ASTM B545–83	Specification for Electro-deposited Coatings of Tin
ASTM B579–73 (1982)	Specification for Electro-deposited Coatings of Tin-Lead Alloy
ANSI/ASTM B605–75 (1982)	Specification for Electro deposited Coatings of Tin-Nickel Alloy
ASTM A623–83	Standard Specification for Tinmill Products, General Requirements
ASTM A623M–83	Metric Version of A623
ASTM A657–81	Standard Specification for Tin Mill Products, Blackplate, Electrolytic Chromium Coated, single and double reduced
ASTM A599–84	Standard Specification for Steel Sheet, Cold Rolled, Tin-Coated by Electrodeposition

Information available from institutional organizations

Table 8.1. Institutes, Societies and Associations

Organization	*Activities and main publications*
Cornish Mines Development Association (Truro)	Annual Report
States of Malaysia Chamber of Mines	Yearbook
Tin Industry (R & D) Board (Kuala Lumpur)	R & D. on tin mining techniques
British Geological Survey (London)	*UK Mineral Statistics* (yearly)
American Bureau of Metal Statistics (Secaueus NJ)	*Non-ferrous Metals Data Yearbook*
Financial Times (London)	*FT Mining International Yearbook*
Australia Bureau of Mineral Resources (Canberra)	*Australian Mineral Industry Review*
Camborne School of Mines (Redruth, Cornwall)	1. Courses on Mining, Extractive Metallurgy, and Applied Geology 2. Library Facilities
Redruth Public Local Studies Library, Cornwall	Specializes in the History of Cornwall, with a section on the history of tin mining in the country.
Southeast Asia Tin Research and Development Centre (SEATRAD) – Ipoh, Malaysia	Research work on the prospecting, mining and tin ore treatment
International Tin Council (ITC – London)	1. *ITC Annual Report* 2. *Quarterly Statistical Bulletin* 3. *Monthly Statistical Summary* 4. *Notes on Tin (monthly)* 5. *World-Tin in Tinplate (1986)*
International Tin Research Council (Greenford Middlesex – Brunel University Science Park)	1. Technical enquiry and consultancy service 2. Library and Information Service 3. Numerous publications on all aspects of tin and its uses and tinmill products (see references) 4. *Tin Snips* free abstracts bulletin
Institution of Mining and Metallurgy (IMM) – London	1. Library and Information Service 2. Abstract and journal publications
BNF Metals Technology Centre (Wantage, Oxfordshire)	1. Library and Information Service 2. R & D. on tin, tin alloys and tinmill products
American Iron and Steel Institute (AISI, Washington DC)	1. *Tinmill Products Manual* 2. *Steel Can Newsletter* 3. *Annual Statistical report* (includes statistics for all tinmill products)
Institute of Metals (London)	1. Library and Information Service 2. Numerous specialist books, journals, conference proceedings and educational aids produced
American Society for Metals (ASM) Metals Park, Ohio, USA	Extensive publisher of technical books and journals on tin and tinmill products
Materials Information	An Inst. of Metals/ASM joint secondary information service

Table 8.1. Institutes, Societies and Associations

Organization	*Activities and main publications*
Centre Recherches de Fer Blanc (CRFB) – Thionville, France	1. R & D projects carried out for the French tinmill products industry 2. *CRFB Technical Bulletin* (annual)
Chambre Syndicate des Producteurs de Fer-Blanc et Fer Noir (Paris)	1. Sponsored by the French tinmill products manufacturers 2. *Le Fer-Blanc en France et dans le Monde* (annual)
Campden Food Preservation Research Association (Chipping Campden, Gloucestershire)	1. Has an interest in food canning and tinplate and can corrosion 2. Library and Information Service 3. *Annual Report* 4. *Newsletter*
Packaging Industries Research (Pira) – Leatherhead, Surrey	*Statistical and Economic Review of Packaging in the UK* by Mills (annual)
Can Manufacturers Institute (CMI) – Washington DC	1. *Monthly Metal Can Shipments Reports* 2. *Can Shipments Report* (annual)
Metal Packaging Manufacturers Association (MPMA) – Cippenham, Slough	1. *MPMA Annual Report* 2. Occasional publications on can-making 3. Promotes Metal packaging in the UK
Canmakers Institute of Australia (Melbourne)	Similar role to the MPMA.
Syndicate National des Fabricants de Boites Emballages et Boucharge Metalliques (Paris)	Promotes metal packaging in France
The Canmakers (London)	1. Represents major UK producers of cans and their raw material suppliers and specialist detinners 2. *Can Makers Report* (annual) 3. *Can Makers Recycling Update* (annual)
Secrétariat Européen des Fabricants d'Emballages Métalliques Légers (Brussels)	Co-ordinating body for the European Community canmakers
Information – Zentrum Weissblech e.v. (Düsseldorf)	1. Promotes the tinmill products industry in West Germany 2. *Weissblech Reflexionen* (quarterly)
Centro Sperimentale Metallurgico SpA (Rome)	R & D on tinmill products and cans
Stazione Sperimentale per l'Industria delle conserve Alimentari (Parma, Italy)	R & D on tinmill products and cans
Instituo de Agroquimica y Tecnologia de Alimentos (Valencia, Spain)	R & D on tinmill products and cans
Steel Can Recycling Association (SCRA) – Pittsburgh, PA	Promotes steel can recycling in the USA
Institute of Scrap Iron and Steel Inc. (Washington DC)	Promotes steel can recycling in the USA
Steel Products Bureau (New York)	Associated with the AISI

Table 8.2. Information available from manufacturing and commercial trading companies

Company	*Activities and main publications*
British Steel plc – BS Tinplate. Port Talbot, West Glamorgan	1. R & D facilities 2. Library and information services 3. Product range brochures on tinmill products and terne 4. Bibliographies and Literature Surveys 5. Occasional publications on tinmill products manufacture and specifications 6. Current awareness abstracts bulletin – *the Tinplate Bulletin* (bi-monthly) 7. *Annual statistics of British Steel plc*
Metal Box plc. (Wantage, Oxfordshire)	1. R & D on can making 2. Library and information services 3. Occasional promotional publications on their range of packaging products 4. *Metal Box News* (monthly)
Shearson Lehman/American Express	1. Commodity Brokers 2. *Annual Review of the World Tin Industry* 3. *Quarterly, Mid-Year and Annual Reviews of Metal Markets* (including tin)

Note: The numerous tin mining refining and using companies invariably produce annual reports, as do the tinmill products manufacturers, canmakers, and can users. These reports represent a valuable source of financial and operation data on these companies.

Information available from government sources

The UK Government through their publishers Her Majesty's Stationery Office (HMSO) produce a number of useful publications. *British Business* (weekly) regularly lists producer price indices (formerly wholesale prices) for 'tinplate and other tinned sheet' and 'blackplate'. The *Monthly Digest of Statistics* presents statistics on the 'home consumption, imports, exports and stocks of tin metal' and the 'production of finished steel products – tinplate'. The *Annual Abstract of Statistics* gives the same information on an annual basis plus details of 'production and stocks of tin ore'. The *Mineral Resources Consultative Committee Mineral Dossiers* on minerals of economic importance to the UK include dossier No. 9 on tin. The *UK Mineral Statistics* (annual) present summaries on UK tin production consumption and overseas trade, in quantity and value terms. The following

Business Monitors: Production Series (quarterly) should also be consulted:

PQ 2396 – Miscellaneous Minerals
PQ 2512 – Basic Organic Chemicals
PQ 3163 – Metal Storage Vessels (mainly non-industrial)
PQ 3164 – Packaging Products of Metal
PQ 2247 – Miscellaneous Non-ferrous Metals and Their Alloys

The Statistical Office of Her Majesty's Customs and Excise (Southend-on-Sea) publish regular quarterly and annual statistics on exports of 'tinplate' and 'tinned and printed canstock'. The United States *Department of the Interior-US Bureau of Mines* has an interest in tin availability, and publishes a *Mineral Yearbook*, which includes a useful chapter listing US tin and tin consumption for five year periods. They also produce a series of reports – *Bureau of Mines Report of Investigations*, many of which are concerned with the recycling and detinning of tinplate recovered from urban or municipal refuse.

Conferences, congresses, symposia, seminars and meetings

Organizations which regularly host these events, and publish the proceedings thereof, for tin, include the ITC, the ITRI, the IMM, the Institute of Metals, Tin International, Metal Bulletin Conferences and SEATRAD. For tinmill products again the ITC, ITRI and the Institute of Metals can be included plus the Comité International Permanent de la Conserve (CIPC), Paris, and the US Food Processor Institute (Washington DC).

Other organizations hold predominantly steel processing and finishing events, which sometimes include papers on tinmill products include the AISI, the American Society for Metals, the Metallurgical Society of the American Institute of Mechanical Engineers (Met. Soc. AIME), the American Electroplaters' Society (ACS), the Institute of Metal Finishing and the Iron and Steel Institute of Japan (ISIJ).

Many of the technical journals already listed will provide information on forthcoming events, and also reference to announcement services such as *Forthcoming Internat. Scientific and Technical Conferences* (ASLIB, London) and the *World Calendar* (Inst. Metals, London).

The best source of abstracts of papers presented at various events worldwide is probably in *Metal Abstracts* or the online version *Metadex*.

Some examples of recent events are:

The 8th International Congress on Canned Foods, Thessaloniki, Greece, April 1983 (CIPC).
The A–Z of Container Corrosion – February 1982, Chicago, Food Processors Institute of the USA.
The 1st Tin Symposium 'The Market for Tin', December 1985, London, Tin International.
The 5th World Conference on Tin, October 1981, Kuala Lumpur, International Tin Council.
The Seminar on 'Beneficiation of Tin and Associated Minerals', October 1982, Bangkok, SEATRAD.
The 3rd International Tinplate Conference, October 1984, London, International Tin Research Institute.

References

Allen, B. M. (1969), *Soldering Handbook* (Iliffe).
American Iron and Steel Institute (1982), *Tin Mill Products – Steel Products Manual Series* (Washington DC, AISI).
American Iron and Steel Institute (1985), *Steel Cans – No. 1 in Packaging Quality and Integrity* (Washington DC, AISI).
American Welding Society (1979), *Soldering Manual* (New York, AWS)
Association of Iron and Steel Engineers (1978), *The Making, Shaping and Treating of Steel* (Pittsburgh, PA, AISE).
Atkinson, R. L. (1985), *Tin and Tin Mining* (Princes Risborough, Buckinghamshire, Shire Publications Ltd.).
Baldwin, W. L. (1983), *The World Tin Market: Political Pricing and Economic Competition* (Durham, North Carolina Duke UP).
Barry, B. T. K. and Thwaite, G. J. (1983), *Tin and Its Alloys and Compounds* (Chichester, Ellis Horwood).
Belyayer, D. V. (1963), *A Handbook of the Metallurgy of Tin* (Oxford, Pergamon Press).
Blunden, S. J., Cusak, P. A. and Hill, R. (1985), *The Industrial Uses of Tin Chemicals* (London, Royal Society of Chemistry).
Boustead, I and Hancock, G. F. (1981), *Energy and Packaging* (Chichester, Ellis Horwood).
Brett, V. (1982), *Phaidon Guide to Pewter* (Oxford, Phaidon Press Ltd.).
Brooke, E. H. (1932), *Monograph on the Tinplate Works in Great Britain* (Swansea, Ernest Davies and Co.).
Brooke, E. H. (1944), *Chronology of the Tinplate Works of Great Britain* (Cardiff, W. Lewis Ltd.).
Brooke, E. H. (1949), *Appendix to the Chronology of the Tinplate Works of Great Britain 1665–1949* (Cardiff, W. Lewis Ltd.).
Cornish Chamber of Mines (1978), *Mining in Cornwall Today* (Truro, Cornish Chamber of Mines).
Davies, R. L. (1977), *The Corrosion of Tinplate and Tinplate Containers – A Bibliography of Published Literature* (1972–77), Report No. WL/LI/Bib.911/77C, 66 pp. (Port Talbot, British Steel Corporation, Welsh Laboratory).
Davies, R. L. (1985), *A World Listing of Electrolytic Tinplating Lines, Electrolytic Chromium/Chromium Oxide Coating Lines and Dual ETL/ECCS Lines* (Port Talbot, British Steel Corporation, Welsh Laboratory).

Davies, R. L. (1985), *A Technical Guide to Tinmill Products*, Report No. WL/FP/1382/85/C (Port Talbot, British Steel Corporation, Welsh Laboratory).
Evans, C. J. and Karpel, S. (1985), *Organotin Compounds in Modern Technology* (Elsevier).
Fox, W. (1974), *The Working of a Commodity Agreement* (London, Mining Journal Publications).
Geological Sciences Institute (1976), *Geology of the Tin Belt in Peninsular Thailand Around Phuket, Phangua and Takua Pa* (London, HMSO).
Govett, M. H. and Robinson, H. A. (1981), *The World Tin Industry: Supply and Demand* (Sydney NSW, Australia, Australian Mineral Economics Pty Ltd.).
Hall, C. and Murrell, B. T. (1985), *The Techniques of Pewtersmithing* (London, B. T. Batsford Ltd.).
Hanlon, J. F. (1984), *Handbook of Package Engineering*, 2nd edn (New York, McGraw-Hill).
Hedges, E. S. (1960), *Tin and Its Alloys* (London, Edward Arnold).
Hedges, E. S. (1964), *Tin in Social and Economic History* (London, Edward Arnold).
Hoare, W. E. and Hedges, E. S. (1945), *Tinplate* (London, Edward Arnold).
Hoare, W. E., Hedges, E. S. and Barry, B. T. K. (1965), *The Technology of Tinplate* (London, Edward Arnold).
International Tin Council (1981), *European Economic Community – Tin in Tinplate* (London, ITC).
International Tin Council (1981), *United Kingdom – Tin in Tinplate* (London, ITC).
International Tin Council (1983), *United States of America – Tin in Tinplate* (London, ITC).
International Tin Council (1984), *Japan – Tin in Tinplate* (London, ITC).
International Tin Council (1986), *World Tin Mining – Operations, Exploration and Developments* (London, ITC).
International Tin Council (1986), *World – Tin in Tinplate* (London, ITC).
International Tin Research Institute (1975), *Tin versus Corrosion*, Publication No. 510 (London, ITRI).
International Tin Research Institute (1977), *A Guide to Tin*, Publication No. 540 (London, ITRI).
International Tin Research Institute (1977), *Soft Soldering Handbook*, Publication No. 533 (London, ITRI).
International Tin Research Institute (1980), *Metallography of Tin and Tin Alloys*, Publication No. 580 (London, ITRI).
International Tin Research Institute (1983), *Guide to Tinplate*, Publication No. 622 (London, ITRI).
Johnson, M. A. (1985), *The Aerosol Handbook*, 2nd edn (Mendham, New Jersey, USA, Wayne Dorland Co.).
Jones, E. R. (1959), *Toilers of the Hills* (Pontypool, Hughes and Sons Ltd., Griffin Press).
Jones, J. H. (1914), *The Tinplate Industry – A Study in Economic Organisation* (London, P. S. King and Son).
Lankford, W. T. *et al.* (ed.) (1985), 'Manufacture of Tin-Mill Products'. Ch 36. In *Making, Shaping and Treating of Steel* (10th edn) (Pittsburgh, Association of Iron and Steel Engineers).
Leidheiser, H. (1971), *The Corrosion of Copper, Tin and their Alloys* (Wiley).
Litwinow, N. W. (1968), *The Metallurgy of Tin* (Boston Spa, Nat. Lending Library).
Lowenheim, F. A. (ed.) (1974), *Modern Electroplating* (Wiley).
McDonald, E. H. (1983), *Alluvial Mining* (London, Chapman and Hall).

Mantell, C. L. (1970), *Tin: Its Mining, Production, Technology and Applications* (New York, Hafner Publ. Co.).
Metal Bulletin (1983), *Wolfe's Guide to the London Metal Exchange*, 3rd edn (London, Metal Bulletin Books).
Mills, R. (1984), *Statistical Review of Packaging 1979–1983. Update United Kingdom* (Leatherhead, Pira).
Mills, R. (1983), *Statistical Review of Packaging 1978–1982. Outlook 1983–1984 United Kingdom* (Leatherhead, Pira).
Minchinton, W. E. (1957), *The British Tinplate Industry, A History* (Oxford, Clarendon).
Morgan, E. (1985), *Tinplate and Modern Canmaking Technology* (Oxford, Pergamon Press).
Morrison, T. A. (1980), *Cornwall's Central Mines: The Northern District 1810–1895* (Penzance, Cornwall TR18 4AJ, Alison Hodge).
Morrison, T. A. (1984), Cornwall's Central Mines: The Southern District 1810–1895 (Penzance, Cornwall TR18 4AJ, Alison Hodge).
Murach, N. N. et al. (1967), *Metallurgy of Tin, Volumes 1 and 2* (Boston Spa, National Lending Library).
Neumann, W. P. (1967), *The Organic Chemistry of Tin* (London, John Wiley).
Open University (1979), *The Metal Can* (Milton Keynes, Open University Press).
Paine, F. A. and H. Y. (1983), *A Handbook of Food Packaging* (Glasgow, Leonard Hill).
Penhallurick, R. D. (1986), *Tin in Antiquity* (London, Institute of Metals).
Price, J. W. (1983), *Tin and Tin Alloy Plating* (Ayr, Strathclyde, Electrochemical Publications Ltd).
Read, H. H. (ed.) (1970), *Rutley's Elements of Mineralogy* (Hemel Hempstead, Thomas Murby and Co.).
Rees, W. J. and Collins, G. (1973), *Corrosion of Tinplate and Tinplate Containers, Parts 1 and 2*, Bibliographies Nos. 900 and 905 (Port Talbot, British Steel Corporation, Welsh Laboratory).
Roberts, W. L. (1978), *Cold Rolling of Steel* (New York, Marcel Dekker).
Roberts, W. L. (1983), *Hot Rolling of Steel* (New York, Marcel Dekker).
Robertson, W. (1982), *Tin: Its Production and Marketing* (London, Croom Helm).
Samuel, J. R. (1924), *A Short History of Tin and Tinplate* (Newport, Gwent, Williams Press Ltd.).
Sanders, P. A. (1979), *Handbook of Aerosol Technology*, 2nd edn (New York, Van Nostrand Reinhold).
Schmidz (1979), *World Non-Ferrous Metal Production and Prices* (London, Cass).
Spence Thomas, H. (1913), *The Tinplate Trade – Some Recent Developments* (Cardiff, Spottiswoode and Co.).
Taylor, R. G. (1979), *Geology of Tin Deposits* (Amsterdam, Elsevier).
Thoburn, J. T. (1951), *A Study of the Tin Industry: Multinationals Mining and Development* (Farnborough, Gower Pubs.).
Thwaites, C. J. and Barry, B. T. K. (1975), *Soldering* (Oxford University Press).
Tilton, J. E. (1983), *Material Substitution: Lessons from the Tin-Using Industries* (Washington DC, Resources for the Future Inc.).
Tin Industry (R & D) Board (1985), *Tin Mining in Malaysia* (PO Box 12560 Kuala Lumpur, TI (R & D) B).
Trouson, J. H. (1980), *Mining in Cornwall Vol. 1* (Ashorne Derbyshire DE6 1DZ, Moorland Publishing Co.).
Trouson, J. H. (1982), *Mining in Cornwall Vol. 2* (Ashorne Derbyshire DE6 1DZ, Moorland Publishing Co.).
Tylecote, R. F. (1986), *The Prehistory of Metallurgy in the British Isles* (London, Institute of Metals).
United Nations (1983), *Sixth International Tin Agreement* (Geneva 10, UN).

Warren, K. (1964), 'The Steel Industry (in South Wales)' in *South Wales in the 60s – Studies in Industrial Geography*, ed. G. Manners, pp. 103–27 (London, Pergamon Press).

West, E. G. (1982), *Copper and its Alloys* (Chichester, Ellis Horwood).

Wilkins, C. (1933), *The History of Iron, Steel, Tinplate and other Trades of Wales* (Merthyr Tydfil, Joseph Williams).

World Health Organization (1980), *Tin and Organotin Compounds: A Preliminary Review* (London, HMSO).

Wright, P. A. (1982), *Extractive Metallurgy of Tin* (Elsevier).

Yeap, C. H. (1980), *Geology of Tin Deposits* (Kuala Lumpur, 22–11, Dept. of Geology, University of Malaya).

Zuckermann, J. J. (ed.) (1977), *Organotin Compounds: New Chemistry and Applications* (Amer. Chem. Soc.).

CHAPTER NINE

Zinc

M. J. CONWAY

Introduction

It is likely that zinc was known in ancient times, though accounts of its production and use are rare, and the nature of the metal was not much understood. The oldest known artifact is an idol from the prehistoric Dacian site of Dordosch, Transylvania, and two bracelets filled with zinc were discovered in Cameros, which was destroyed in 500 BC. Pompeii, destroyed in AD 79, had a fountain with its upper portion covered in zinc. The Romans used brass at least by 200 BC, and later the production of brass was well known to medieval alchemists. It was believed then that the use of zinc or zinc bearing materials might transmute copper to gold. The word 'zinck' first appears in Europe around the fifth century. Zinc was known in India between 1000 and AD 1300, and was commercially smelted as early as the fourteenth century. During the seventeenth and eighteenth centuries large amounts of slab zinc were imported to Europe from the East. The metal was called by many names, such as Indian tin, calaaem, tutaneg or spiauter. This latter name gave rise to the term spelter, which for many years was the commercial designation of slab zinc.

Around 1730 the knowledge of zinc smelting was brought to England from China, and in 1739 a patent was granted for the process of downward distillation. Between 1740 and 1743 a smelter was erected near Bristol which produced some 200 tons per year. Up to 1758 zinc oxides were used as ore, but in that year a patent for making zinc from blende (ZnS) by roasting the ore and mixing it with charcoal, then smelting it, was granted. Zinc distillation also began in Silesia towards the end of the century, and over the next 100 years various improvements and alterations to furnace design were carried out in Europe. In England, it was discovered at Sheffield that zinc could be rolled into sheets at a temperature of 100–150°C. The first commercial sheet rolling mill was built in Liège, Belgium in 1812, and the first sheet in the USA was produced in Philadelphia in 1857, though the first industrial sheet zinc was not rolled until 1866, at LaSalle, by Matthiessen and Hegler. Several process improvements took place over the next few decades, and by the First World War the electrolytic process, first developed in the 1890s, became widely used. Although there are still several processes in use, more than 75 per cent of the World's zinc is produced by the electrolytic process, and much of the remainder by the Imperial Smelting blast furnace process.

Zinc is primarily used for galvanizing iron and steel products and in zinc-based alloys for diecastings. It is also used in large quantities for brass making, and in smaller amounts for batteries, photoengraving and lithographic plates, flashings on roofs, in pigments, in desilverizing lead, as a stabilizer in rubber and plastics, in ceramics, and in cosmetics, biocides, floor coverings and other textiles.

General information

As with most metals, zinc is a prime commodity and therefore there is a wide range of material containing information on it. Books, periodicals, patents, standards, reports, trade literature and conference proceedings are all standard sources. The majority of the sources quoted below are books, periodicals, reports and conference proceedings. Information on standards is obtainable from many national centres, but those most often encountered are from the British Standards Institution in the UK, from two sources in the USA, the American National Standards Institute (ANSI), or the American Society for Testing and Materials (ASTM), and in West Germany the Deutsches Institut für Normung e.V. (DIN). Details of other countries can be found in *Use of Engineering Literature* edited by A. J. Anthony (Butterworths, 1986).

Patents are best obtained from the appropriate national Patents Office, which will have records of the most recent, relevant documents. The most important sources are the British Patent Office, the US Patent Office and the German Patent Office. Patents may also be searched for online, the major source being Derwent's World Patents Index, available through DIALOG and Orbit. Anthony contains a section dealing with other sources for patents, and these are equally relevant to zinc.

Books, reports, proceedings and periodicals

All inclusive works

The best, and most comprehensive single-volume work on the chemistry and metallurgy of zinc and its alloys is *Zinc: the Science and Technology of the Metal Its Alloys and Compounds*, by C. H. Mathewson (Reinhold/Chapman and Hall, 1959). This provides an account of the metallurgy of zinc, with as much attention paid to economic, scientific and general technological and adaptive aspects of the metal as possible. The chapters are arranged under broad titles, namely: Historical Background, Economics and Statistics, Geology of Zinc Deposits, Chemistry and Physics of Zinc Technology, Treatment of Ore Concentrate, Metallurgical Extraction, Refining, Manufacture of Zinc Oxide, Processing and Uses of Metallic Zinc and Zinc-Base Alloys, Zinc as an Alloying Agent, Use of Zinc in Various Forms for the Extraction of Other Metals, Industrial Zinc Oxide, Sulphide and Other Compounds, the Biological Significance of Zinc, and Use of Zinc in Wood Preservation and in Agriculture. The chapters are divided into their principal subordinate subjects, each division being contributed by an appropriate expert in the field. As the chapters or subchapters are presented as technical papers, each contains a set of references. Although there have been substantial technological changes, particularly in zinc uses, Mathewson remains a most valuable source. Another primary source worthy of note is Gmelin's *Handbuch der Anorganischen Chemie* (Springer Verlag, 1974) – already mentioned in the introductory chapter.

For a historical overview of the development of the technology of metals, *Metallurgy in Antiquity* by R. J. Forbes (R. J. Brill, 1950) is worthy of note, and also Gmelin Vol. 32.

The Zinc Development Association (ZDA) publishes several booklets, conference reports and notes on the general aspects of zinc. *Eurozinc 80: Zinc in Europe* is a report on a seminar held on 28–9 October 1980, which reviews the outlook for zinc production in Europe and how the markets could be expanded, and includes discussions on the sources of zinc concentrates and current trends in zinc using industries. *Eurozinc 79: Zinc in Europe* reports a similar seminar held in April 1979 which reviewed developments and production and uses of zinc. Discussions on the economic outlook for major zinc using industries and current trends in zinc coatings, sheet die casting and brass are included. *Lead and Zinc into the 80s* is a report of an International Conference held in June 1977, and reviews zinc and lead technology and the critical factors governing the demand for zinc and lead in the 1980s. *Lead, Zinc and Cadmium: Retrospect and Prospect* reports a seminar held in

New Delhi in November 1981. Ninety technical papers are presented and discussed. *Technical Notes: Zinc Production, Properties and Uses*, reviews major uses of zinc, and the properties of purer zinc, alloys for die casting and zinc coatings. *ILZRO: Zinc Research Digest* is a yearly review of research carried out for ILZRO (International Lead Zinc Research Organization), which describes programmes on die cast and wrought zinc, surface finishes, architectural applications, zinc for corrosion protection, and zinc compounds. *Zinc Wastes and their Utilization* contains papers from a seminar held by the Indian Lead Zinc Information Centre in October 1980. It deals with zinc sulphate, zinc oxide, recycling of zinc from galvanizer's ash and dross, dechlorination of zinc ash, and other aspects of zinc recovery.

Information sources arranged by major area of interest or uses

Analysis

Zinc by J. H. Kanzelmeyer, pp. 95–169, in *Treatise on Analytical Chemistry of the Elements* Vol. 3 edited by I. M. Kolthoff, P. J. Elving and E. B. Sandell (Interscience, 1964) gives full details of the wide range of conventional analytical techniques available for zinc. *Chemical Analysis in Extractive Metallurgy* (Charles Griffin and Co., 1971) and *Chemical Phase Analysis* (Charles Griffin and Co., 1974) both by R. S. Young, also cover the subject, as does Gmelin, Vol. 32.

Periodicals dealing with this topic are many and varied, but the following are most often encountered. *Analyst*, *Analytica Chimica Acta*, *Analytical Chemistry*, *Atomic Spectroscopy*, *Current Topics in Materials Science*, *Electrochimica Acta*, *Environmental Technology Letters*, *Journal of Applied Physics*, *Journal of the Association of Official Analytical Chemists*, *Journal of Crystal Growth*, *Journal of the Electrochemical Society*, *Journal of Materials Science*, *Journal of Power Sources*, *Metall*, *Oppervlaktetechnieken*, *Physica Status Solidii Parts A and B*, *Plating and Surface Finishing*, and *Scripta Metallurgica*.

Batteries

The Primary Battery, in two volumes, contains a wealth of information on zinc cells, and covers the science involved in primary batteries. Volume 1, edited by W. Heise and N. C. Cahoon (John Wiley, 1971), covers the historic development and chemistry of primary cells, and has the following chapters of interest: 'The alkaline copper oxide: zinc cell' (4) by E. A. Schumacher, 'Mercuric oxide: zinc cell' (5) by S. Ruben, 'Silver oxide: zinc system' (6) by T. P. Dirksen, 'Alkaline manganese dioxide: zinc' (7) N. C. Cahoon and H. W. Holland, and 'Zinc: oxygen cells with alkaline electrolyte' (8) by E. A. Schumacher. Volume 2 by N. C. Cahoon and G. W. Heise (John Wiley, 1976) deals with the Leclanche cell and its related types, together with the zinc-chloride cell. Another book, *Alkaline Storage Batteries* by S. U. Falk and A. J. Salkind (John Wiley, 1969), gives useful information on silver-zinc, manganese-zinc and mercury-zinc batteries. Rechargeable secondary batteries, e.g. silver-zinc, nickel-zinc and zinc-chloride types, are covered in *Small Batteries Vol. 1. Secondary Cells* by T. R. Crompton (Macmillan Press, 1982). Silver-zinc batteries are also dealt with in 'Zinc-silver oxide batteries', edited by A. Fleischer (John Wiley, 1971). ZDA has produced a bibliography, *Zinc and Zinc alloys in Electric Cells*, which covers the period 1929–64, and refers to books, journal articles, patents and trade literature on the use and behaviour of zinc and zinc alloys for primary and secondary cells. Later references are contained in *Zincscan* (q.v.).

Periodicals include *Batteries Today*, *Battery Man*, *Electrical Review*, *Electrochimica Acta*, *Journal of the Electrochemical Society*, and *Journal of Power Sources*.

Building

Roof Design by P. von Flotow (Karl Kramer Verlag, 1964) details zinc uses in various buildings. The book is produced in English, German and Italian. ZDA has published jointly with other organizations, *Eurozinc '82*, and *Eurozinc '83: Zinc in Building*, in which outstanding applications of zinc in building, mainly roofing in sheet zinc, but with examples of continuously galvanized steel cladding and galvanized steel roofing structures presented. *Working Zinc Sheet and Strip* gives specifications for zinc sheet in buildings, for roofing, wall cladding, guttering, flashings and weatherings. These notes are intended for operatives who need practical advice on working zinc sheet. A series of leaflets, *Galvanized Steel Case Histories*, have many applications of galvanizing in buildings, titles are: *Dome for the Imperial War Museum*, *Swimming Pool near Paris*, *Steel Lifeboat*, *Houses of Parliament* (galvanized iron roof tiles on Big Ben's clock tower), *Bayer Cross*, *Roll-on/Roll-off Facilities for the Holyhead Car Ferry*, *Steel Bridge Beams*, *Whitbread Brewery*, *Edinburgh Military Tattoo* (grandstands), and *Scheveningen Pier*. *Rust Prevention by Hot Dip Galvanizing*, an English translation of a booklet by R. Thomas of the Nordic

Galvanizers Association, contains much on steel structures and their protection by galvanizing. *Hot Galvanized Steel in Building, and Galvanizing for Civil Engineering* also deal with building. *Fe + Zn 14, 15, 16* and *17* all concern themselves with various structures, and are reviews of new and interesting applications of hot galvanizing to protect structures in industrial or semi-industrial sites. A bibliography, *Rolled Zinc and Rolled Zinc alloys*, covers the period 1929–64, and refers to books, technical articles, patents and selected trade literature on the production, properties and main uses of zinc, zinc alloy sheet and strip. Later references are covered by *Zincscan*.

Information on zinc in building in periodicals is widely scattered throughout the journals concerned with architecture, building and construction.

Chemicals

Gmelin-Handbuch Vol. 32 gives the most complete survey of information on zinc chemicals. The best English survey of the properties and industrial applications of zinc chemicals is *Zinc Chemicals* by M. Farnsworth and C. H. Kline, ILZRO (International Lead Zinc Research Organization, 1973) with a chapter on organozinc compounds by J. G. Noltes. There are two other multi-volume series which include good sections on zinc chemicals. *Nouveau traite de chimie minérale. Tome V. Zinc, cadmium, mercure*, edited by P. Pascal (Messon et Cie, 1963), contains a bibliography giving a complete survey of the literature to 1957. *Zinc Oxide Rediscovered* by E. Brown (New Jersey Zinc Co., 1957) provides a comprehensive overview of its use both as a pigment and in semiconductors together with information on its properties and further applications. A bibliography of 163 references is included. Most of the journals mentioned in the other sections contain information about chemicals and chemistry involving zinc.

Ceramics

As zinc is not widely used in ceramics there are few sources. *Ceramic Glazes* by F. Singer and W. L. German (Borax Consolidated Ltd., 1960) and *Ceramic Glazes* by K. Shaw (Elsevier, 1971) both contain sections dealing with zinc compounds. Modern uses of ceramics in semiconductors are adequately covered in books dealing with electronics, though zinc is not a predominant metal in such devices. *American Ceramic Society Bulletin*, *Journal of Materials Science* and *Journal of Non-Crystalline Solids* are the main periodicals, though other publications dealing with materials in general will also have occasional articles.

Corrosion

As zinc is mainly used as an anti-corrosion agent, this section overlaps greatly with Galvanizing below. So, listed here are the sources mainly concerned with corrosion and corrosion resistance *per se*, rather than the details or properties of the coatings that prevent it.

There are a large number of reference works, among which the following are most useful. *Corrosion* in 2 volumes edited by L. L. Shreier (Newnes-Butterworths, 1976), *Korrosionverhalten von Zink Vol. 2. Verhalten von Zinc in Wassern* by W. Weiderholt (Metall-Verlag GmbH, 1965). An English translation, *The Corrosion Behaviour of Zinc*, is available. *Corrosion and Corrosion Control* by H. H. Uhlig (John Wiley, 1963) and *An Introduction to Metallic Corrosion* by U. R. Evans (Arnold, 1963) both give good introductions to the subject. The American Society for Testing and Materials produces *Underground Corrosion*, a symposium volume of a November 1979 meeting in Williamsburg, edited by E. Escalente (ASTM, 1981). Cathodic protection is covered in *Cathodic Protection: Its Theory and Practice in the Prevention of Corrosion* by J. A. Morgan (Leonard Hill Books, 1959) and in *Handbook of Cathodic Protection* by W. v. Baeckmann and W. Shenck (Portcullis Press, 1975). *Protection against Atmospheric Corrosion; Theories and Methods* by K. Barton (John Wiley, 1976) disusses this aspect of corrosion, and E. Rabald, in *Corrosion Guide* 2nd revised edition (Elsevier Publishing Co., 1968), subdivides the book by the corrosive agent, and each subdivision deals with the materials affected. *Corrosion Resistance of Metals and Alloys*, 2nd edn edited by F. L. LaQue and H. R. Copson (Reinhold Publishing Corp., 1963), summarizes knowledge on corrosion rates and processes. Mathewson likewise deals with corrosion in detail.

The Zinc Development Association (ZDA) also publishes much material on corrosion as such, and the following are currently available. *Zinc – Its Corrosion Resistance* is a comprehensive data compilation of works on the corrosion resistance of zinc in air, water, soils and other media, prepared by the Battelle Institute for ILZRO, and published by ZDA and other associations. *Atmospheric Corrosion Resistance of Zinc* is a critical review by Prof. G. Shikorr of the corrosion behaviour of zinc, zinc alloys and zinc coatings under atmospheric exposure. This was specially commissioned through the Zinkberatung eV, for the European Zinc Producers Technical Committee and was first published in German. A companion volume *Aqueous Corrosion Resistance of Zinc* by Dr W. Wiederholt has also been published in Germany, but an abridged translation is available on loan from the ZDA/LDA

Library. *Technical Notes on Zinc – Zinc for Cathodic Protection* gives introductory information for those not familiar with the process.

Periodicals for this section are many and varied. The major sources are *Corrosion Prevention and Control*, *Korrosion (GDR)*, *Metal Finishing*, *Metall*, *Metals Progress*, *Neue Hutte*, *Oppervlaktetechnieken*, *Surface Technology*, and *Werkstoffe Korrosion*. There are further periodicals detailed under Galvanizing and Hot Dip Galvanizing below.

Diecasting

Most sources for diecasting are in technical papers, but there are a few noteworthy books. *The Diecasting Book*, edited by A. V. Street (Portcullis Press, 1977) is the first comprehensive book on diecasting in English since H. H. Doehler's book, *Die Casting* (McGraw-Hill, 1951). It contains chapters of special interest to zinc diecasters which cover die casting machines and their maintenance, controlling and monitoring the process, zinc alloys, designing for die casting and die design, die steels, handling metal and castings, trimming operations, finishing processes and quality control. A chapter specifically on zinc describes the properties and uses of zinc alloys for die casting, and lists many national specifications. The book has many references and has an extensive bibliography. *Diecasting Metallurgy* by A. Kaye and A. Street (Butterworths, 1982) comprehensively covers significant developments in die casting over the past few years, and emphasizes the importance of metallurgical factors in efficient and economic production. *Pressure Diecasting Part 1: The Technology of the Casting and the Die* by D. F. Allsop and D. Kennedy (Pergamon Press, 1983) and *Pressure Diecasting Part 2: Metals – Machines – Furnaces* by B. Upton (Pergamon Press, 1982) both deal with high-pressure diecasting. Volume 1 covers furnaces, alloys and machines used. The equipment needed and the calculations involved to obtain performance data are detailed, and many examples demonstrate methods of parameter measurement and calculation. Volume 2 gives background information and explains the diecasting process, while focusing on die design and the effects of the materials and processes on the quality of the product.

As this is one of the main areas in which zinc is used, the Zinc Development Association (ZDA) has published many works. There follows a list of the majority of them: *Zinc Die Casting Guide*; *Zinc Die Castings: Design Hints and Directory* (to members of the Zinc Alloy Die Casters Association); *Engineering Standards for Zinc Die Castings to BS 1004*; *Technical Notes on*

Zinc: Zinc Alloy Die Castings; *Safety in Pressure Die Casting*, a handbook on guarding pressure die casting machines, avoiding fire and electrical hazards and other risks. *Technical Information Sheets*, which are a series of four-page leaflets. *Machining and Forming Zinc Die Castings*; *Corrosion Resistance of Zinc Alloy Diecastings*; *Joining and Fixing Zinc Alloy Die Castings*; *Zinc Alloy Diecastings: Prototypes and Short Runs*; *Zinc Alloy Die Cast Gears and Gearboxes*; *Training Die Casting Machine Operators* (includes 24 slides). *Code of Practice for Heat Treatment of 5% Chromium Steels (H13 type)*; *Guidelines for Die Casting Die Makers*; *Die Casting Dies: Designing* by E. A. Herman, a book published by the Society of Die Casting Engineers, USA, which gives a very detailed account of pressure die design. *Developments in Zinc Die Casting Technology*; *Zinc Die Casting Die Design Aid*; *Die Casting Die Enquiry Forms*, which give die makers all the additional detailed information necessary for them to quote for pressure die casting dies. *Eurozinc – Zinc Die Castings: Better Quality at Lower Cost*, a series of 4-language bulletins which describe and illustrate up-to-date applications. *Engineering Properties of Zinc Alloys*; *Designing in Zinc*; *Glossary of Die Casting Terms*, in 5-languages (English, French, German, Italian and Spanish). *Finishing and Electroplating Die Cast and Wrought Zinc*; *International Pressure Die Casting Conferences*, which are major conferences on developments in pressure die casting held every 3 years from 1954 to date. Edited proceeding are available for several of them. *Markets for Zinc Die Castings in the UK*; *Zinc Die Casting: A New Film for Designers and Engineers*, which is a brief description of a joint US/UK film on up-to-date zinc die casting, plus a booking form for those wishing to borrow a copy. Periodicals include *Die Casting Engineer*, *Diecasting Bulletin*, *Erzmetall*, *Foundry Trade Journal*, *Journal of Metals*, *Metall*, *Modern Metals* and *Precision Metal* as the major sources, but such journals as *Casting Engineering and Foundry World*, *Design Engineering*, *Fonderie Fondeur d'Aujourd'hui*, *Materials Engineering*, *Materials Performance*, and *Metalworking and Production* also contain relevant material.

Electrochemistry and electronics

Gmelin Vol. 32 deals in detail with this, while T. F. Sharpe in the *Encyclopedia of Electrochemistry of the Elements* Vol. 1 (Marcel Dekker, 1973) concerns himself more with the theoretical aspects. Periodicals dealing with electrochemistry overlap greatly with the Batteries literature. *Electrochimica Acta*, *Journal of Applied Physics* and the *Journal of the Electrochemistry Society*, *Journal of*

Materials Science, *Journal of Power Sources*, *Metalloberflache*, *Oppervlaktetechnieken*, *Physica Status Solidii A*, *Physical Review B*, *Scripta Metallurgica*, and *Surface Technology* are the most useful.

Environment

Zinc is not toxic, and is recognized as being a key element in our health, and as a beneficial micronutrient in other organisms. Because of this, there is relatively little on the environmental aspects of its use. The major work is *Zinc in the Environment* by J. O. Nriagu, in 2 volumes (Vol. 1 Ecological cycles; Vol. 2 Health effects) (Wiley Interscience, 1980). It is a comprehensive review of the current knowledge about the role of zinc in the environment, and concentrates on the chemical behaviour and biological effects of zinc in the biosphere, hydrosphere and geosphere. *Zinc*, a report of the Subcommittee on Zinc of the Committee on Medical and Biologic Effects of Environmental Pollutants of the National Research Council (University Park Press, 1979), is an attempt to place in perspective the role of zinc in the environment. Periodicals dealing with the environment are many and varied, and may contain occasional references to zinc. The most important ones are, *Archives of Environmental Health*, *Bulletin of Environmental Contamination and Toxicology*, *Chemical Engineering*, *Chemistry in Industry*, *Environmental Health Perspectives*, *Environmental Pollution Parts A and B*, *Environmental Science and Technology*, *Environmental Technology Letters*, *Journal of Water Pollution Control Federation* and *Science in the Total Environment*.

Zinc coatings – general

Zinc coatings may be applied in several different ways, hot dip galvanizing (or true galvanizing), zinc spraying, zinc plating (electrogalvanizing), sherardizing and other mechanical plating processes. The sources mentioned here are concerned with processes other than hot dip galvanizing, which will be covered in detail in the following section.

The Canning Handbook on Electroplating contains a chapter on Zinc (Ch. 22, pp. 540–63, in the 20th edition, 1966). The ZDA produces several publications dealing with these processes. *Inspection of Zinc Sprayed Coatings* is a detailed guide to inspection of zinc sprayed steel; *Technical Notes on Zinc – Zinc Spraying*, outlines the advantages of spraying over other methods; *Painting Sprayed Zinc Coatings* gives details of paint treatments and the results of exposure trials carried out by the Paint Research Association on behalf of ILZRO. A bibliography, *Zinc Spraying*, covers the period 1913–68 and deals with the process, properties and uses of the resulting coatings. Technical notes on *Zinc Plating* and *Sherardizing* are also available. *Galvanizing, Sherardizing and Other Diffusion Coatings*, a translation by D. E. Hayler of E. V. Proskurkin and N. S. Gorbunov's book (Technicopy Ltd. and ZDA, 1975) is an integration of Russian and Western knowledge of the title processes.

Periodicals for generalized coatings are the same as for hot dip coating, so refer to the next section for details of these.

Hot dip galvanizing

The majority of works in this area are produced under the auspices of either the Zinc Development Association, the International Lead Zinc Research Organization (ILZRO), or other similar national organizations worldwide. The greater part of ZDA publications on zinc coatings refer to the hot dip process. There follows a list of those presently available.

Galvanizing (Hot-Dip) by H. Bablik, and translated by C. A. Bentley (3rd edn, 1950, available from ZDA), was the first book to undertake a scientific treatment of hot-dip galvanizing, and remains a key work on the subject. *Galvanizing, Sherardizing and Other Diffusion Coatings*, a translation by D. E. Hayler of E. V. Proskurkin and N. S. Gorbunov's book (Technicopy Ltd. and ZDA, 1975) is an integration of Russian and Western knowledge of the title processes. *Technical Notes on Zinc – Zinc Coatings* gives a brief account of the methods by which zinc coatings can be applied; *Zinc Coatings* is a comprehensive survey covering corrosion theory, production methods, uses, finishes etc; *Welding Zinc Coated Steel* gives technical details of all forms of welding, quality, safety and health, and is published by the American Welding Society by arrangement by ILZRO. *Galvanized Guide* is information required by the specifier of galvanizing for fabricated products in a quick reference form; *1981 Directory of General Galvanizers (UK)* is self explanatory. *Technical Notes on Zinc – Galvanizing* gives a brief description of the procedures used in general galvanizing; *Galvanizing for Structural Steelwork*, a manual giving advice on design, joining and painting galvanized steel. *Galvanizing for Structural Steelwork*, a reprint of an article by J. Wilcock in the Nov. 1980 issue of *Building with Steel*. *Galvanized Steel Case Histories*, a one-page, colour series with the following titles: *Dome for the Imperial War Museum*, *Swimming Pool near Paris*, *Steel Lifeboat*, *Houses of Parliament* (galvanized iron roof tiles for Big Ben's clock tower), *Bayer Cross*, *Roll-on/Roll-off Facilities for the Holyhead Car Ferry*, *Steel Bridge*

Beams, *Whitbread Brewery*, *Edinburgh Military Tattoo* (grandstands), and *Scheveningen Pier*. *Intergalva News Sheet*, periodically issued by the European General Galvanizers Association. *Galvanizing in Sport and Transport* (English, Italian and Spanish edition). *Galvanizing for Civil Engineering*, a reprint article by F. C. Porter and P. W. Toseland covering specification of galvanizing and uses in chemical engineering, river works, marinas, docks, piers, reinforced earth and reinforced concrete. *Galvanizing Steel into Low-Cost Corrosion Resistance* gives a brief account of galvanized products corrosion resistance, cost and practical considerations. *Galvanizing of Steel Reinforcement*, a case history of 30 years' use in Bermuda. *Galvanized Reinforcement for Concrete II*, is a critical review of the literature, plus papers from conferences. *Guidance on the Use of Galvanized Tanks and Cisterns*; *Galvanized Pipes for Drinking Water*, which covers corrosion and hygiene aspects of galvanized steel piping for work at Mannesmann AG. *Galvanizing of Steel: Modern Techniques, Developments and Performance*, a reprint article by F. C. Porter from *Metallurgia*. *Non-ferrous Waste as a Source of Zinc for Electrogalvanizing* by E. R. Cole was produced by the US Bureau of Mines in 1987. *Reactions between Iron and Molten Zinc*, a comprehensive review by Dr Horstmann translated into English. *Galvanized Fasteners*, a leaflet on the use of hot dip galvanized fasteners; *Galvanized Steel in Friction Grip Connections* gives engineering data on galvanized bolts and on galvanizing of faying surfaces in friction grip connections. *Welding of Galvanized Steel* is a collection of five papers given at a 1972 symposium; *Fusion Welding Galvanized Steel* covers arc, carbon dioxide and shielded metal arc processes, and cracking behaviour of shielded metal-arc welds; *Rust is not a Must*, a summary of information from the 1980 Galvaforum seminars. *Fe + Zn European General Galvanizers Bulletins* review and illustrate new and interesting applications of hot dip galvanizing to protect steelwork in a variety of industrial, semi-industrial and other fields, and are also published in French, German, Italian and Spanish. The following are available: *Fe + Zn 4*, on mining; *Fe + Zn 7*, on exhibitions in Europe; *Fe + Zn 14*, galvanized steel in chemical engineering; *Fe + Zn 15*, on tubular structures; *Fe + Zn 16*, galvanized steel reinforcement of concrete; *Fe + Zn 17*, galvanized steel in agriculture. General Recommendations for Painting Galvanized Steelwork which is a folder published jointly with the Paintmakers Association and gives simple recommendations for painting steel which has been galvanized after fabrication; *General Galvanizing Practice* is a guide to practices and techniques of proved worth; *Preserving Galvanized Steel and Zinc from Wet Storage* deals with the occurrence of wet storage stain, its prevention, and restoration of surfaces where damage has occurred; *Zinc and Its Role in Coil Coated Products* is an account of zinc production and its use in coatings, together with worldwide statistics; *Galvanizing for Profit* is a manual for costing practice covering the principles of costing both by-the-job costing and by-the-bath hour rate; *General Galvanizing – A Manual of Safety* is a revision of *A Manual of Good Housekeeping and Safety*, which takes into account changes in law and recent statistics, and stresses the importance of safety training. *Trends in General Galvanizing* is a comprehensive survey of end uses of general galvanized products in the UK, carried out each year by ZDA; *International Glossary of Galvanizing Terms* gives equivalents of 531 terms in English, Danish, Dutch, French, German, Italian, Spanish and Swedish. *The Galvanized Sheet Industry* gives a list of continuous sheet galvanizing facilities worldwide as in 1976. *World Continuous Electrozinc Lines* is a list of electroplating lines for sheet with capacity, size, type, date and trade name. *Rust Prevention by Hot Dip Galvanizing*, an English translation of a booklet by R. Thomas of the Nordic Galvanizers Association which compares zinc coatings, reactions between iron and steel, design and fabrication of structures, testing, corrosion resistance, painting and welding. *Hot Galvanized Steel in Buildings* shows the wide range of uses of galvanized steel in buildings. Sections are on main structures, windows, cladding and roofing, exterior steelwork, interior steelwork and services. *Training for Operators in Hot Dip Galvanizing* is a set of training elements prepared to cover operations in the galvanizing process. These documents are published by the Engineering Industry Training Board, and are available from ZDA. A bibliography, *Hot Dip Galvanizing*, covers articles published in *Zincscan* between January 1955 and June 1975, together with a comprehensive subject index. Later references are published in *Zinc Abstracts*. Another bibliography is *Wire Galvanizing*, covering the period 1943–65, on the process of hot dip galvanizing wire, and the properties and applications of the resulting coating. It does not include material of general interest, but only contains documents specifically concerned with wire. *Galvanizing Conference Proceedings* are collections of papers given at conferences held every three years by the European General Galvanizers Association, of which ZDA holds the Secretariat. They are available as follows. The edited proceedings of the 1st (Copenhagen), 2nd (Düsseldorf), 3rd (Oxford), 5th (Benelux) and 7th (Paris) are now out of print, but details of papers contained are available on request, and photocopies can be supplied at cost. The 10th conference had no volume issued, being a discussion only. Edited proceedings are available for the 4th (Milan 1956), 6th (Interlaken 1961), 8th (London 1967), 9th (Düsseldorf 1970), 11th (Madrid 1976), 12th (Paris 1979) and 13th (London 1982).

Two ILZRO seminar proceedings are also of interest. *Galvanizing of Silicon Containing Steel* are papers from a seminar held in Liège, Belgium on 21–22 May 1975. *The Hot Dip Galvanizing Seminar* was held in St Louis Missouri on 9–10 June 1976.

A number of films are also available, either from the Central Film Library (CFL) or the ZDA Library, as noted. *No Rust Here* (CFL), general galvanizing process and applications, 16mm colour/sound, 20 mins; *Too Late for Wuppertal* (CFL), German film with English commentary on protection of steel structures and goods by galvanizing, 16mm colour/sound, 20 mins; *Think About It* (CFL), shows the wide range of zinc coatings, how protection is obtained and the applications of zinc coated steel, 16mm colour/sound, 20 mins; *Hot Dip Galvanizing after Fabrication* (ZDA) shows how zinc protects iron and steel from rust and reduces maintenance, 16mm colour/sound, 15 mins; *Steel in Colourful Life* (ZDA), a Japanese film with English commentary dealing with galvanized and painted steel sheet, 16mm colour/sound, 20 mins; *Engineering Craft Studies – Control of Corrosion* (ZDA), a BBC film which includes a sequence on galvanizing, 16mm colour/optical sound; 30 mins; and *Street Furniture* (ZDA), filmed slides showing the need for good protection of street furniture, for use with La Belle Courier 16mm projector.

There are also many journals in this field. The most often encountered are: *Corrosion Prevention and Control*, *Finishing Industries*, *Electrochimica Acta*, *Finishing*, *Galvano Organo*, *Galvanotechnik*, *Industrial Corrosion*, *Jahrbuch Oberflachentechnik*, *Journal of Coatings Technology*, *Journal of the Electrochemical Society*, *Metal Finishing*, *Metall*, *Metalloberflache*, *Modern Metals*, *Neue Hutte*, *Oppervlaktetechnieken*, *Plating and Surface Finishing*, *Precision Metal*, *Product Finishing (UK)* and *Products Finishing (USA)*, *Surface Technology*, *Transactions of the Institute of Metal Finishers*, *Wire*, and *Zeitschrift für Metallkunde*. The *International Iron and Steel Institute Reports* are also a valuable source.

Joining

The Soldering Handbook (B. M. Allen, Iliffe Books, 1969) and *Solders and Soldering Materials, Design, Production and Analysis for Reliable Bonding* (K. H. Manko, McGraw-Hill, 1964) give good surveys of solders and soldering, the latter reflecting American practice.

Periodicals include *Brazing and Soldering*, *Metall*, *Plating and Surface Finishing*, *Tin and Its Uses*, and the *Welding Journal*.

Metal working

Mathewson deals with the subject in detail, while J. T. Cairns and P. T. Gilbert in *The Technology of Heavy Non-ferrous Metals and Alloys* (George Newnes, 1967) cover only some aspects. Periodicals would be much the same as for Building and Joining above. In addition, *Acta Metallurgia*, *The Engineers Digest*, *Machinery and Production Engineering*, *Materials Engineering*, and *Metallurgical Transactions A* would be useful.

Phase diagrams and thermodynamics

Good, annotated phase diagrams can be found in Mathewson and in the series *Constitution of Binary Alloys*, 2nd edn (McGraw-Hill, 1958) by M. Hansen; 1st supplement 1965, by R. P. Elliott; 2nd supplement 1969, by F. A. Shunk. They are also contained in two ASM books, *Metals Handbook Vol. 8. Metallography, Structures and Phase Diagrams* (1973), and *Selected Values of the Thermodynamic Properties of Binary Alloys* by R. Hultgren *et al.* (1973). Gmelin Vol. 32 also includes critical reviews of thermodynamic data. Microstructural and metallographic information is contained in the ASM Metals Handbook, and phase diagrams for ceramic systems have been compiled by the American Ceramic Society in *Phase Diagrams for Ceramists* (1964).

Periodicals for this aspect of the metal are: *Analytical Chemistry*, *Applied Physics Letters*, *Erzmetall*, *Journal of Applied Physics*, *Journal of the Electrochemistry Society*, *Journal of Materials Science*, *Metallurgical Transactions A*, *Physica Status Solidi A*, *Scripta Metallurgica*, *Transactions of the Institute of Mining and Metallurgy*, *Zeischrift für Metallkunde*, and many more, which also deal with physics or chemistry or other properties of materials.

Pigments

The primary source is the *Pigment Handbook* edited by T. C. Patton, 3 vols (John Wiley, 1973), though this reflects American usage and terminology. Individual pigments are considered in detail, including an historical introduction, methods of manufacture and applications. An introduction to the subject of pigments can be found in *Pigments* by J. S. Remington and W. Francis (Leonard Hill Ltd., 1954), and *Zinc Oxide Rediscovered*, described in the Chemicals section above.

The ZDA publishes several works covering pigments, as

detailed below. *Technical Notes on Zinc – Zinc Dust*, which is basic information for engineers, designers and teachers; *Protecting Steel with Zinc Dust Paints/2* are papers and discussions from a seminar held in May 1972, which covers worldwide developments particularly in the UK, Netherlands, USA and Japan. *Protecting Steel with Zinc Dust Paints/3* reports a seminar held in May 1975. *Zinc Dust in Paints* outlines dust grades, formulation of paints, applications and uses. *Zinc Dust and Powder*, an ILZRO publication, reviews information on production, properties and applications. Zinc Oxide Properties and Applications is a comprehensive review and commentary on the literature, also produced by ILZRO. Two bibliographies, *Zinc Dust in Paints* (1931–68), and *Zinc Dust in Paints (1969–76)*, from the ZDA, cover the literature on research and development in the field. Subjects include surface preparation, formulation of inorganic and organic zinc dust paints, and their uses in a wide range of environments. The latter volume contains papers from 3 seminars organized by the ZDA in 1969, 1972 and 1975.

Periodicals include *American Paint and Coatings Journal*, *Anti-Corrosion Methods and Materials*, *Corrosion Prevention and Control*, *Finishing Industry*, *Journal of Coatings Technology*, *Journal of the Oil Colour Chemists Association*, *Korrosion*, *Metall*, *Metal Finishing*, *Paint and Resin*, *Pigment and Resin Technology*, and *Plating and Surface Finishing*.

Statistics

Zinc mining, production and end-use statistics are collated by many organizations. The most reliable sources are monthly bulletins published by the International Lead Zinc Study Group (ILZSG), which produces *Lead and Zinc Statistics*, and the World Bureau of Metal Statistics, producing *World Metal Statistics*. The American Bureau of Metal Statistics produces the *Yearbook of the ABMS*, which gives lists of major smelting and refining plants, and includes mine production figures for individual US mining companies. Metallgesellschaft AG also produces a yearbook called *Metal Statistics*, which contains a historical graph showing price changes of zinc since 1850.

Periodicals for this section cover a wide field, and include most mining and engineering journals, such as the *Canadian Mining Journal*, *Engineering and Mining Journal*, *Erzmetall*, the *Journal of Metals*, *Metall*, the *Mining Magazine*, *Transactions of the Institute of Mining and Metallurgy*, and the *US Bureau of Mines Mineral Commodity Profiles and Reports*.

Rubber and plastics

Zinc Oxide and Rubber (Zinc Institute Inc., 1983) describes the American and French processes, and chemical precipitation method for manufacture of zinc oxide. Size measurement methods, the effects of impurities, and the positive effects of zinc oxide on polymers are all covered in detail, as are the pigments contribution to thermal properties, mould shrinkage and cleanliness, tyre abrasion wear resistance etc. *Zinc Oxide: A Weathering Stabilizer for Plastics* by B. Baum and D. S. Carr (ILZRO, 1981) summarizes laboratory tests, outdoor exposure results and commercial applications of zinc oxide in preventing degradation by UV light. The manual is aimed at chemists and engineers. Periodicals include *Metal Progress*, *Modern Plastics International*, and *Plastics Engineering*.

Miscellaneous

There are a wide variety of books which deal with non-ferrous metals or chemical compounds and include sections on zinc, and these may be found through any specialized library. ZDA publishes a bibliography, *Zinc and Zinc Alloy Printing Plates* which reviews the literature from 1929–63, and deals with the manufacture and properties of rolled zinc and zinc alloy plates for photo-engraving, lithography and allied printing processes.

There are a large number of journals which contain information on zinc and related subjects, in addition to those mentioned.

A full list of all the journals monitored for *Zincscan* is available from the Zinc Development Association.

Organizations, research associations etc.

The *Zinc Development Association (ZDA)* is engaged in promoting the use of zinc and its compounds and derived products. To this end it provides a technical service and an information service, run from 42 Weymouth Street, London. It has close links with similar organizations throughout the world, and is actively engaged in organizing conferences and meetings for and between producers and potential end-users of the metal. As part of the information service an abstracts journal, *Zincscan*, is published quarterly.

Zincscan stemmed from the original ZDA Abstracts which began in 1943. Lead and zinc abstracts were first published as separate journals in 1962. The major areas covered by the journal are batteries (mainly dry cell), battery-powered vehicles, coatings with the metal or its compounds (including paint), diecasting, galvanizing, hot dip galvanizing, zinc in buildings, joining and metal working, and corrosion and cathodic protection. There are many other smaller uses of zinc such as in pigments, as stabilizers in rubber and plastics, and as semiconductor material. Regional Information centres are listed below.

Affiliated with the ZDA are the Galvanizers Association, the Diecasting Society, the Zinc Alloy Diecasters Association, and the Zinc Pigment Development Association. The ZDA also serves as the Secretariat for the European General Galvanizers Association.

Information centres

Asia

Zinc and Lead Asian Service, 95 Collins Street, Melbourne, Victoria 3000, Australia.

Australia and New Zealand

Australian Zinc Development Association, 95 Collins Street, Melbourne, Victoria 3000, Australia.

Brazil

Instituto Brasileiro de Informacao do Chumbo, Niquel e Zinco, Avenida Nove del Julho, 4015, 01407 Sao Paulo.

India

Indian Lead Zinc Information Centre, B–6/7, Shopping Centre, Safdarjung Enclave, New Delhi 110029.

Japan

Japan Lead Zinc Development Association, New Hibiya Building, 3–6 Uchisaiwaicho, 1-chome, Chiyoda-ku, Tokyo.

Mexico

Instituto Mexicano del Zinc, Plomo y Coproductos, AC Av. Sonora No. 166–1er Piso, Colonia Hippodromo Delegacion Cuauhtemoc, 06100 Mexico DF

Scandinavia

Nordic Galvanizers Association, Kungsgatan 37, 4tr., S–111 56 Stockholm, Sweden.

Scandinavian Lead Zinc Association, Sturegatan 22, S–114 36 Stockholm, Sweden.

United Kingdom

Zinc Development Association, 42 Weymouth Street, London W1N 3LQ.

USA

Zinc Institute Inc., 292 Madison Avenue, New York, NY 10117.
International Lead Zinc Research Organization, 292 Madison Avenue, New York, NY 10017.

Other areas

Contact ZDA, 42 Weymouth Street, London W1N 3LQ, UK or nearest Regional Centre.

CHAPTER TEN

Other metals and some related materials

M. N. PATTEN

Introduction

The title of this chapter is not meant to give the impression that the metals considered within it are any less worthy than those which have been singled out for individual attention. Somewhere along the way, a line had to be drawn and the remainder have been grouped together in a single chapter because of space limitations as much as anything else. A few elements which would not be classed as metals but which play an important part in metallurgy have been included.

For convenience, the metals have been dealt with in alphabetical order. While some of these fall into natural groups, there would always be a residue requiring classification, so the simple route has been chosen. The exceptions are the platinum group and rare earth groups which are not normally separated in the literature.

Some key sources such as Gmelin's *Handbuch* and Hampel's *Rare Metals Handbook* have already been mentioned in Chapter 1 but a detailed bibliographical guide is worthy of fuller mention here: K. Boodson's *Non-ferrous Metals: A Bibliographical Guide* (Macdonald, 1972) contains over 4,000 annotated entries for material published largely in the previous five years. Coverage ranges from metal physics, crystallography, physical metallurgy and physical, mechanical and chemical properties through analysis and testing to mining and metal production, followed by entries for specific metals.

The most comprehensive and authoritative source to which the reader's attention should be drawn is the Kirk-Othmer *Encyclopedia of Chemical Technology*, 3rd edn, a mammoth work of 24 volumes (Wiley). It is now also available online (see section on online databases in Chapter 1). Most large public reference libraries or appropriate academic or industrial libraries will house this work. Most of the metals dealt with in this chapter plus their compounds are dealt with at length and all the articles are well referenced. Occurrence, production methods, properties and applications are all covered and the work should be looked upon as the most important source of an encyclopaedic nature. A more general encyclopaedia with good introductory articles on most metals is the *McGraw-Hill Encyclopedia of Science and Technology*, 5th edn (1982), while sources mentioned in detail elsewhere such as *ASM Metals Handbook* and *Smithells Metals Reference Book* should not be overlooked. A major new source of information is the eight-volume *Encyclopedia of Material Science and Engineering*, edited by M. Bever and published in 1984 by Pergamon. Its contents cover production, processing and properties of metals and other materials, materials for specific applications and subjects such as materials engineering and materials resources. Its lists of data sources in the chapter by J. H. Westbrook are particularly useful. A first supplement to this encyclopaedia was published in 1988 and has been described in Chapter 1.

Online databases such as Metadex, BNF Abstracts (although this is no longer available outside BNF itself), Electronic Materials Information Service (IEE), Chemical Abstracts and Compendex are all relevant sources and a number of metals have their own databases. These are all detailed elsewhere. A comprehensive directory of textual and numeric databases which lists all those known to relate to metals is the *EUSIDIC Database Guide* published every three to four years by Learned Information (Oxford and NY). A new publication is Brit-line – a directory of British databases which aims to produce a six-monthly update and which is published by EDI Ltd., Lingfield, Surrey.

Data sources are more plentiful than many people think and in addition to the *Encyclopedia* from Pergamon mentioned above one should look at the *Handbook of Tables in Science and Technology*, edited by R. H. Powell, 2nd edn (The Library Association, 1983). This lists 3,400 handbooks and tables and metals are extensively covered. The National Physical Laboratory has an Alloydata database dealing with phase equilibria of alloys and the Metals Datafile from Materials Information, although new and comparatively untried, is attempting to build in as much retrospective data as it can. Specific handbooks worth mentioning are *Engineering Alloys* from Van Nostrand and R. B. Ross *Metallic Materials Specifications Handbook* (Spon). Data journals such as the *Bulletin of Alloy Phase Diagrams* and the *Journal of*

Chemical and Engineering Data are valuable current sources of information. Vladimir Sedacek's *Non-ferrous Metals and Alloys* was published by Elsevier in 1986 and should be mentioned. Handbooks such as Gmelin's and those emanating from bodies such as ASM and SAE have been described in the introductory chapter and where appropriate in other chapters. Conference proceedings should not be overlooked, and typical of these is *Light Metals 1986: Proceedings of the Technical Sessions Sponsored by the TMS Light Metals Committee, 1986*, edited by R. E. Miller (Metallurgical Society, 1986).

Antimony

This is usually classified as a non-metal or metalloid although it has in its trivalent state certain metal characteristics. It is mainly used as an alloying ingredient for applications such as lead-acid storage batteries and as an addition agent for copper-base alloys and ductile iron. A detailed treatment can be found in Kirk-Othmer, Vol. 3, pp. 96–128 and a small but very well referenced volume by C. Y. Wang entitled *Antimony, Its Geology, Metallurgy, Industrial Uses and Economics* is well worth perusing despite the fact that the latest edition is the third and this appeared as long ago as 1952. The US Dept of the Interior, Bureau of Mines produces *Antimony Quarterly* as well as *Antimony Materials Surveys*.

Arsenic

This too is classed as a non-metal but it can be found in over 50 minerals. It is commonly obtained as a by-product from the treatment of copper, lead, cobalt and gold ores and is extensively used in agriculture, glass manufacture and the semiconductor industry. Again, there is good coverage in Kirk-Othmer, Vol. 3, pp. 243–66. The US Bureau of Mines cover arsenic in their commodity reports and the Arsenic Development Committee has published a number of papers on applications. More recently, W. H. Lederer and R. J. Fensterheim edited *Arsenic: Industrial, Environmental and Biomedical Perspectives* (Van Nostrand Reinhold, 1982).

Barium

This is classed as an alkaline earth metal, occurring in compound form and widely distributed in igneous rocks, sandstone and the like. Probably its better known use is as a getter to remove gas traces from vacuum and TV tubes. Other applications include thin films, lead-acid batteries, alloying additions and oxygen reduction of molten steel and copper. There is a sound treatment in Kirk-Othmer, Vol. 3, pp. 457–79. C. A. Hampel's *Rare Metals Handbook*, 2nd edn (Reinhold, 1961), contains a chapter by C. L. Mantell on the alkaline earth metals – calcium, barium and strontium. The Food Machinery and Chemical Corporation, New York Mineral Products Division produced a useful barium bibliography in 1961.

Beryllium

Described as the only light metal with a high melting point, most of the production is used in alloys, notably copper-beryllium. The metal has found extensive application in X-ray tubes and as a heat sink material for aircraft brakes, as well as in areas such as inertial guidance components, space optics and microelectronics. A well-referenced article appears in Kirk-Othmer Volume 3, pp. 603–29 and useful early works include D. W. White and J. E. Burke, *The Metal Beryllium* (ASM, 1955); G. E. Darwin and J. H. Buddery, *Beryllium* (Butterworth, 1960); R. G. Bellamy and N. A. Hill, *Extraction and Metallurgy of Uranium, Thorium and Beryllium* (Macmillan, 1963); H. H. Hausner (ed.), *Beryllium, Its Metallurgy and Properties* (University of California Press, 1965); and A. J. Breslin, *Beryllium: Its Industrial Hygiene Prospects* (Academic Press, 1966). More recently, a two-volume work entitled *Beryllium Science and Technology* has been published by Plenum Press (1977–79). Volume 1, edited by D. Webster and G. J. London, contains some of the material presented at the Fourth International Conference on Beryllium held under the auspices of the Royal Society, London, 1977. Volume 2, edited by D. R. Floyd and J. N. Lowe, contains a four-page bibliography. A volume on *Engineering Applications of Beryllium* by P. Greenfield was published by Mills and Boon in their Mechanical Engineering Monographs series (1971) and both the US Bureau of Mines and the US Geological Society have produced numerous reports on the metal.

Bismuth

This falls in the same group as arsenic and antimony and is mainly used in pharmaceuticals and as a metallurgical addition agent. There was some interest at one time in the use of bismuth in lead bath annealing by the British Iron and Steel Research Association, and it is interesting to see from the literature that similar research is still going on elsewhere. The Institute of Metals Conference on Clean Steel, held in Hungary in 1986 referred to the alloying of steel with bismuth to improve machinability. In its liquid form, it has also been used in liquid metal fission reactors. Other applications include irradiation in medicine and thermoelectrics. Kirk-Othmer gives the metal detailed treatment in Volume 3, pp. 912–37 and a report by N. P. Sajin and P. Y. Dulkina on the *Production of High Purity Metallic Bismuth* appeared in the serial *UN Peaceful Uses of Atomic Energy*, No. 9 (1955). The Bismuth Institute has produced a number of publications.

Boron

Usually appearing in alkali form, boron's uses range from abrasive wheels to reinforcing materials for composites. Other applications are in nuclear technology, fibreglass and borosilicate glass. The Kirk-Othmer article in Volume 4, pp. 62–110 lists 150 references. Useful works include *The Chemistry of Boron and Its Compounds*, edited by E. L. Muetterties (Wiley, 1967); *Boron, Metallo-boron Compounds and Boranes*, edited by R. M. Adams (Interscience, 1964) and W. A. Gales's *History and Technology of the Borax Industry* (US Borax and Chemical Corp.). More recently, A. G. Von Malushka produced *Boronizing* (Wiley, 1981).

Cadmium

This is a soft ductile element, usually recovered as a byproduct of the processing of lead, zinc and copper ores. A significant use is in corrosion protection due largely to its ability to deposit rapidly and uniformly on intricate surfaces, either by plating from a cyanide bath or by vacuum deposition, dipping, spraying or powder coating. Cadmium is extensively used in rechargeable batteries and has found application as a copper hardener, as a semiconductor and in soldering and brazing. Kirk-Othmer deals with the metal in Vol. 4, pp. 387–411 and the US Bureau of Mines produces reports from time to time. The NTIS produced a report entitled *Technical and Microeconomic Analysis of Cadmium and its Compounds* (1975) and a useful early work is *Cadmium* by D. M. Chizhikov (Pergamon, 1966). International Cadmium Conferences are held regularly under the auspices of the Cadmium Association, the

Cadmium Council and the International Lead Zinc Research Association.

Much of the recent literature has been concerned with the toxicity of cadmium, e.g. L. Friberg and others, *Cadmium and Health: A Toxicological and Epidemiological Appraisal Effects Response* (CRC Press, Florida, 1986). The Cadmium Association has published a reort on a seminar held in Brussels in 1985 entitled *Cadmium Today* (Cadmium Assn., London, 1986).

Calcium

Classed as an alkaline earth metal and the fifth most abundant in the Earth's crust, calcium is used as a reducing agent in the production of the less common metals and as an anode material in thermal batteries. It is widely used in the steel industry for deoxidizing, desulphurizing or degassing steel and cast iron and to control non-metallic inclusions. It is also effective in reducing the bismuth content of lead and in improving the electrical and mechanical properties of aluminium alloys. Kirk-Othmer covers calcium in Vol. 4, pp. 412–48. An early work was C. L. Mantell and C. Hardy's *Calcium Metallurgy and Technology* (Reinhold, 1945). F. J. Shortsleeve and D. C. Hilty wrote on 'Calcium in iron and steel' in their work *Boron, Calcium, Columbium and Zirconium in Iron and Steel* (Wiley, 1957). A recent publication in Springer Verlag's Materials Research and Engineering Series is *Calcium Clean Steel*, by T. Ototani (1986).

Carbon

This can occur in amorphous form in coal or lignite or in crystalline form in graphite or diamond and is used in an extremely wide range of products with exceptional electrical, thermal and physical properties. The Kirk-Othmer article in Vol. 4, pp. 556–614 has extensive bibliographies on carbon and carbides. Pergamon Press has been publishing the journal *Carbon* since 1963 and Japan has had its own journal *Tanso* since 1949. Proceeding of conferences on carbon are regularly published by Pergamon and the American Carbon Committee holds biennial conferences. Another active body is the European Carbon Black Centre. Works worthy of examination include G. M. Jenkins and K. Kawamura *Polymeric Carbons – Carbon Fibre, Glass and Charcoal* (Cambridge UP); *Carbon Fibres in Engineering*, edited by M. Langley (McGraw-

Hill, 1973); A. A. Linari-Linholm, *Occurrence, Mining and Recovery of Diamonds* (De Beers Ltd., Kenion Press, Slough, UK); *Akron Rubber Group Technical Symposium* (April 1977) covering the carbon black process; J. Delmonte's *Technology of Carbon and Graphite Fibre Composites* (T-C Pubns, California, 1981); and *Chemistry and Physics of Carbon*, edited by P. A. Thrower (Dekker, 1987).

Cesium

An alkali metal which is generally preferred to other alkali metals when maximum reactivity, atomic weight or vapour pressure is desired. Its chief use is in research on power generation while the main commercial use is in the manufacture of vacuum tubes as a getter to remove oxygen traces and as a means of inducing electron emission from the cathode. Kirk-Othmer deals with cesium in Vol. 5, pp. 327–35 and the US Dept of the Interior *Minerals Yearbook* usually contains up-to-date information on the metal. A major producer in the US is Penn Rare Metals Division of Kawecki Berylco Industries Inc., Revere, Pa. A useful pamphlet by D. B. Robinson entitled *Characteristics of Cesium* was published in 1978 by Gemfield Association, Florida.

Chromium

Chromite ores are frequently used in intermediate form in refractory bricks, in steelmaking and in metal-finishing, in wood preservatives and as catalysts. For chromium itself, electroplating and chromizing, stainless and other alloy steels are important uses. Kirk-Othmer gives the metal detailed coverage in Vol. 6, pp. 54–82. Other works of note are M. J. Udy's *Chromium: Metallurgy of Chromium and Its Compounds* (Reinhold, 1956); *Chromium*, by A. H. Sully and E. A. Brandes, 2nd edn (Butterworth, 1967); *The Superalloys*, edited by C. Sims and W. Hagel (Wiley, 1972) pp. 87–151; *Modern Electroplating*, edited by F. A. Lowenheim, 3rd edn (Wiley, 1974); and *Nickel and Chromium Plating*, by J. K. Dennis and T. E. Such, 2nd edn (Butterworth, 1986). Chromium is extremely well-referenced throughout the metal-finishing and steelmaking literature and the interested reader should also examine the chapter on stainless steels.

Cobalt

Uses are largely in magnetic alloys, superalloys and wear-resistant alloys as a metal and in plating baths and in plant drying as a salt. Kirk-Othmer coverage is in Vol. 6, pp. 451–94. Useful sources of information include the journal *Cobalt* and the Cobalt Information Centre in Brussels, which has produced a number of monographs on the metal in collaboration with the Battelle Memorial Institute, and R. S. Young's *Cobalt* (American Chemical Society, no. 149, Reinhold, 1960). This latter work contains over 1,500 references. More recently, W. Betteridge published *Cobalt and Its Alloys* (Horwood, Chicherster, 1982); N. Aldyne's *Cobalt* (Avon, 1982) and *Cobalt – A Market Appraisal* by M. Hale appeared in the Occasional Papers series of the Institution of Mining and Metallurgy in 1984. Nickel and cobalt extraction using organic compounds appeared in the European Patent Office series (Pergamon, 1985). In 1986, the US Bureau of Mines produced *Processing Technologies for Extracting Cobalt from Domestic Resources* by C. E. Jordan.

Germanium

Generally recovered as a byproduct of other metals by sintering and leaching, germanium is a semiconductor element with an electrical resistivity midway between that of metallic conductors and good electrical insulators. As would be expected, the main application is in semiconductors, particularly as a substrate in wafer technology. Kirk-Othmer deals with the metal in Vol. 11, pp. 791–802. Works of interest include F. Glocking's *The Chemistry of Germanium* (Academic Press, 1969) and V. A. Nazarenko's *The Analytical Chemistry of Germanium* (Wiley, 1974). World reserves of the metal are estimated at 4,000 metric tons and this has led to considerable interest in supply and demand. Roskill Information Services have produced reports on production, consumption, prices and future trends in the material and a paper on the *Supply of Germanium for Future World Demands* was presented at the Fourth European Electro-Optics Conference, Society of Photo-Optical Instrumentation Engineers (October 1978), pp. 216–22. A work which is well worth examining is *Semiconductors and Semimetals*, edited by R. K. Willardson and others. More than twenty volumes of this work have been published to date by Academic Press. The Materials

Research Society Symposia Proceedings series and Metallurgical Society Conferences are frequent publishers of papers on the metallurgy of semiconductor materials.

Gold

Credited as being the first metal known to man, gold occurs naturally as a metal of high purity and is the noblest of the noble metals. After its uses in the money world, gold finds its largest industrial use in jewellery, followed by the electronics industry and a host of competing uses from dentistry through temperature measurement to catalysts for medicinal purposes. Kirk-Othmer coverage is in Vol. 11, pp. 972–95. A regular source of up-to-the-minute information on gold research and industrial applications is the *Gold Bulletin* published by the International Gold Corporation of South Africa. Another periodical source is the *Engelhard Industries Technical Bulletin* published by the Engelhard Minerals and Chemical Corporation of New Jersey. Consolidated Gold Fields of London produce annual reports on the metal. The Gold Institute is a good source of current data and the International Precious Metals Institute holds regular conferences and seminars while the National Association of Recycling Industries, New York has produced a report on *Recycling Precious Metals* in 1979. Useful works in recent years include *Gold*, by C. Glynn *et al.* (Consolidated Goldfields, 1979); *Gold Usage*, by W. S. Rapson and T. Groenewald (Academic Press, 1978); *Anaylsis of Noble Metals*, by F. E. Beamish (Academic Press, 1977); *The Chemistry of Gold*, by R. J. Puddephatt (Elsevier, 1978); *Guide to Precious Metals and Their Markets*, by P. Robbins (Nichols, New York, 1979); *Gold* (Graham and Trotman, 1982), by B. Bettel (contains a good bibliography); *Gold and Other Precious Metals*, by C. I. Coombs (Morrow, 1981); and *Recovery and Refining of Precious Metals*, by C. W. Ammen (Van Nostrand/Reinhold, 1984). A useful history is Alexander del Mar's *History of the Precious Metals* (A. M. Kelley, New York, 1969) and an early comprehensive work which should not be passed over is *Gold: Recovery, Properties and Applications*, edited by E. M. Wise (Van Nostrand, 1964).

Hafnium

Usually linked with zirconium, hafnium finds use in nuclear reactors as an absorber of neutrons, in alloys as a strengthening agent and in flashbulbs and tool bits. In general, the literature does not separate the two. Kirk-Othmer treatment is in Vol. 12, pp. 67–80 and useful works are D. E. Thomas and E. J. Hayes, *The Metallurgy of Hafnium* (US Govt Printing Office, 1960); R. J. H. Clark *et al.*, *The Chemistry of Titanium, Zirconium and Hafnium* (Pergamon, 1973); and D. R. McClintock and G. A. Tracy, *Physical and Mechanical Properties of Hafnium* (Westinghouse Elec. Corp., 1958) (Report no. WAPD–SFR–Met 653). An Atomic Energy Review Special Issue entitled *Hafnium: Physico-chemical Properties of Its Compounds and Alloys* was published for the IAEA by Unipub in 1981.

Indium

A soft, silvery metal with very good plastic properties, indium has been used extensively as a bearing material in aircraft, automobile and diesel engines. Its largest single use is as a solder or alloy although its semiconductor characteristics ensure that it still retains a high level of use in research and development. Other uses include nuclear control rods, neutron monitoring badges, corrosion protection and, in the oxide form, in sodium-vapour lamps and as conductive coatings on glass and ceramics. It is dealt with by Kirk-Othmer in Vol. 13, pp. 207–12. There is a useful chapter in C. A. Hampel's *Encyclopedia of the Chemical Elements* (Reinhold, 1968) and there is a comprehensive bibliography in M. T. Ludwick's *Indium* (Indium Corporation of America, 1959). Other companies active in the field include Cominco and American Smelting and Refining.

Lime and limestone

As an important material which is heavily used in the metallurgical industries, limestone ought to be mentioned in this chapter. Limestone is widely distributed around the world and is used predominantly as crushed or broken stone. Uses abound in concrete aggregates, road construction, agriculture, metallurgy, water treatment and air pollution control. Kirk-Othmer treats the material very thoroughly in Vol. 14, pp. 343–82. Regular sources of information include *Pit and Quarry Magazine* and *Rock Products Magazine*, the proceedings of the National Crushed Stone Association, publications of the National Lime Association and the International Lime Congresses. Useful book sources are R. S. Boynton's *Chemistry and Technology of Lime and Lime-*

stone, 2nd edn (Wiley, 1979) and the *Pit and Quarry Handbook*, published annually by P. & Q. Publications, Chicago. A new publication from ASTM is STP 931 entitled *Lime for Environmental Uses* (1987) which includes a chapter on the thiosorbic lime wet scrubbing process for flue gas desulphurization.

Lithium

Trace amounts of lithium are present in many minerals and precipitation with lime is one of the methods used to recover lithium from brines. Uses range from metallurgical ones such as hardeners or reactive agents to high energy batteries and mental health. Kirk-Othmer deals with lithium in Vol. 14, pp. 448–76. A number of publications have emerged from the Lithium Corporation of America and S. S. Penner's *Lithium Needs and Resources* (Pergamon, 1978) is still in print. Two recent publications are *Lithium: Current Applications in Science, Medicine and Technology* (Wiley, 1985) and *Aluminium-lithium Alloys*, edited by C. Baker (Institute of Metals, 1986).

Magnesium

In its primary form, magnesium lacks strength and is generally used in alloyed form with other metals, particularly in aluminium and zinc die-casting alloys. Other uses include corrosion protection, batteries and applications where lightness and structural strength achieved through suitable alloys are desirable. Kirk-Othmer coverage is in Vol. 14, pp. 570–615. Important publications are those of the Dow Chemical Co., Esso Research and Engineering and the International Magnesium Association whose *Proceedings* are particularly useful as they contain annual supply and demand reports. An International Conference on Energy Conservation in Production and Utilization of Magnesium was held at the Massachusetts Institute of Technology in 1977. Two early works are *The Story of Magnesium* by W. H. Gross (ASM, 1959) and *The Physical Metallurgy of Magnesium and Its Alloys* by G. V. Raynor (Pergamon, 1959), which is very well-referenced. In 1986, the Institute of Metals held a Magnesium Technology Conference in London and the proceedings of this were published a year later. *Magnesium Products Design* by R. S. Busk and sponsored by the International Magnesium Association was published by Dekker in 1987.

Magnetic materials

Most commercially important magnetic materials comprise ferromagnets and ferrimagnets and generally fall into the categories of soft magnetic materials with high permeability and low coercivity and hard or permanent magnet materials which are characterized by high coercivity and high energy product. A useful summary appears in Kirk-Othmer Vol. 14, pp. 646–707 and this article also covers thin-film materials. There is also a good introductory article in the McGraw-Hill *Encyclopedia of Science and Technology*. Regular sources of information include *Magnetism Letters*, *Journal of Magnetism and Materials*, *IEEE Transactions on Magnetics* and the *Annual Conferences on Magnetism and Magnetic Materials*. Useful works to note are C. Heck's *Magnetic Materials and Their Applications* (Crane, Russack, 1974); *Permanent Magnets in Theory and Practice*, by M. McCaig (Wiley, 1977); *Magnetic Materials* by R. Ball (Heyden, 1979); *Ferromagnetic Materials,* edited by E. P. Wohlfarth and published in two volumes by Elsevier (1980); and *Critical Materials in the Electrical and Electronics Industry*, by D. M. Jacobson and D. S. Evans (Institute of Metals, 1984) including a select bibliography which lists sources of information and statistics for a wide range of metals.

Recent publications of note include *Magnetism and Metallurgy of Soft Magnetic Materials* by Chih-wen Chen (Dover Publications, 1986), which was originally published in 1977 and contains a fourteen-page bibliography. *Metallic Magnetism* by H. Capelmann was published by Springer Verlag (1986) and this too is well-referenced. Other works with substantial bibliographic content are *Metal Semiconductor Contacts and Devices* by S. S. Cohen (Academic Press, 1986) and P. L. Rossiter's *The Electrical Resistivity of Metals and Alloys* (Cambridge University Press, 1987). The Proceedings of the ASM's 1984 Materials Science Seminar have been published as *Advances in Electronic Materials*, edited by B. R. Wessels (ASM, 1986) and discusses chemical composition, defect structure and electronic properties of semiconductors in the light of recent developments in electronic materials.

Manganese

Manganese ore can be reduced by the addition of carbon to produce material of a quality suitable for use in ferrous metallurgy. The largest use is in steel manufacture, followed by cast iron and aluminium, stainless and other alloy steels and the general non-ferrous industry. Kirk-Othmer covers manganese in Vol. 14, pp. 824–43. The Manganese Centre, Paris, has produced a number of useful publications such as *Manganese in Ferrous Metallurgy*. An important work is A. H. Sully's *Manganese* (Butterworth, 1955) and a good paper by J. P. Faunce and J. Y. Welsh on *The Production of Manganese Metal* can be found in AIME 105th Annual Meeting Proceedings, Las Vegas, 1976. The UNCTAD Secretariat produced a study entitled *The Processing and Marketing of Manganese: Areas for International Cooperation* (United Nations, 1985).

Mercury

Also known as quicksilver, mercury occurs naturally, often in combination with sulphur, to form some dozen or more minerals. Its main use is in electrical applications but it is also used in fungicides, pesticides and pharmaceuticals, as a catalyst in the production of urethane foams and as a cathode in the preparation of chlorine and caustic soda. Mercury is dealt with by Kirk-Othmer in Vol. 15, pp. 143–56. A good annotated bibliography can be found in *Mercury in the Environment*, by G. M. Caton *et al.*, Oak Ridge National Laboratory, 1972 and Roskill Information Services have published a number of surveys dealing specifically with mercury. A recent volume in Pergamon's *Solubility Data* series is *Metals in Mercury* by G. and C. Gurninsky (1986). In 1987 the UK's Ministry of Agriculture produced its second supplementary report on its *Survey of Mercury in Food*. This is available from the British Library (DSC3983.70 MAFF–FSP–17).

Molybdenum

This metal's potential as an additive in armour plating and tool steels was first exploited in the 1880s but its main use today is in alloy and stainless steels, as a furnace resistance element among a number of furnace-related uses, in the glass industry and in the machine tool industry. Kirk-Othmer coverage of molybdenum is in Vol. 15, pp. 670–82. Companies such as Climax Molybdenum Co., Ann Arbor, Michigan and Amax Specialty Metals Division, Greenwich have produced a number of booklets dealing with the metal and the Second International Ferroalloys Conference in Copenhagen in 1979 contained a paper by J. W. Goth on *Molybdenum in the Eighties*. An early authoritative work was L.

Northcutt's *Molybdenum* (Butterworth, 1956) and the most important work in recent years is A Suthlov's *International Molybdenum Encylopedia*, 2 vols (International Publications, Santiago, 1979). Other works to note are G. A. Parker's *Analytical Chemistry of Molybdenum* (Springer Verlag, 1983) and the *Molybdenum Resources Guidebook* (Minobras, CA, 1980). Climax also produce a quarterly journal entitled *Molybdenum Mosaic* dealing with the technology of the metal and a monthly journal, *Moly Corrosion Inhibitors*. A Metal Bulletin conference on Nickel/Molybdenum was held in Paris in December, 1986.

Nickel

As an element, nickel has been known to man for two hundred years but it was produced as an alloy in China probably a thousand years ago. Its properties of good ductility and high melting point have found it many applications in the superalloys category, particularly in gas turbines.

Combined with chromium, copper or molybdenum, or iron and chromium, nickel has achieved a high reputation as an alloy. The latter alloys, typified in the Inconel alloy series, are well known for their mechanical properties at cryogenic through to elevated temperature ranges and for their corrosion resistance. Almost half of the use of nickel worldwide is in stainless steel and other alloy steels. Nickel ores which can be economically mined are either of the sulphide or lateritic type, with the former accounting for most of the world's production. Ore reduction is usually carried out by crushing, froth flotation or magnetic separation, followed by roasting, smelting and converting before casting into anodes for electrolytic refining. Very pure nickel can be produced by the carbonyl method.

Nickel and its alloys is dealt with very extensively in the literature and, before it ceased publication in 1968, the *Nickel Bulletin* was a key current awareness source for nickel users. Despite its demise, it still has value as a retrospective research tool and has retained its place on the shelves of many libraries. Many of the standard sources mentioned elsewhere such as Gmelin and Smithell deal with nickel and its alloys at length and the Kirk-Othmer coverage is in Vol. 15, pp. 787–800.

Probably the best known introductory text to the metal is W. Betteridge's *Nickel and Its Alloys*, in the Macdonald and Evans Industrial Metals Series (1977). This author, with J. Heslop, also produced *The Nimonic Alloys and Other Nickel-Base Alloys*

(Arnold, 1974). B. O. Holland's *Extractive Metallurgy of Nickel* (BHP Co.'s Central Research Laboratories, 1967) contains a fifty page bibliography and this aspect has been updated in *Extractive Metallurgy of Nickel*, edited by A. R. Burkin and published by Wiley on behalf of the Society of Chemical Industry (1987). *Corrosion of Nickel-Base Alloys*, edited by R. C. Scarberry (ASM, 1985), is useful for information on industrial corrosion tests and topics such as influence of high temperatures and corrosive environments. A related volume is *Nickel and Chromium Plating* by J. K. Dennis and T. E. Such, 2nd edn (Butterworth, 1986), which deals at length with electroplating baths and anodes for industrial nickel deposition which has seen a significant increase in engineering applications in recent years. It also deals with plating plastics, nickel alloys and difficult to plate metals.

The US Bureau of Mines regularly covers nickel in its *Mineral Commodity Summaries* and the Huntingdon Alloy Products Division of the International Nickel Co. Inc. has published a wide range of booklets such as *Nickel Alloys* (1972); *Inconel Alloy 600*; and *High Temperature High Strength Nickel Base Alloys* (1977). Other active companies include Cabot Corporation, Special Metals Corporation and in the UK, Henry Wiggin Ltd – now Wiggin Alloys Ltd., Hereford.

Niobium

Also known as columbium, this is used in ferroniobium form as an additive in the manufacture of high-strength low-alloy steels and carbon steels. Niobium also finds important uses as an additive to stainless steels as a corrosion inhibitor and as a strengthening agent in a number of nonferrous metals and superalloys. Its high corrosion resistance makes it a valuable material in a wide range of chemical and metallurgical activities. Kirk-Othmer treatment is in Vol. 15, pp. 820–40. Works of note include G. L. Miller, *Tantalum and Niobium* (Metallurgy of the Rarer Metals series) (Butterworths, 1959); D. L. Douglass and F. W. Kunz (eds) *Columbium Metallurgy* (symposium proceedings, New York 1960; Interscience, 1961); F. T. Sisco and E. E.Premian, *Columbium and Tantalum* (Wiley, 1963); R. J. H. Clark and D. Brown *The Chemistry of Vanadium, Niobium and Tantalum* (Pergamon, 1975). A good review paper is 'Niobium in superalloys – a perspective', by C. T. Sims, which appeared in *High Temperature Technology* **2** (4) Nov. 1984, pp. 185–201.

Platinum and the platinum group

This metal occurs naturally and is usually paired with palladium within a group which also contains ruthenium, rhodium, osmium and iridium. Platinum finds significant applications in the dental and medical areas, in jewellery, as catalysts in the chemical industry and in a wide range of electrical and electronic fields. Iridium is the only metal which can be used in air at temperatures up to 2,300°C and it finds a ready use in platinum alloys. Combined with platinum, it makes an excellent electrical contact material as well as having a number of medical applications. Osmium alloys are extremely hard and are well known for their use in pen nibs. Palladium has a unique property in that it can absorb and retain over 800 times its volume of hydrogen. Uses, like platinum, are in the dental, medical, electrical, chemical and aeronautical industries as well as in jewellery. Rhodium also alloys readily with platinum and is widely used in jewellery, as a reflective material and for electrical contacts. Ruthenium also finds a wide use in electrical contacts and is used as a hardener in platinum and palladium alloys.

Kirk-Othmer deals with the group in Vol. 18, pp. 228–53 and a useful *History of Platinum and Its Alloys*, by D. McDonald and L. B. Hunt, was published by Johnson, Matthey (1982). A regular source of information is *Platinum Metals Review* published quarterly by the same company. The National Academy of Sciences published *Supply and Use Patterns* for the platinum group metals in 1980 (Publication no. NMAB–359) and a useful text is *Physical Metallurgy of Platinum Metals* by E. Savitsky *et al.* (MIR Publications, Moscow, 1979).

Plutonium

Discovered as recently as 1940, plutonium's most significant property is the enormous amount of energy which can be released from just one gram (equivalent to burning three tons of coal). Nuclear reactors can produce plutonium by the irradiation of uranium fuels and its radioactivity, nuclear potential and chemical reactivity are well known. Kirk-Othmer covers the metal comprehensively in Vol. 18, pp. 278–301. Significant works include *Plutonium Handbook*, edited by O. J. Wick (Gordon & Breach, 2nd edn, 1980); M. Taube's *Plutonium* (Macmillan, 1964); W. D. Wilkinson's *Extractive and Physical Metallurgy of Plutonium and Its Alloys* (Interscience, 1960); and publications of the American

Nuclear Society such as J. M. Cleveland's *The Chemistry of Plutonium* (1979). A similar work is *Plutonium Chemistry*, edited by W. T. Carnall and G. R. Choppin (American Chemical Society, 1983). Important periodical sources include *Radiochimica Acta*, *Progress in Nuclear Energy* and *Atomic Energy Review*. See also the section dealing with uranium.

Potassium

The primary use of potassium is as a heat transfer fluid for reactors. It readily forms alloys with other alkali metals and potassium compounds are used in the manufacture of glass, as reactants, and in food and medicine. Kirk-Othmer coverage is in Vol. 18 pp. 912–20 and a useful volume by O. J. Foust entitled *Sodium and Sodium Potassium Engineering Handbook* was published by Gordon & Breach (1972). *A World Survey of Potash Resources* was published by the British Sulphur Corporation (1979). The International Minerals and Metals Corporation has published a number of papers dealing with potassium.

Radium

Radium was normally recovered along with barium in the processing of pitchblende for uranium although commercial production from this route has now virtually ceased. Nowadays most of the radium is a recycled product from users who have switched to other radioactive sources. Radium has been superseded by other radiation sources in many medical and industrial applications although it still finds occasional specialized use. The chapter in Vol. 19 of Kirk-Othmer on Radioactivity (pp. 639–81) gives a useful background to developments and is well referenced. A relevant work is *The Radiochemistry of Radium* (US National Academy of Sciences and National Research Council, NAS–NS 3057).

Rare earths

This group of seventeen elements comprises scandium, yttrium, lanthanum, cerium, praseodymium, neodymium, promethium, samarium, europium, gadolinium, terbium, dysprosium, holmium, erbium, thulium, ytterbium and lutetium. Rare earths can be reduced to the metal by reaction with calcium, sodium or potassium but the properties of rare earth metals can be significantly affected by very small amounts of impurities which, while not critical for many industrial uses, can certainly be so when required for research purposes. Rare earths find significant use in petroleum cracking plants and in iron and steelmaking to remove undesirable elements. Another major use is in optics and there are growing uses in electronics and magnetics, in artificial diamond manufacture and in nuclear research. Kirk-Othmer discusses them in Vol. 19, pp. 833–54. Useful works include A. H. Sully, *Metallurgy of the Rarer Metals* (Butterworths, 1967); E. V. Kleber and B. Love, *The Technology of Scandium, Yttrium and Rare Earth Metals* (Pergamon, 1963); C. A. Hampel, *Rare Metals Handbook* (Reinhold, 1961); K. A. Gschneidner and L. Eyring (eds), *Handbook on the Physics and Chemistry of the Rare Earths* (North Holland, 1979); R. J. Elliott (ed.), *Magnetic Properties of the Rare Earth Metals* (Plenum Press, 1972); E. J. McCarthy *et al.*, *The Rare Earths in Modern Science and Technology* (Plenum Press, 1980); K. A. Gschneidner, *Industrial Applications of Rare Earth Elements* (Symposium) (American Chemical Society, 1981); Proceedings of the Rare Earth Research Conference (published at 18 month intervals usually in the *Journal of the Less Common Metals*); and the Rare Earth Information Centre, Ames Laboratory of the US DOE, Ames, Iowa 50011 deals with enquiries on rare earths and can provide detailed bibliographies.

Rhenium

A significant use of this comparatively rare element is in filaments in light bulbs and electron tubes, in mass spectrometers and magneto contacts, but the largest use is in petroleum-refining catalysts. Kirk-Othmer coverage is in Vol. 20, pp. 249–58 and useful references include J. G. F. Druce's *Rhenium* (Cambridge UP, 1948); R. Colton's *The Chemistry of Rhenium and Technetium* (Elsevier, 1965); and R. D. Peacock's *The Chemistry of Technetium and Rhenium* (Elsevier, 1966). E. M. Savitsky's *Rhenium Alloys* was published by Coronet Books in 1970 and more recent material has been largely confined to the periodical literature.

Rubidium

An alkali metal with similar properties to cesium, rubidium finds many uses in research and pharmaceutics. Kirk-Othmer discusses it in Vol. 20, pp. 492–9. It receives treatment in C. A. Hampel's *Rare Metals Handbook*. An old but useful work is F. M. Perelman's *Rubidium and Cesium* (Macmillan, 1965) but a more recent publication is *Current Trends in Lithium and Rubidium*, edited by G. U. Gorsini (MTP Press, UK, 1984).

Selenium

Usually obtained as a byproduct of precious metals recovered from copper refineries, selenium's best known use is in xerography. There is a wide range of optical and electronic uses as well as applications in glass and ceramics, pigments, lubricants, pharmaceuticals and plating. Kirk-Othmer treatment is in Vol. 20, pp. 575–601. A useful translated work is that of D. M. Chizhikov and V. P. Suchastlivyi on *Selenium and Selenides* (Collets, 1968). *The Chemistry of Sulphur, Selenium, Tellurium and Polonium*, by M. Schmidt *et al.* was published by Pergamon (1973) and in 1980, a Symposium on Industrial Uses of Selenium and Tellurium was held in Toronto. A useful source of information is the Selenium Tellurium Development Association, Connecticut, USA. *The Physics of Selenium and Tellurium*, edited by E. Gerlach was published by Springer Verlag (1979) as Vol. 13 of the Springer Series in Solid State Sciences and the *Proceedings of the Third International Symposium on Selenium in Biology and Medicine*, edited by G. F. Combs, Jr. and J. E. Spallholz was published by AVI (1986).

Silver

One of the oldest metals known to man and now much better appreciated for its industrial applications (it has the highest electrical and thermal conductivity of all the metals), silver finds significant application in areas such as photography, electrical contacts, catalysts, brazing solders, mirrors and dental fillings. The metal is covered by Kirk-Othmer in Vol. 21, pp. 1–15 and also by A Butts and C. D. Coxe in *Silver – Economics, Metallurgy and Use* (Van Nostrand, 1967). The Silver Institute in Washington DC produces regular statistics on the silver market as do Messrs Handy and Harman of New York. *Silver: The Restless Metal*, by R. W. Jastram, was published by Wiley-Interscience (1981).

Silicon

Although silicon makes up about 25 per cent of the Earth's crust, the element does not occur naturally but rather as a constituent of various minerals such as silica. It is widely used in metallurgy, particularly in steelmaking but it is best known for its use today in the semiconductor industry. Kirk-Othmer discusses it in Vol. 20, pp. 826–45. An early work is A. S. Berezhnoi's *Silicon and Its Binary Systems* (Consultants Bureau, NY, 1960) and more recently H. R. Huff and E. S. Sirtl produced *Semiconductor Silicon* (Electrochemical Society, 1977). *The Journal of Crystal Growth* is frequently cited. See also the section on magnetic materials for recent sources in the electronics field and Chapter 12 on ceramics.

Sodium

Its metallic form is produced by thermal reduction of sodium compounds and its use as an anti-knocking agent has been a major application in the petroleum industry. Important applications include refractory materials, catalysts, descaling, reducing agents, batteries, heat-transfer mediums in nuclear plants as well as a number of metallurgical uses such as deoxidation. It is dealt with in Kirk-Othmer Vol. 21, pp. 181–204 and a useful work is M. Sittig's *Sodium, its Manufacture, Properties and Uses* (Reinhold, 1956). A number of inorganic chemistry works deal with sodium, notably J. W. Mellor's *Comprehensive Treatise on Inorganic and Theoretical Chemistry* (Wiley). O. J. Foust's *Sodium-Nak Engineering Handbook* appeared between 1972 and 1979 in five volumes as part of the US Atomic Energy Commission Monograph series published by Gordon and Breach.

Tantalum

Probably best known as a refractory metal, tantalum finds significant application in capacitors, in cutting tools and as a steel hardening alloying agent. Kirk-Othmer treatment is in Vol. 22, pp. 541–64 and useful works include G. L. Miller's *Tantalum and Niobium* (Butterworths, 1959); F. T. Sisco and E. Epremian, *Columbium and Tantalum* (Wiley, 1963); M. Schussler's *Fansteel Corrosion Data Survey on Tantalum* (Fansteel Inc., Chicago, 1972); and the First International Symposium on Tantalum held in Brussels in 1978 under the auspices of the Tantalum Producers International Study Centre. A useful periodical source before it ceased publication was the *Murex Review*. *Tantalum: Physico-chemical Properties of its Compounds and Alloys* was published in 1973 for the IAEA by Unipub as Atomic Energy Review Series No. 3.

Tellurium

Usually found in association with gold, silver and copper deposits, its uses include additives to plain carbon and leaded steels to improve machinability, chilled iron castings and oxygen steelmaking additives, copper and lead alloys, catalysts, lubricants, medical, photographic and semiconductor applications. Kirk-Othmer deals with it in Vol. 22, pp. 658–79 and notable works include *Tellurium* edited by W. C. Cooper (Van Nostrand, 1971) and A. A. Kudryavtsev's *The Chemistry and Technology of Selenium and Tellurium* (Collets, 1974). See also the Selenium entry.

Thorium

This was formerly grouped with titanium and zirconium but it is now considered to be part of the rare earth group. The dominant source is monazite of which large deposits exist in, for instance, India, Malaysia, Australia and North and South America. It is usually separated from rare earth by precipitation, leaving a residual thorium hydroxide which can be further refined by dissolution in a suitable acid. It yields a bright, silvery metal with a very high melting point and good chemical reactivity. Thorium alloys readily with a wide range of metals and also has a number of compound forms which are classified as source materials for nuclear energy and are thus subject to government regulations. Since it can be produced as a byproduct of uranium it is frequently discussed in the literature of the latter. Kirk-Othmer coverage is in Vol. 22, pp. 989–1002. An early work was *Thorium Production Technology*, by F. L. Cuthbert (Addison-Wesley, 1958) and, in the same year, the ASM published *The Metal Thorium*, edited by H. W. Wilhelm and containing some useful bibliographies. More recently J. F. Smith *et al.* produced *Thorium: Preparation and Properties* (Iowa State University Press, 1975). The US Geological Survey and the US Atomic Energy Commission have produced a number of reports on the metal and there have been frequent references in the chemical journals such as the *Journal of the American Chemical Society*.

Titanium

Although discovered two hundred years ago, titanium came into its own during the Second World War and because of its high strength-to-weight ratio it has found significant use in aircraft engines and airframes. Its high corrosion resistance makes it particularly suitable for use in corrosive environments for heat exchanger pipes and tubing. Kirk-Othmer coverage is in Vol. 23, pp. 98–130. Works of note include A. D. McQuillan and M. K. McQuillan's *Titanium* (Butterworths, 1956); B. Jelk's *Titanium: Its Occurrence, Chemistry and Technology* (Ronald Press, NY, 1966); R. I. Jaffee and N. E. Promisel's *The Science, Technology and Application of Titanium* (Pergamon, 1970); M. J. Donachie's *Titanium and Titanium Alloys: Source Book/A Collection of Outstanding Articles from the Technical Literature* (American Society for Metals, 1982). Two volumes by E. W. Collings are worth noting: *Applied Superconductivity, Metallurgy and Physics of Titanium Alloys, Vol. 1 Fundamentals* (Plenum Press, 1985) and *A Sourcebook of Titanium Alloy Conductivity* (Plenum Press, 1983). Earlier Plenum had published *Titanium Science and Technology* in four volumes, edited by R. I. Jaffee (1973) and a work with the same title and also in four volumes and edited by G. Luetjering *et al.* was published in Germany by DGM Metallurgy. The metal has figured prominently in international conference proceedings such as those of the AIME Metallurgical Society, in NASA and NTIS reports and in the publications of the Titanium Metal Corp. of America, Pittsburgh, Pa. The Titanium Development Association organized an International Titanium Conference in San Francisco in October 1986. *Industrial Applications of Titanium and Zirconium*, edited by C. S. Young and J. C. Durham arose from a symposium sponsored by ASTM Committee B–10 on Reactive and Refractory Metals and Alloys in Philadelphia (October 1984). The reader should also examine the section in the Aerospace chapter (16) relating to titanium alloys. *Bureau of Mines Development of Titanium Production Technology* by J. L. Henry was published by the US Dept of the Interior in 1987.

Tungsten

Also known as wolfram and characterized by having the highest melting point of all the metals, tungsten finds significant use in tool steels and in lamp filaments, X-ray tubes, heat shields, furnace elements and glass-to-metal seals – to mention but a few of the metal's numerous applications. In tungsten carbide form, it is used for cutting and drilling tools and a number of applications where wear-resistance is critical. Kirk-Othmer treatment is in Vol. 23, pp. 413–25. Useful works include C. J. Smithell's *Tungsten* (Chapman and Hall, 1953); G. D. Rieck *Tungsten and Its*

Compounds (Pergamon, 1967); S. W. H. Yih and C. T. Wang, *Tungsten Sources, Metallurgy, Properties and Applications* (Plenum Press, 1979); P. M. Harris and D. S. C. Humphreys, *Tungsten: A Review* (Institution of Mining and Metallurgy, 1983); and the *First and Second International Tungsten Symposiums* held in Stockholm and San Francisco respectively under the auspices of the Primary Tungsten Association and the Consumer Reporting Group and edited and published by Mining Journal Books Ltd, 1979 and 1982. For the electronics industry, a new volume edited by R. S. Blewer entitled *Tungsten and Other Refractory Metals for VSLI Applications* was published in 1986 by Materials Research. The Primary Tungsten Association is a good source of current information.

Uranium

Uranium resources are mostly concentrated in North America, Africa and Australia in the Western World and although the metal's radioactivity was discovered almost a century ago it was another fifty years before its potential was unleashed with the discovery of the fissionable natural isotope. This, coupled with the need for feed material for nuclear power generation, has created a strong demand for high purity uranium, although in the present anti-nuclear climate, the potential for the metal is somewhat uncertain. Kirk-Othmer discusses it in Vol. 23, pp. 502–47. Notable works include A. N. Holden *Physical Metallurgy of Uranium* (a USAEC contracted publication with an appendix of binary uranium-alloy systems) (Addison-Wesley, 1958); C. D. Harrington and A. E. Ruehle, *Uranium Production Technology* (Van Nostrand, 1959); W. D. Wilkinson *Uranium Metallurgy* (Wiley, 1962); J. H. Gittus *Uranium* (a systematic appraisal of the world's literature) (Butterworth, 1963); R. C. Merritt, *The Extractive Metallurgy of Uranium* (again prepared under USAEC contract and containing extensive bibliographies) (Colorado School of Mines Research Institute, 1971); and *Metallurgical Technology of Uranium and Uranium Alloys: Transcripts of a State-of-the-art Seminar* (May 1981), sponsored by the Academy for Metals and Materials (American Society for Metals, 1982). A series of state-of-the-art seminars sponsored by the Academy for Metals and Materials and published by the ASM in 1982 dealt with Extractive and Process Metallurgy; Manufacturing; and Physical Metallurgy of Uranium and Uranium Alloys. A number of bibliographies have been produced by the US DOE Technical Information Centre, Oak Ridge, Tennessee. *Annual Uranium Seminar Proceedings* are published by the Society of Mining Engineers. A volume entitled *Uranium and Nuclear Energy* edited by staff of the Uranium Institute was published by Butterworths (1983) and the DOE Office of Scientific and Technical Information produced an update to its bibliography on *Radioactive Waste Management: Uranium Mill Tailings*, in 1985. A number of bibliographies have also been produced by the IAEA.

Vanadium

Its chief application is as an alloying agent in steels where grain refinement and hardenability are sought. Vanadium has found use in a variety of ferrous products such as constructional steels, heavy iron and steel castings, forgings, bearing, gears and springs. A further use is in jet aircraft engines as a high temperature alloy. Fast breeder reactors have emerged as a significant user due to vanadium's corrosion resistance and high-temperature strength in preference to stainless steel as a fuel cladding. Another developing use is in the field of superconductivity (Tohoku University Research Institute for Iron, Steel and Other Metals, for example, has been publishing in the *Journal of the Physical Society of Japan*). Kirk-Othmer treatment is in Vol. 23, pp. 673–87. Useful early works are R. Rostoker, *The Metallurgy of Vanadium* (Wiley, 1965); C. A. Hampel, *Rare Metals Handbook* (Reinhold, 1961); and publications of the Vanadium Corporation of America and US Dept of Commerce (e.g. *Economic Analysis of the Vanadium Industries*, PB–176471, 1967). The Argonne National Laboratory has been a prolific publisher of papers in recent years and a well-referenced paper published under its Fusion Power Programme by D. L. Smith and others entitled 'Vanadium-base alloys for fusion reactor applications' appeared in *Journal of Nuclear Materials*, v. 135 (2–3) Oct. 1985, pp. 125–39.

Zirconium

Classified alongside titanium and hafnium, zircon has been in use for thousands of years as a gem mineral but its largest use today is in foundry sand as the basic mould material, in mould cores and ram mixes. The oxide forms are heavily used in ceramic glazes. Zirconium metal finds use in military ordnance, titanium and copper alloys and niobium superconductors. It has good corrosion resistant properties and is finding increasing use in pumps, valves, pipes and heat exchangers. Hafnium-free zirconium has high ductility and excellent oxidation resistance and has found several applications in nuclear reactors. Kirk-Othmer deals extensively with it in Vol. 24, pp. 863–902 (the bibliography contains 266 referrences). Useful works include D. L. Douglass, *Metallurgy of Zirconium* (IAEA, Vienna, 1976); J. Schemel, *ASTM Manual on Zirconium and Hafnium* STP 639 (ASTM, 1977); D. G. Franklin (ed.) *Zirconium in the Nuclear Industry* STP 754 (ASTM, 1982); and publications of the Zirconium Corporation of America. See also Thomas's book on hafnium. Recent publications include the *Symposium on Industrial Applications of Titanium and Zirconium* (ASTM, 1984), and *Industrial Applications of Zirconium*, edited by C. S. Young and J. C. Durham, STP No. 917 (ASTM, 1986).

PART II

Applications

CHAPTER ELEVEN

Powder metallurgy

Dr D. S. COLEMAN

Introduction

Historical background

The knowledge that the compaction of powders[1] to form a shape, and the subsequent heating of this mass to give a hard product, is probably as old as civilized man. The first step in civilization was made when man began to change his environment and shape materials for his own use. The first materials used were, most probably, animal skins and bones, along with wood and clay. Of these simple materials, clay was the only one which required a secondary process, that of heating, to complete its formation into a useful product for man's benefit.[2]

History records that the pressing of metal powders is not new. The Egyptian civilizations certainly knew of the technique and they produced a number of iron implements long before the smelting of iron ore was discovered. The Incas of the Americas knew how to produce articles from powders of precious metals, such as gold and silver ornaments.

The Delhi iron pillar is reported to have been made some thousand years ago by Indian craftsmen forging iron particles together to give a solid mass. Compaction of gold dust is reported in 1542[3] to have been used to fill teeth. The discovery of platinum in the eighteenth century meant that new methods were needed to fabricate it, since it could not then be melted and was only available in powder form. Rochen[4] in 1786 produced solid platinum by hammering hot platinum particles together while held between two sheets of this metal. This method of hot forging sponge platinum raw material was used up to the beginning of this century in the fabrication of platinum crucibles.[5] The compaction of platinum powders to make blanks of this metal was reported in 1829 by Wollaston,[6] and in 1834 Sobolevskyi[7] pressed this powdered metal to produce coinage blanks; examples of three rouble coins made this way can be seen in the Lavra Museum at Kiev. At about the same time (1832) Marshall[8] carried out similar experiments. This process is still used today by the Sherritt Gordon plant in Edmonton, Canada, where powdered nickel is rolled into sheets from which nickel coin blanks are stamped from sintered strips.

The development of the electric incandescent lamp in the last century demanded the production of tungsten bars for drawing into wires for the filaments. Coolidge developed this process, which became the first large-scale production route for the manufacture of a metal from the compaction of a powder raw material. This process is still used, where wire bars are made by the cold compaction of tungsten powder, which are subsequently heat treated and drawn into filament wire.

Osann[9] in Germany in 1830 also found that cold-impacted metal powders could have a high strength after heating or sintering. Although these powder compacts decreased in size on sintering, this shrinkage was uniform and the original shapes were retained without distortion. He used copper and silver powders to produce medals and suggested that more complex shapes could be possible. Nearly 100 years later this was proved to be true in the early 1920s, when the first engineering components were produced from powders.

The early products of powder metallurgy were those composed of materials which were difficult to fabricate by normal metallurgical techniques. The first cemented tungsten carbides were made by the cold compaction of the carbide with a metal powder and the sintering of these compacts to form mining and cutting tool tips. This process was also suitable for the production of porous bronze bushes and bearings for machines. The inherent porosity produced in these components was ideally suitable for self-lubricating bearings, since the pores could be filled under a vacuum with a lubricant, such as an oil or wax.

The demands of the Second World War in 1940 saw the rapid growth of powder metallurgy in many countries. This technology was extended beyond the precious metals, the hard metals, and the refractory metals; to include the copper based alloys (mainly brasses and bronzes), other non-ferrous metals (aluminium and beryllium), nuclear metals (uranium and plutonium) and ferrous alloys. Powder metallurgy has expanded rapidly since that time and now includes the use of large tonnages of metal powders, mainly ferrous alloys.

The powder metallurgy process

The process is a relatively simple one with many advantages over the established metallurgical routes used for the fabrication of a product.[10] There are three main stages, which are as follows:

(1) *Mixing and blending* of the powders constituting the final product, this usually includes the addition of a bond and a lubricant.
(2) *Consolidation of this powder blend* by any of a number of processes. These include: die-compaction, isostatic compaction (both hot and cold processes), extrusion, rolling into sheets, slip-casting, loose filling of a mould, cold and hot forging. These are traditional metallurgical techniques which have been adapted to the use of a powder raw material. Other processes are now being studied in order to extend the range of consolidation processes and these include, injection moulding of powders,[11] spray-forming on a mould,[12] and flame or plasma spraying on to a shape. Ideas from both the metals and ceramics industries therefore have been developed and adapted for this technology.
(3) *Heating of the powder compact* The sintering of the consolidated compact is carried out in a reducing gas atmosphere to protect the metals from oxidation. In some cases (cermets) a vacuum or an inert gas atmosphere (nitrogen or argon) is used.
(4) *Post-sintering operations* Because the products are mainly used in the engineering industries, other operations sometimes need to be carried out on the product before delivery to a customer. These include: *repressing* or 'coining' to increase the final density or realign parts of the product; *machining*: drilling holes, lathe work, milling slots and other processes; *coating* for protection: painting, electroplating, and impregnation with sealant or lubricant; *final fabrication*: assembly of parts to form one component by welding, brazing or other assembly operations.

Advantages of powder metallurgy

The competition of this technology with other established metallurgical processes is based on the following advantages.[13-15]

(i) This technology is suitable for processing materials with high melting points, such as the refractory metals:

tungsten, tantalum, molybdenum etc., which cannot be produced by normal casting methods.

(ii) Alloys can be produced easily since intimate mixtures of alloying elements in powder form can be made. This is especially useful where normal melting and casting processes would have problems with the immiscibility of phases and lack of solution. This is particularly useful for the production of superalloys, special steels and cermets.

(iii) Controlled microstructures can be produced. Required levels of porosity and permeability can be manufactured which make possible the production of self-lubricating bearings and filters. Powder metallurgy also enables the development of special microstructures with specific features, e.g. grain orientation for permanent magnet production.

(iv) Large numbers of small engineering components can be produced rapidly and economically, if die-compaction processes are used.

(v) This technology is versatile since it can make parts of widely differing shapes and sizes ranging from a few grammes to several kilogrammes. If all the consolidation processes, such as isostatic compaction are considered, there are no real restrictions on the shape or size of an engineering component. Large structural girders, for example, can be fabricated from aluminium powders for aerospace vehicles.[16] Fully dense parts, as large as 400 kg, are now possible using hot isostatic pressing techniques.

(vi) Components can be made to very close tolerances with a range of finishes.

(vii) There are distinct economic advantages in the efficient use of raw materials, energy, capital equipment and space required to produce an engineering component, when compared with the traditional metallurgical processes.

(viii) The mechanical properties in most components are now as good as, or in some cases better, than the conventionally manufactured product. Powder forged components, for example, have properties that are better than the normally cast and forged product.

Powder metallurgy production

This technology has seen almost continuous growth from the 1950s, with some slowing down with the worldwide recession which started in the late 1970s. The number of companies involved in the powder metallurgy industries had risen in 1984 to a world total of over 1,500, which included 279 companies in Great Britain, and 443 companies in Western Europe. In Eastern Europe (including Finland) the number of industries involved in powder metallurgy was 49. This number compares with the United

Table 11.1. Companies in the world involved in powder metallurgy in 1984

Country	Companies	Country	Companies
Western Europe		*Eastern Europe*	
Austria	15	Czechoslovakia	5
Belgium	27	Finland	4
Denmark	3	E. Germany	8
France	94	Hungary	2
W. Germany	145	Poland	5
Great Britain	279	USSR	22
Greece	1	Yugoslavia	3
Holland	17		
Ireland	4	TOTAL:	49
Italy	45		
Luxembourg	2		
Norway	2	*Middle East Area*	
Portugal	5	Israel	8
Spain	3	Iran	2
Sweden	40	Turkey	5
Switzerland	40	Egypt	1
TOTAL:	443	TOTAL:	16
North America		*Africa*	
Canada	4	South Africa	20
USA	742	Zaire	1
		Zambia	1
TOTAL:	746	Zimbabwe	1
		TOTAL:	23
Central and South America			
Argentine	5	*Far East*	
Bolivia	1	Japan	160
Brazil	20	Korea	16
Colombia	2	China	3
Mexico	11	Taiwan	10
Peru	1	Hong Kong	3
Venezuela	7	India	40
TOTAL:	54	Philippines	2
		Singapore	6
		Australia	24
		New Zealand	1
		TOTAL:	265

Table 11.2. Companies in the world involved in powder metallurgy in 1986

Europe:	Great Britain	358
	Western Europe	1026
	Eastern Europe (including Finland)	53
Middle East:	Israel	10
	Turkey	6
The Americas:	USA and Canada	1035
	Central and South Americas	66
Africa:	South Africa	24
	Rest of Africa	3
Asia:	India	54
	Japan	235
	China, Taiwan and Hong Kong	40
	Rest of Asia	33
Others:	Australia and New Zealand	25
	WORLD TOTAL:	2968

Companies are involved in one of the following activities:
(a) Production and selling of powders.
(b) Manufacture and distribution of powder metal/ceramic engineering products.
(c) Equipment for the powder metallurgy industries, e.g. presses, furnaces etc.
Source: International Powder Metallurgy Directory, 1983–84.[16]

States of America and Canada, who had a total of 746 companies (only 4 in Canada). Elsewhere, the largest number of powder metallurgy companies was in Japan with 160. The rest of Asia had 26 companies if we include Iran (2), and to this number can be added 40 for India, with China 3, Taiwan 10 and Hong Kong 3. See Table 11.1. By 1986 world numbers had doubled as Table 11.2 shows.

Products for structural or mechanical applications

This is the largest group of products. The main powder raw material is iron powder, for which there is at present a world production capacity of over half a million tonnes. The products are distributed as in Table 11.3.

Table 11.3. Distribution of ferrous product applications

65–70%:	Automotive and constructional parts
10–15%:	Business machines, domestic applicances and agricultural applications
Rest:	Sports equipment, hobbies industries, garden equipment and others

There are significant tonnages of copper, brass, bronze and aluminium used for the production of non-ferrous structural and mechanical applications

Special high duty alloys

This is a rapidly growing category. Products with high mechanical strength, such as the high-speed sintered steels and the so-called superalloys (based on nickel or cobalt), can now be manufactured with properties which are superior to those produced by casting or forging.

The powder route has the advantage of a higher yield of usable material from the blank or billet made from powder. The microstructure is usually more uniform and can have a finer grain structure; these impart improved mechanical properties to the product. This route has also allowed the use of the newer powder grades of rapidly solidified metals and alloys which have a 'glass' type microstructure. The products from these materials have high tensile strengths, good ductilities and good thermal stabilities. See Table 11.4.

Table 11.4. Range of products produced by powder metallurgy

Structural components:	Ferrous parts; parts based on: copper, nickel, aluminium, stainless steels; powder forgings; high speed steels
Bearings:	Self-lubricating types based on copper or iron; steel backed types
Magnets:	Soft types; hard types; rare earth types
Electrical materials:	Electrical contacts; electric motor brushes
Wrought products based on powders:	Parts made from: aluminium, titanium, high speed steels, refractory metals, beryllium; structural ceramics
Composite materials:	Friction materials: brakes, clutches
Filters:	Bronze types; stainless steel types; others
Hard metals or cemented carbides:	Indexible tool tips; mining and drilling bits; drills; milling cutters; saws; wear parts; forming tools: dies, punches; high stress bearings
Diamond components:	Bonded diamond tools

A useful volume for those interested in equipment has recently been published. *Powder Metallurgy Equipment Manual*, edited by S. Bradbury, was issued by the Powder Metallurgy Equipment Association division of Metal Powder Industries Federation in 1986.

Reference literature

1 Fischmeister, H. F., John Player lecture on 'Powder compaction – fundamentals and recent developments', *Proceedings of the Institution of Mechanical Engineers*, **196** (18) (1982), 105.
2 Kingery, W. D., *Introduction to Ceramics* (J. Wiley and Sons Inc., New York and London, 1977).
3 Jones, W. D., *Fundamental Principles of Powder Metallurgy* (E Arnold, London, 1960).
4 Rochen, A, *J. Phys. Chem. Hist. Nat. Acta.*, **37**, 33 (1978).
5 Smith, C. S., in *Powder Metallurgy* by J. Wulff (Am. Soc. Metals, Cleveland, Ohio, 1942), Ch 2.
6 Wollaston, W. H., *Trans. Roy. Soc. London*, **119**, 1 (1829).
7 Sobolevsky, P, *Ann. Physik Chemie*, **109**, 99 (1834).
8 Marshall, W, *Phil. Mag.*, **11**, 321 (1832).
9 Osann, G, *ibid.*, **128**, 406 (1841).
10 'Powder metallurgy', *Metals Handbook*, Vol. 7 (American Technical Publishers Ltd, Hitchin, Herts SG4 0TP, England).
11 Quackenbush, C. L. *et al.*, *Proc. Ceramic Eng. and Sci. Am. Ceram. Soc.*, Jan/Feb. 20 (1982).
12 Singer, A. R. E., *Powder Metallurgy*, **25** (4) (1982) 195.
13 James, P. J., *The Production Engineer*, 32 (Sept. 1980).
14 James, P. J., *Chartered Mechanical Engineer*, **26** (7) (1979), 67.
15 Haynes, R, *The Metallurgist and Materials Technologist*, **13** (12) (1981), 355.
16 *International Powder Metallurgy Directory* (Metal Powder Report Publishing Services Ltd, Shrewsbury, England).

Information sources

Reference journals

Metal powders

Powder Metallurgy. Quarterly journal published by the Institute of Metals, 1 Carlton House Terrace, London SW1Y 5DB.
Metal Powder Report. Monthly journal published by MPR Publishing Services, Old Bank Buildings, Bellstone, Shrewsbury.
Journal of Powder Metallurgy and Powder Technology. Published quarterly by the American Powder Metallurgy Institute, 105 College Road East, Princeton, NJ 08540, USA.
International Powder Metallurgy. Monthly journal published in English by Verlag Schmid GmbH, D-78 Freiburg, PO Box 1722, W. Germany.
Soviet Powder Metallurgy and Metal Ceramics. Journal of English translations published monthly by Consultants Bureau, 227 West 17th Street, New York 10011, USA.

Other powders

Powder Technology. Published six times a year by Elsevier Sequoia SA, PO Box 1001, Switzerland. An international journal on the science and technology of wet and dry particulate systems.
Journal of Powders and Bulk Solids. Published quarterly by the International Powder Institute, 222 West Adams Street, Chicago, Illinois 60606, USA.

International Journal of Refractory and Hard Metals. Published quarterly by MPR Publishing Services, Old Bank Buildings, Bellstone, Shrewsbury.

Ceramic powders

Transactions and Journal of the British Ceramic Society. Published quarterly by the British Ceramic Society, Shelton House, Stoke-on-Trent ST4 2DR, Staffordshire.
Journal of the American Ceramic Society and *Ceramic Bulletin*. Published monthly by the American Ceramic Society Inc., 65 Ceramic Drive, Columbus, Ohio 43214.
Journal of the Institute of Refractories Engineers. Published by the Institute of Refractories Engineers, 15 St Benedict's Road, Wombourne, West Midlands WV5 9HP.

Reference source books and reviews

Source Book on Powder Metallurgy, S. Bradbury (ed.) (American Society for Metals, Metals Park, Ohio 44073, USA, 1979).
Powder Metallurgy Literature Reference Guide (Metal Powder Industries Federation, Princeton, New Jersey 08540, USA, 1974–75).
Sintering by F. Thümmler and W. Thomma. Metallurgical Reviews No. 115 (Institute of Metals, 1967).
Powder Metallurgy Guide for Engineers (Metals Society (now the Institute of Metals), 1 Carlton House Terrace, London SW1Y 5DB, 1978).
Reviews on Powder Metallurgy and Physical Ceramics. Periodic reviews published by Freund Publishing House Ltd., Chesham House, 150 Regent Street, London W1R 5FA.
Engineering Properties of Selected Ceramic Materials, J. F. Lynch *et al.* (eds) (American Ceramic Society Inc., 4055 N. High Street, Columbus, Ohio 43214, USA, 1966).
Growth Opportunities in Powder Metallurgy (Wooltech Reports, 44 Woodbine Avenue, Northpost, New York 11768, USA, 1985).
Metals Information: Computer Search Service. The Institute of Metals, 1 Carlton House Terrace, London SW1Y 5DB.
Treatise on Powder Metallurgy, C. G. Goetzel. Volume 3 (1952) and 4 (1963): Classified and annotated bibliography. Interscience, New York.

Reference books

Powder Metallurgy, Metals Handbook (American Society of Metals, 1984). Volume 7; American Technical Publishers Ltd, Hitchin, Herts SG4 0TP.
Introduction to Powder Metallurgy, J. S. Hirschhorn (American Powder Metallurgy Institute, Princeton USA, 1969).
Powder Metallurgy: Principles and Applications, F. V. Level (American Metal Powder Industries Federation, Princeton USA, 1980).
The Mechanical Behaviour of Sintered Metals, R. Haynes (Freund Publishing House Ltd, London, 1981).
General Principles of Powder Metallurgy, M. Yu Balshin and S. S. Kiparisov (in English) (Mir Publishers, Moscow, USSR, 1980).
Granulation, P. J. Sherrington and R. Oliver (1981).
The Production of Metal Powders by Atomisation, J. K. Beddow (1978).
Sintering, M. B. Waldron and B. L. Daniell (1978).*
Isostatic Pressing, P. Popper (1976).*
Roll Pressing, W. Pietsch (1976).*

Testing and Characterisation of Powders and Fine Particles, J. K. Beddow and T. P. Meloy (eds) (1979).*
Powder Metallurgy Processing: New Techniques and Analyses, H. A. Kuhn and A Lawley (Academic Press, New York and London, 1978).
Powder Metallurgy Equipment Manual (Metal Powder Industries Federation, Princeton, New Jersey 08540, USA, 1977).
Introduction to Ceramics, W. D. Kingery (J. Wiley and Sons Inc., New York and London, 1977).
The Refractory Carbides, E. K. Storms (Academic Press, New York and London, 1967).
Perspectives in Powder Metallurgy, H. H. Hausner *et al.*, Vol. 1: New Methods for the Consolidation of Metal Powders; Vol. 2: Vibratory Compaction, Principles and Methods (Plenum Press, New York, 1967).
Handbook of Metal Powders, A. R. Poster (ed.) (Reinhold, New York, 1966).
Researches in Powder Metallurgy, B. A. Borok (ed.) (Consultants Bureau, New York, 1966).
Powder Metallurgy: Practice and Applications, R. L. Sands and C. R. Shakespeare (Newnes, London, 1966).
Powder Metallury, S. A. Tsukerman. Trans. from the Russian by R. E. Hunt and H. S. H. Masset, ed. A. R. Entwhistle (Pergamon Press, Oxford, 1965).
Borides, Silicides and Phosphides, C. Aronsson *et al.* (Methuen, London, 1965).
Ultrafine Particles, W. E. Kuhn *et al.* (Wiley, New York and London, 1963).
New Types of Metal Powders, H. H. Hausner (ed.) (Gordon and Breach, Science Publishers, London and New York, 1964) (Proceedings of a Symposium held in Cleveland, Ohio, USA, 24 October 1963).
A Practical Course in Powder Metallurgy, D. Yarnton and M. Argyle (Cassell, London, 1962).
Powder Metallurgy, W. Lezynski (Interscience Publishers, London and New York, 1961) (Proceedings of an International Conference, New York, 13–17 June 1960 held by the American Metal Powder Industries Federation).
Cemented Carbides, P. Schwarzkopf and R. Kieffer (Macmillan, New York and London, 1960).
Fundamental Principles of Powder Metallurgy, W. D. Jones (Arnold, London, 1960).
Tooling for Metal Powder Parts, J. A. de Groat (McGraw-Hill, New York and London, 1958).
Powder Metallurgy, Selected government research reports, vol. 9 (HMSO, Ministry of Supply, 1951).
Treatise on Powder Metallurgy, C. G. Goetzel (Interscience, New York).
Vol. 1.:. Technology of Metal Powders and their Products (1949).
Vol. 2.:. Applied Physical Powder Metallurgy (1950).
Vol. 3.:. Classified and Annotated Bibliography (1952).
Vol. 4.:. Classified and Annotated Bibliography 1950–1960 (1963)
Part 1: Literature Survey
Part 2: Patent Survey
A Course on Powder Metallurgy, W. J. Baeza (Reinhold, New York, 1943).

*All published by Heyden and Son Ltd., Spectrum House, Hillview Gardens, London NW4 2JQ.

Proceedings of conferences, symposia and other meetings

Proceedings of Annual Conference of Powder Metallurgy Group of the Institute of Metals (Previously the Metals Society). Held yearly since 1954. Institute of Metals, 1 Carlton House Terrace, London SW1Y 5DB.

East European International Conferences Held in rotation with a conference every four years in each host country, for example:

German Democratic Republic: 8th International Conference in Dresden 1989. Languages: English, German and Russian. *Contact*: Zentralinstitut für Festkärperphysik und werkstofforschung, DDR-8027 Dresden, Helmholzstrasse 8.
Czechoslovakia: 7th International Conference in Pardubice 1987. *Contact:* Výzkumný ustav pro práškovou metalurgii (Research Institute for Powder Metallurgy), Žerotinova 60, 787–63 Šumpark, ČSSR.
Other host countries are Poland (1988) and Russia.

International Plansee Seminar. Held every four years in Reutte, Tirol, Austria; the 11th International Seminar was held 20–24 May 1985 and the 12th Seminar on 8–12 May 1989. Contact: Metallwerk Plansee GmbH, A–6600 Reutte, Tirol, Austria.

Western European International Conferences. Four yearly conferences organized by the European Powder Metallurgy Foundation. Recent conferences are:

8th International Conference: July 1990: London.
7th International Conference: July 1986: Düsseldorf.
6th International Conference: June 1982: Florence.
5th International Conference: June 1978: Stockholm.
4th International Conference: May 1975: Grenoble.
Contact: Institute of Metals, 1 Carlton House Terrace, London, SW1Y 5DB.

International Conferences on Isostatic Pressing. 1st Conference at Loughborough University of Technology, 19–21 September 1978 run jointly by MPR Publishing Services Ltd. and the Powder Metallurgy Research Group of the Department of Materials Engineering and Design, 1978.
2nd Conference at Stratford-upon-Avon, 21–3 September 1982, organized by MPR Publishing Services Ltd., Old Bank Buildings, Bellstone, Shrewsbury.

International Conference in Recent Advances in Hard Metals. Held at Loughborough University of Technology, 17–19 September 1979, run jointly by MPR Publishing Services Ltd. and Powder Metallurgy Research Group of the Department of Materials Engineering and Design.

International Conference on Advances in Hard Metal Production. Lucerne, Switzerland, 7–9 November 1983. Published by organizers: MPR Publishing Services Ltd., Old Bank Buildings, Bellstone, Shrewsbury.

International Conference on PM Aerospace Materials. Berne, Switzerland, 12–14 November 1984. Published by organizers: MPR Publishing Services Ltd., Old Bank Buildings, Bellstone, Shrewsbury.

Conferences on the Science of Hard Materials.
First International Conference: Moran, Wyoming, USA, September 1982.
Second International Conference: Rhodes, Greece, September 1984.
Organizers: National Physical Laboratory, Teddington, Middlesex TW11 0LW.

American International Conferences on Powder Metallurgy. Four yearly conferences, June 1988, Orlando, Florida; June 1984, Toronto, Canada and June 1980, Washington, USA, with others in 1970s. Organized by Metal Powder Industries Federation, American Powder Metallurgy Institute, 105 College Road East, Princeton, New Jersey 08540, USA.

International Conference on Powder Metallurgy. Modern Developments in Powder Metallurgy (3 vols). H. H. Hausner (ed.) (Plenum Press, New York, 1966).

Symposium: New Types of Metal Powders. Cleveland, USA, 1963. H. H. Hausner (ed.) (Gordon and Breach, New York, 1964).

Symposium: Fundamental Phenomena in the Material Sciences, H. H. Hausner (ed.) (Plenum Press, New York, 1964).

Conference: Handling of Solids, London 1962, P. A. Rottenburg (ed.) (Institution of Chemical Engineers).
Proceedings of International Conferences on the Compaction and Consolidation of Particulate Matter. First Conference: Brighton 3–5 October 1972. Second Conference: Brighton 2–4 September 1975. Published by the organizers: Powder Advisory Centre, 10 St Johns Road, Golders Green, London NW11 0PG.
Conference: Powder Metallurgy in Atomic Energy, Philadelphia, USA. 1955. H. H. Hausner (ed.) (American Society of Metals, 1958).
Symposium: Powder Metallurgy, London, 1954. Iron and Steel Institute, special reports No. 58, now Institute of Metals, 1 Carlton House Terrace, London SW1Y 5DB.
Symposium: Powder Metallurgy, London, 1947. Ibid. Special report No. 38.

Commercial sources

There are over 3,000 powder metallurgy companies scattered throughout the industrialized areas of the World. Therefore contact with a local powder metallurgy firm is one way to gain information in this area. Factories with their addresses are listed in: *International Powder Metallurgy Directory* (MPR Publishing Services Ltd., Shrewsbury). There are also many companies involved in carbide and cement manufacture. This can be found in the following publication:

World Directory and Handbook of Hard Metals (3rd edn). Published by the Engineers Digest, Swan House, 32 Swan Court, Leatherhead, Surrey KT22 8AH.

USA

Metal Powder Industries Federation. American Powder Metallurgy Institute, 105 College Road East, Princeton, New Jersey 08540, USA. Telephone: 609 452 7700. Telex: 510 685 2516.
American Society of Metals. Metals Park, Ohio 44073, USA. Telephone: 216 358 5151, Telex. 980619 ASMINT.

Japan

Japan Powder Metallurgy Association. 2–16 Iwamoto-cho 2-chome, Chiyoda-ku, Tokyo 101, Japan. Telephone: (03) 862 6646.

India

Powder Metallurgy Association of India. P26 Lakminagar, PO Saidabad, Hyderabad 500659, India. Telephone: 222579. Telex: 014457.

Australia

Powder Metal Industries Association. GPO Box 817, Canberra, ACT 2600, Australia. Telephone: 06259 6360.

South Africa

Powder Metal Association. c/o CSP PO Box 395, Pretoria 0001, South Africa. Telephone: 012 869211. Telex: 3–630 SA.

Research laboratory sources in Britain

National Physical Laboratory. Queens Road, Teddington, Middlesex TW11 0LW. Telephone: 01 977 3222. Telex: 262344.
National Engineering Laboratory. East Kilbride, Glasgow G75 0DU, Scotland. Telephone: 03552 20222. Telex: 777888.
Powder Metallurgy Research Group. Institute of Polymer Technology and Materials Engineering, University of Technology, Loughborough, Leicestershire LE11 3TU. Telephone: 0509 223152. Telex: 34319.
BNF Technology Centre. Grove Laboratories, Denworth Road, Wantage, Oxfordshire. Telephone: 02357 2992. Telex: 837166.
GKN Technology Ltd. Birmingham New Road, Wolverhampton WV4 6BW. Telephone: 0902 334361. Telex: 339724.
Production Engineering Research Association (PERA). Melton Mowbray, Leicestershire LE13 0PB. Telephone: 0664 64133. Telex: 34684.

Other countries

Research laboratories and institutes involved in powder metallurgy can be found in Reference 16, p. 224.

Manufacturing associations and associations in powder metallurgy

Great Britain:

British Powder Metal Federation. Contact: Secretary, 124a Compton Road, Wolverhampton WV3 9QZ. Telephone: 0902 28987.
British Hard Metal Association. Light Trades House, Melbourne Avenue, Sheffield S10 2QJ. Telephone: 0742 663084.
Powder Metallurgy Group. Institute of Metals, 1 Carlton House Terrace, London SW1Y 5DB. Telephone: 01 839 4071. Telex: 8814813.

Europe:

European Powder Metallurgy Federation. Contact: The Institute of Metals, 1 Carlton House Terrace, London SW1Y 5DB. Telephone: 01 839 4071. Telex: 8814813.
Fachverband Pulvermetallurgie. Goldene Pforte 1, 5800 Hagen-Emst, Postfach 921, West Germany. Telephone: 02331 51041 45. Telex: 0823806.
Fédération des Chambres Syndicales des Minerais et des Métaux Non-ferreux. 30 Avenue de Messine, 75008 Paris, France. Telephone: 145630266. Telex: 650 438 CSMETO.
International Society for Powder Metallurgy. PB 74, A-6600 Reutte, Tirol, Austria. Telephone: 0567270. Telex: 55505 PLANA.
Jernkontoret, Powder Metallurgy Group. Box 17721, S-11187 Stockholm, Sweden. Telephone: 0822 4670. Telex: 10165 JERNGRCS.

International Standards on powder metallurgy

*ISO, 1973:*2737-	Code B-	Permeable sintered metal materials – Determination of oil content
2738-	Code C-	Permeable sintered metal materials – Determination of density and open porosity
2939-	Code B-	Sintered metal bushes – Determination of radial crushing strength
2740-	Code B-	Sintered metal materials (excluding hard metal) – Tensile test pieces
*ISO, 1975:*3252-	Code G-	Powder metallurgy – Vocabulary Bilingual edition
3312;	Code B-	Sintered metal materials and hard metals – Determination of Young's modulus
3325	Code B-	Sintered metal materials, excluding hard metals – Determination of Young's modulus
3326-	Code B-	Hard metals – Determination of (the magnetizational) coercivity
3327-	Code B-	Hard metals – Determination of transverse rupture strength
3369	Code C-	Impermeable sintered metal materials and hard metals – Determination of density
*ISO, 1976:*3878-	Code C-	Hard metals – Vickers hardness test
3908-	Code B-	Hard metals – Determination of insoluble (free) carbon – Gravimetric method
3909-	Code C	Hard metals – Determination of cobalt – Potentiometric method
3923/1-	Code C-	Metallic powders – Determination of apparent density – Part 1: Funnel method
*ISO, 1977:*3907-	Code C-	Hard metals – Determination of total carbon – Gravimetric method
3927-	CodeC-	Metallic powders, excluding powders for hard metals
3927-	Code C-	Determination of compactibility (compressibility) in uniaxial compression
3928-	Code C-	Sintered metal materials, excluding hard metals – Fatigue test pieces
3953-	Code C-	Metallic powders – Determination of tap density
3954-	Code D-	Powders for powder metallurgical purposes – Sampling
3955-	Code D-	Sintered metal materials, excluding hard metals – Sampling
3995-	Code D-	Metallic powders – Determination of green strength by transverse rupture of rectangular compacts
4003-	Code C-	Permeable sintered metal materials – Determination of bubble test pore size
4002-	Code E-	Permeable sintered metal materials – Determination of fluid permeability
4505-	Code D-	Hard metals – Metallographic determination of porosity and uncombined carbon
*ISO, 1978:*4489-	Code B-	Sintered hard metals – Sampling and testing
4490-	Code C-	Metallic powders – Determination of flowability by means of a calibrated funnel (Hall flowmeter)
4491-	Code C-	Metallic powders – Determination of loss of mass on Hydrogen reduction (hydrogen loss)

4482- Code D- Metallic powders, excluding powders for hard metals – Determination of dimensional changes associated with compacting and sintering
4495- Code B- Lubricated metallic powders – Determination of lubricant content – Soxhlet extraction method
4496- Code C- Metallic powders – Determination of acid insoluble content in iron, copper, tin and bronze powders
4498/1- Code C- Sintered metal materials, excluding hard metals – Determination of apparent hardness – Part 1: Materials of essentially uniform section hardness
4499- Code C- Hard metals – Metallographic determination of microstructure
4501- Code C- Hard metals – Determination of titanium – Photometric peroxide method
4503- Code C- Hard metals – Determination of contents of metallic elements by X-ray fluorescence – Fusion method
4507- Code C- Sintered ferrous materials, carburized or carbonitrided – Determination and verification of effective case depth by the Vickers microhardness testing method
4883- Code B- Hard metals – Determination of contents of metallic elements by X-ray fluorescence – Solution method
4884- Code B- Hard metals – Sampling and testing of powders using sintered test pieces
5754- Code B- Sintered metal materials, excluding hard metals – Unnotched impact test piece
*ISO, 1979:*4506- Code C- Hard metals – Compression test

The above standards are available for purchase from all national standard organizations.

CHAPTER TWELVE

Ceramics

D. MURFIN

It seems best to begin a discussion of information on ceramics by defining the materials with which we are concerned. While this might seem obvious and straightforward, for materials which have been made in one form or another for many thousands of years, it has in fact proved difficult to arrive at a general definition which includes all the materials the various commercial and academic interests would wish to include, while also excluding materials which no-one would wish to consider as ceramics.

Perhaps the best definition is that of ASTM C-242, which states that ceramics are products made by a ceramic process. Lest this be thought circular, a ceramic process is defined as:

> The production of articles or coatings from essentially inorganic, non-metallic materials, the article or coating being made permanent and suitable for utilitarian and decorative purposes by the action of heat at temperatures sufficient to cause sintering, solid-state reactions, bonding or conversion partially or wholly to the glassy state.

This definition is extremely wide, embracing the products of the traditional pottery industries, clay bricks and pipes, glass products, modern engineering ceramics and many others. Even so, the heat treatment must be allowed to take place under geological conditions if materials such as naturally occurring silica (used as a refractory) are to be included, while the reader must judge whether the definition encompasses so-called ceramic powders (manufactured powder of inorganic compounds such as silicon carbide or boron nitride) or autoclaved compounds such as silicate bricks, or indeed cement and concrete, materials which the

American Ceramic Society and now others (including the Institute of Ceramics in the UK) certainly regard as products of the ceramics industry.

Basically, then, we are concerned with inorganic, non-metallic materials for which heat treatment plays a significant role at some stage in their production (and normally we are thinking of red-heat at least). By extension, we are also concerned with the equipment and processes used to make articles or starting powders from such materials (extruders, presses, kilns); or to finish them by machining or decorating purposes. We are concerned with their raw materials and with a range of auxiliary materials used to make these into products (plaster for moulds in which to cast shapes; gums and resins to bind them until hardened by firing). We need to consider their applications (cups and saucers; bricks, tiles and pipes; furnace linings; wear resistant parts; electrical components and electronic sensors); the measurement of their properties and the specialized test methods particularly applicable to them (thermal shock tests, the crazing resistance of glazes, impact and chipping tests, and many more). Their uses may be governed by design codes (for wall-tile fixing or load-bearing brickwork, for example); by detailed standards (for the composition and properties of refractories or electrical ceramics); by legislation (limiting the release of toxic metals from decorated ceramics, glassware or vitreous enamelled ware).

To this might be added, though in the context of this volume we shall not do so, a vast literature on the historical, aesthetic and antiquarian aspects of ceramics, as products of artistic endeavour.

Thus it will be seen that proper coverage of the information relating to ceramic materials is a wide-ranging task, spreading over physics, chemistry, engineering, economics, marketing and so on. It is the purpose of this chapter to chart a path to these sources of information which are specific to the ceramics industry or to ceramic products; to try to give some indication of their completeness and value and to indicate where no such sources exist and reliance must be placed on those general sources common to information work in other disciplines.

Dictionaries and glossaries

It may seem odd to begin at this point, but ceramics is bedevilled by an arcane and inconsistent nomenclature stemming from its roots in several disparate craft industries, combined with differences in meaning between British and American practice, and difficulties in translation when the terms refer to equivalent materials, but materials which are not identical materials – the differences arising from variations in industrial practices and from differences in the indigenous raw materials available. Anyone coming fresh to the ceramics literature will soon find the need for some brief but clear explanation of terms such as slip casting, spalling, sintering, frit, gunning refractory, fibre anchor and many more. And what is the difference between bone china, porcelain, earthenware, hotelware and stoneware? Or between RBSN (reaction-bonded silicon nitride), HIPed (hot isostatically pressed) silicon nitride and hot-pressed silicon nitride? What is partially stabilized zirconia, and what is a sialon?

Unfortunately, there is no one publication which can be recommended to answer all these questions.

The American Ceramic Society's *Ceramic Glossary*[1] is a convenient size, accurate and not expensive but, although reasonably up-to-date in other respects, includes few terms relating to modern engineering ceramics.

O'Bannon's contemporary *Dictionary of Ceramic Science and Engineering*[2] is seriously flawed in the proportion of entries which relate to basic electrical engineering rather than ceramics, and has duplicated but inconsistent entries betraying lack of final editorial control.

Both of the above reflect American practice. For Britain, there is as yet no replacement for Dodd's classic *Dictionary of Ceramics*,[3] though this now 20-years-old and out-of-print.

British Standard BS 3446 *A Glossary for the Refractories Industry*[4] is useful and a new edition is at an advanced stage, with a glossary for engineering ceramics to follow. The trilingual (English, French, Russian) international standard ISO/R836[5] is however still based on the first edition of BS 3446. Other multilingual glossaries exist for the refractories industry. The PRE Glossary[6] covers French, English, German and Italian, while its ALAFAR counterpart[7] has English, French, Spanish and Portuguese.

The CEC Ceramic Tile Dictionary[8] is multilingual and in fact provides terminology more generally applicable in the ceramics industries than its title suggests. In general, for translation purposes, reliance must be placed on good general dictionaries, large enough to include specialized terms – Harrap's French–English dictionary[9] is particularly worth mentioning for the number and accuracy of its entries related to ceramics.

At this point some words of warning about vocabulary, nomenclature and translation are pertinent. First, there are some straight oddities: in the USA, drain tile are pipes, and other structural tiles in the UK would be called hollow blocks, while the Americans would conversely understand clearly our use of the words floor tile and wall tile. This brings us to the second point – translations often choose to render words into their equivalent in other languages. It must be remembered that to translate faience as earthenware or gres as stoneware does not make the compositions of these materials identical. Finally, if one becomes involved with tariff nomenclature, one must stick carefully to the appropriate definitions and classifications. Quite often these do not accord fully with definitions to be found in technical glossaries or textbooks. In this context it is worth remarking that as well as there being definitions of terms in tariff nomenclature, many standards also define terms, and these definitions are by no means all consistent with one another, nor are they always easy to find. For example, definitions of china, bone china and porcelain are to be found in British Standard BS 5416:1976 Water Absorption and Translucency of China or Porcelain Tableware. The American Ceramic Society's glossary has brought together the ASTM definitions, and Dodd brought together these and others, but there is no single source for British or European terminology of the last twenty years, though the *Silikat Lexicon* (W Hing, Akademic-Verlag, Berlin, 1985) provides for those who read German a detailed and extensive (900 pp.) illustrated dictionary of ceramics technology (whose entries average 2 column-inches).

The entries are further listed separately in German, with Russian, English and French equivalents, though this is of limited use if approached through these other languages.

Sources

1 Perkins, W. W. (ed.), *Ceramic Glossary 1984* (Columbus, Ohio, American Ceramic Society, 1984).
2 O'Bannon, L. S., *Dictionary of Ceramic Science and Engineering* (New York, Plenum, 1984).
3 Dodd, A. E., *Dictionary of Ceramics* (2nd edn) (London, Newnes, 1967).
4 British Standards Institution BS 3446 (1962). A glossary of terms relating to refractory materials.
5 International Organization for Standardization, *Vocabulary for the Refractories Industry*. Trilingual edition. ISO/R836 – 1968.
6 Fédération Européenne des Fabricants de Produits Réfractaires, *PRE Glossary* – French, German, English and Italian (Zurich, PRE, 1973).
7 Asociacion Latinoamericano de Fabricantes de Refractarios, *Terminologia* ALAFAR. Spanish–Portuguese–English–French (AKAFAR, undated).
8 *CEC Ceramic Tile Dictionary* (Genoa, Italy, CEC, 1970).
9 *Harrap's New Standard French and English Dictionary*, J. E. Mansion (London, Harrap, 1977).
10 Savage, G. and Newman, H, *An Illustrated Dictionary of Ceramics* (Thames and Hudson, 1974).

(Though the artistic side of ceramics will receive little attention from here on, G. Savage and H. Newman's *An Illustrated Dictionary of Ceramics*[10] is worth mentioning here. It includes an introductory list of European factories and their marks, which is available in more complete form in J. P. Cushion, *Handbook of Pottery and Porcelain Marks* (London, Faber and Faber, 1980).

Textbooks and monographs

True textbooks in ceramics appear relatively infrequently, so some fairly elderly titles are listed, as well as the newer ones. Of interest are the following. *An Introduction to Ceramics Technology* 2nd edn, W. D. Kingery *et al.* (John Wiley, 1976) contained in its first edition substantial chapters on industrial practice. These were omitted in the second edition in favour of an expansion of the treatment of the basic science of ceramics to cover new materials and new theoretical developments.

An elementary general introduction, providing an easy-to-read, balanced (though now somewhat dated) view of the industry as a whole is M. Chandler, *Ceramics in the Modern World* (Aldus Books, 1967). *Chemistry and Physics of Clay*, 3rd edn, A. B. Searle and R. W. Grimshaw (Benn, 1960) provides detailed coverage of the raw materials for the traditional clay-based side of the industry. *Industrial Ceramics*, F. Singer and S. Singer (eds) (Chapman and Hall, 1963) was a comprehensive account of pottery manufacturing technology which is now rather old.

The *Ceramic Monographs* series, published as regular brief pull-out supplements to the journal *Interceram* goes some way to providing a modern equivalent, but it is multi-author, as yet incomplete, with some parts that are quite old, and not readily accessible to those who have not saved the parts.

A wide-ranging coverage at the introductory textbook level (suitable for, say, first year university students of ceramics) is provided by the *Institute of Ceramics Textbook* series. Titles so far include (some in new, revised editions):

Whitewares: Production, Testing and Quality Control
Ceramics Drying
Health and Safety in Ceramics
Ceramics Glaze Technology
Ceramic Raw Materials
Portland Cement
Vitreous Enamelling
Cement
Rheology of Ceramic Systems
Action of Heat on Ceramics

More specialized books include *An Introduction to the Technology of Pottery*, P. Rado (Pergamon, 1970). This is a readable yet technically accurate and quite complete account of the production of tableware, rather than of pottery in general and a new edition is in preparation. *Pottery Science: Materials, Processes and Products* by A. Dinsdale (Halsted Press, 1986) is a clear and comprehensive account marred only by its relatively high price. *Introduction to Whitewares* by G. Jackson (Maclaren, 1969) in its day provided good coverage of all aspects of pottery manufacture, including sanitaryware and tiles. *Fine Ceramics* by F. H. Norton (McGraw-Hill, 1970) is a rather newer, more detailed equivalent, describing American practice. It is valuable for including flow-charts, and short sections describing how British practice differs from that in the USA. More specific still are *Ceramic Glazes*, 3rd edn, by C. W. Parmelee (Cahners Publishing Co., Boston, 1973) and *Ceramic Screen Printing* by A Kosloff (Sign of the Times Publishing Co., Cincinatti, 1977). The intensive literature in studio pottery is outside the scope of this review.

Books on ceramics for the building industry include *W. Bender and F. Handle* (eds), *Brick and Tilemaking* which is in fact an expensive multi-author compilation describing the best modern practice in brickmaking. Older, but still a useful introduction is F. H. Clews, *Heavy Clay Technology*.

The refractories industry is slightly better served: J. H. Chesters' two-volume work is still of real value: Vol. I *Refractories, Production and Properties* covers the production and properties of the whole range of conventional refractories. The bibliographies at the end of each chapter provide references to the earlier literature. Vol. II *Refractories for Iron and Steelmaking* continues the story, considering the service behaviour of refractories in the various processes from blast furnace to casting-pit. F. H. Norton's *Refractories* is the American equivalent, while a more recent textbook is in French: *Traite pratique sur l'utilisation des produits refractaires*. The Iron and Steel Society of the AIME published *Blast Furnace Refractories and Repairs* in 1987 (edited by George Harry). More specialized is the 1984 translation of the 1979 Japanese edition of *Technology of Monolithic Refractories* by A. Nishikawa (Plibrico Japan Co, 1984. This provides a useful and reasonably up-to-date account of refractory concretes, their properties, installation and uses. The US Bureau of Mines produced *Evaluation of Refractories for Aluminium Recycling Furnaces*, by E. E. Davis in 1988.

Advanced, or engineering ceramics have still to find their chronicles. The best introduction of conference proceedings is R. Morrell, *Handbook of Properties of Technical and Engineering Ceramics*, Part 1. Frequently, modern ceramics must be content with chapters (of very varied comprehensiveness and quality) in books on materials science, aimed at students on general materials science courses in universities and polytechnics. In this category is the eight-volume *Encyclopaedia of Materials Science and Engineering* edited by M. B. Bever (Pergamon, 1986). The entries in this encyclopaedia are of a high standard, each providing a good, and brief, introduction to its field. Similarly, the reprint collection of entries from the Kirk-Othmer *Encyclopaedia of Chemical Technology* provides an excellent starting point.

The reprint *Encyclopaedia of Glass, Ceramics and Cement* is a very reasonable way of obtaining the relevant parts of Kirk-Othmer. Texts on glass, glassmaking and glass-ceramics include *Glass-making Today* edited by P. J. Doyle (Portcullis Press, 1979) and *Glass Ceramics*, P. W. McMillan, 2nd edn (Academic Press, 1979).

A more detailed and extensive treatment of glass science and technology is to be found in the series with that general title published by Elsevier. Titles include:

Vol. 1.:. J. Stanek, *Electric Melting of Glass*
Vol. 2.:. C. R. Bamford, *Colour Generation and Control in Glass*
Vol. 3.:. H. Rawson, *Properties and Applications of Glass*
Vol. 4.:. J. Hlavac, *The Technology of Glass and Ceramics: An Introduction*
Vol. 5.:. I. Fanderlik, *Optical Properties of Glass*
Vol. 6.:. K. L. Loewenstein, *The Manufacturing Technology of Continuous Glass Fibres*
Vol. 7.:. M. B. Volf, *Chemical Approach to Glass*
Vol. 8.:. Z. Strnad, *Glass-Ceramic Materials*.

Conference proceedings

Conference proceedings represent one of the richest sources of technical information in ceramics. In some areas they overshadow the journal literature in providing up-to-date and detailed articles on current technology. The drawbacks are that the proceedings of a given conference do not always appear with the rapidity that might be expected; some of them are hard to locate; their forms are varied. Some masquerade as special issues of journals; some are annual series (*Advances in. . . .*); some have more or less catchy titles and may be mistaken for books (for which indeed they sometimes substitute if the theme of the conference is to provide an up-to-date review of the state-of-the-art in a given field – e.g.

Science of Ceramic Chemical Processing by Hench and Ulrich Wiley, 1986). Some declare themselves formally and accurately.

The papers which appear in them are equally varied. Accounts of recent technical developments predominate, though sometimes one is snared by a lengthy abstract appearing instead of the actual paper. Also useful are review papers and keynote addresses, and where questions and discussions have been reported, these often give some flavour of the reception accorded to particular papers.

Sometimes a little imagination is necessary to identify likely targets. *Materials Science and Engineering* **71** (5) 1985 was in fact the proceedings of the International Symposium on Engineering Ceramics held in Jerusalem in December 1984 – very creditable as a rapid publication. Engineering ceramics show a tendency to dominate the current ceramics literature, and conference proceedings contribute substantially to this trend (the proportion of papers on engineering ceramics included in *British Ceramics Abstracts*, which covers individual papers in conference proceedings, has risen from about 1/10 to 1/3 in the last 10 years).

More traditional materials do have their place, and indeed the total of papers published on ceramic masonry in conferences outweighs in both quantity and quality the relatively few to be found in the regular journals, which often concentrate on more ephemeral topics.

A complete survey of conference proceedings related to ceramics is not practical in this context, but it seems valuable to list, first those regular series specifically concerned with ceramic developments, and then a selection of recently published books and proceedings to give an idea of the range of recent literature of this type.

Institute of Ceramics (formerly British Ceramic Society) – Proceedings series have covered all aspects of ceramic materials and processes, from electronic ceramics to loadbearing brickwork. The Institute now also publishes *Special Ceramics Conference Proceedings* which were formerly published by the then British Ceramic Research Association.

Science of Ceramics conference proceedings are published biennally by the respective organizing committees. The *Materials Science Research Conference Proceedings* are published by Plenum Press. The CIMTEC series (International Meetings in Modern Ceramics technologies) ventured into high technology ceramics at its 6th meeting, whose proceedings were published in three substantial volumes.

International Brick Masonry conferences are published regularly by the various conference organizers while the proceedings of the 14th Conference on Silicate Industry and Silicate Science were published in 1985 and the series on raw materials for the refractories and glass industries published by *Industrial Minerals* are of value, as are Electric Furnace Proceedings and Ironmaking Proceedings published by the Iron and Steel Society of the AIME.

The annual Aschen conference on refractories is published as an *Interceram* special issue. It is also probably appropriate at this point to interject some comment on the recent upsurge in the publishing activities of the American Ceramic Society. As a deliberate policy, the ACS has greatly expanded its range of publications, so that it is now the world's dominant publisher of technical information on ceramics, much of this by way of conference proceedings and reviews.

The first International Symposium on Advances in Refractories for the Metallurgical Industries was held in Winnipeg in Canada in 1987, sponsored jointly by the American and Canadian Ceramic Societies, edited by M. A. J. Rigaud (Pergamon, 1988).

Earlier publications have survived: the *Journal* containing erudite technical papers; the *Bulletin*, providing news of Society affairs, and papers of more general interest, though still of high quality; and *Ceramic Abstracts* providing summaries of ceramics literature from sources worldwide.

To these have been added two further regular publications. *Ceramic Engineering and Science Proceedings* is a bimonthly journal containing papers related to a particular topic, usually the proceedings of a conference. *Advanced Ceramic Materials* is a new quarterly journal concentrating on high-technology ceramics for electronic, electromagnetic, optical and mechanical applications. It reports on the development worldwide of new advanced ceramics and ceramic composites, discussing their properties, fabrication and applications. Technical and scientific reports are complemented by review features and news items. *The Advances in Ceramics* series which, again, are fully edited conference proceedings, provide authoritative reviews and up-to-date descriptions of fields at the forefront of ceramic development. In its series *Ceramics and Civilisation* the ACerS breaks new ground by providing a platform for interdisciplinary studies which correlate artistic, historical and technical perspectives on ceramics.

Primary journal literature

A complete listing of relevant journals is not feasible, and reference to the abstracting service (see below) is necessary. Here we shall provide a general overview of the problems of adequately covering journals containing information on ceramics. Until the last few years, almost all of the relevant journal articles were contained in what might be termed the ceramics core journals – those devoted almost exclusively to ceramics, and published either by the various national ceramics societies, or by commercial publishers. An illustrative though not exhaustive list would include the following current journals:

British Ceramic Transactions (formerly *Transactions of the British Ceramic Society*)
American Ceramic Society Bulletin
Journal of the American Ceramic Society
Ceramic Forum International/Berichte der Deutacher keramischen Gesellschaft
Industrie Ceramique
Journal of the Japanese Ceramic Society
Interceram
Ceramica Informazione
Ceramics Informazione International
Ceramic Industries Journal
Ceramics Japan
Ceramic Industry
Australian National Clay
Canadian Clay and Ceramics
Journal of the Canadian Ceramics Society
Glass and Ceramics (a Consultants Bureau Translation of *Steklo i Ceramica*)
Australian National Clay
Journal of the Australian Ceramic Society
Sprechsaal
Ceramurgia
Ceramurgia International
Silikaty
Sklar a Keramik
Silicates Industrial
Silikattechnik
Indian Ceramics
Klei/Glas/Keramiek
Keramische Zeitschrift
Epitoanyag
Ceramica (Sao Paulo)
Bulletin de la Sociedad Espanola de Ceramica y Vidrio
Central Glass and *Ceramics Research Institute Bulletin*

To these could be added journals with a wider coverage but a substantial ceramic content, such as:

Journal of Materials Science
Science of Sintering
Inorganic Materials (*Neorganichesk Materiali*)
Materials Forum
High Temperature Technology
Industrial Minerals

Further journals deal with more specialized areas of ceramics technology, or with other technologies which are major users of ceramics:

Tableware International
Bagno e Accessori
Baustoffindustrie
Blast Furnace and Steel Plant
Brick and Clay Record
Cement and Concrete Research
Clay Minerals
Drying Technology
Euroclay
Fliessen und Platten
Glass Industry
Glass Technology
Glustechnische Berichte
Industrial Heating
Interbrick
International Journal of High Technology Ceramics
International Journal of Mineral Processing
Journal of the Institute of Refractories Engines
Journal of Materials Research
Prosthetic Dentistry
Magazine of Concrete Research
Materials Chemistry and Physics
Materials and Design
Materials Forum
Metals and Materials
Philip Technical Review
Physics and Chemistry of Glasses
Physics and Chemistry of Minerals
Powder Metallurgy
Refractories (*Ognaupory*)
Refractories (*Taikabutsa*)
Refractories Journal
Refrattari e laterizi
Revue internationale des hautes températures et des refractaires
Steel Times

Stroitelnye Materiali
Structural Engineer
Tableware International
Tiles and Tiling
TIZ Fachberichte
Vitreous Enameller
World Cement
ZI International (formerly *Ziegelindustrie*)

Two general trends are apparent in ceramics journal literature. One, the increasing tendency of journals in foreign languages to publish English versions of their papers, or to have 'International' editions, makes a greater proportion of the ceramics literature more readily available in English.

The other is a recent great increase in the diversity of sources in which papers of real value to the ceramics industries are published infrequently. This trend is particularly noticeable in, though not confined to, the recent developments in advanced engineering ceramics. Basically, materials scientists and development and test engineers whose background relates to the applications for which ceramics are being used (and not directly to the ceramics industry) are increasingly publishing worthwhile papers through the channels familiar to them. Some examples will illustrate this:

Hamid, A. A. and Chukwunenye, A. O., 'Effect of bearing plate properties on the behaviour of block masonry prisms under axial compression', *J. Test Eval*, **14** (3), 156 (1986).
Driscoll, T. J., 'Bottom purging of steel ladles in directional porosity plugs', *J. Inst. Eng.* (Spring), 2 (1986).
Shima, S. and Mimura, K, 'Densification behaviour of ceramic powder', *Int. J. Mech. Sci.*, **28** (1), 53 (1986).
Klima, S. J., 'NDE of advanced ceramics', *Mater Eval.*, **44** (5), 571 (1986).
Matsushita, K, Okamoto, T. and Shimada, M, 'Internal friction and mechanical properties of new ceramics', *Journal de Physique*, Colloque C10, Suppl. to **46** (12), p. C10 (1985).
Stedham, M. E. C., 'Quality assurance in the building brick industry', *Quality Assurance*, **12** (2), 31 (1986).

Thus, far from being something of a closed world, the literature on ceramics is rapidly widening into journals which a few years ago would virtually never contain materials on ceramics. Such a paper is still an infrequent occurrence in any one journal but, overall, the journals required to cover ceramics adequately have widened in numbers and scope very considerably, making further demands on and increasing the need for good abstracting services in the field. This trend is likely to be accelerated rapidly by a torrent of publications on superconducting ceramics in such well-established and prestigious publications as *Nature*, and *Physical Review*, though doubtless specialist journals will soon appear in this area.

Reports, translations and theses

This is a brief section. Specific items must be located using abstracting services or the British Library's *Reports, Translations and Theses*. Particularly prolific sources are: US Government Reports NTIS series (available from Microinfo or on microfiche from the British Library) and abstracted in Materials Sciences, NTIS, US Dept of Commerce; British Ceramic Research Ltd (a list of those of its publications which are openly available is obtainable from Queens Road, Stoke-on-Trent ST4 7LO); Building Research Establishment Digests and Current Papers include reports on the constructional use of ceramics, particularly bricks and tiles; Brick Institute of America publishes relevant reports; and the Institute of Metals publishes monthly lists of published translations on refractories, and latterly on advanced ceramics.

Some years ago the Institute of Geological Sciences published a series of *Mineral Dossiers* which included several of ceramics interest – *Ball Clays*, *Fireclays*, *China Clays*, *Building Clays* and others. Reports on energy audits and energy conservation including those emanating from the DTI or the Department of Energy have included several on particular aspects of the British ceramics industry – for ceramics production is inherently a greater user of energy. Such reports can include useful production figures. The Noyes Data Corp., New Jersey, as well as publishing critical reviews of the US Patent literature, also publishes, in convenient form, papers and reviews from US Government reports and research contracts. Titles include:

Ceramic Heat Exchanger Concepts and Materials Technology;
Corrosion and Chemical Resistance Masonry Materials Handbook;
Fracture in Ceramic Materials; and
Ceramic Containing Systems – Mechanical Aspects of Interface and Surfaces.

The Parthenon Press, Carnforth, Lancs., has published a series *Research Reports in Materials Science* which is a reviewed selection of doctoral theses on topics of current practical significance. Those of ceramic interest include:

Sintering Additive for Zirconia Ceramics
Suxing Wu,
University of Zhejiant, Hangzhou, China;

The Sintering of Nitrogen Ceramics
S. Hampshire,
National Institute for Higher Education, Limerick, Ireland;

High Temperature Creep Crack Growth Models in Ceramics
M. Thouless,
University of California at Berkeley, USA;

Thermo-mechanical Properties of Ceramic Fibres
V. K. Marghussian,
University of Science and Technology, Tehran, Iran;

Silicon Carbide Alloys
M. M. Dobson,
University College, Cardiff, Wales; and

Mechanics of Microcrack Toughening in Ceramics
Yen Fu,
University of California at Berkeley, USA

Patents

Basically, the position with regard to patents is simple. Reliance must be placed on abstracting services to identify relevant new patents. British Patents are covered comprehensively by *British Ceramic Abstracts*. US Patents are included in the American Ceramic Society's *Ceramic Abstracts*. Other national patents are listed in the national ceramic society's journal, or else the major patent abstracting and alerting services must be used – e.g. Derwent, or Auszuge aus den Europaischen Patentanmeldungen for European Patents.

However, there is a 10-year compilation of European Patents in ceramics: *Materials for Refractories and Ceramics – A Study of Patents and Patent Applications* by M. Terpstra (Elsevier, London and New York, 1986). This covers patents published in Federal Germany, France, United Kingdom and European patents since 1976, related to ceramic raw materials, intermediate products and unfired bodies, finished refractories and ceramics, and equipment for their production.

Similar, but reviewing US Patents in specialized areas, are the Noyes Data Corp. Chemical Technology Reviews, of which the following are relevant to ceramics:

No. 6.3 Glass Technology – Recent Developments;
No. 7.6 Refractory Materials;
No. 1.51 New Dental Materials;
No. 1.78 Refractory Materials – Developments since 1977;
No. 1.90 Industrial Abrasive Materials and Composites; and
No. 1.84 Glass Technology – Developments since 1978.

Standards

Standardization is very varied in the various branches of the ceramics industry. In some areas such as traditional raw materials, and in the products of the tableware industry, there are few standards and, those that there are, specify but few properties, usually those which affect performance as seen by the customer.

Engineering ceramics also are not yet the subject of extensive standardization. They are too new, and appropriate test methods have yet to be devised. In between, refractories are the subject of many standards, with BS 1902 running to many parts to specify appropriate test methods and acceptable values of properties for a wide range of refractory materials. German, American, Japanese and Russian standards are similarly prolific. The ISO catalogue lists 14 standards, while the list in the ASTM guide runs to over two pages.

Reference should be made to the appropriate national and international catalogue of standards, not forgetting the test methods listed in PRE Recommendations – Refractory Materials 1978 (Fédération Européenne des Fabricants de Produits Refractaires).

An appropriate selection of foreign standards can be obtained using the BSI Worldwide Standards Information. British Standards include:

BS 1902:1967 Methods of testing refractory material,
BS 1598:1964 Specification for ceramic insulating materials for general electrical purposes,
BS 3921: 1985 Specification for clay bricks,
BS 6466:1984 Code of practice for design and installation of ceramic fibre furnace linings to which might be added: Guide to Good Practice for the Installation of Monolithic Refractories; Institute of Refractory Engineers 1986; and Design Guide for Loadbearing Brickwork Special Publication 56 (British Ceramic Research Ltd, 1988).

Property data

Property data in ceramics must be approached with extreme caution. With the exception of R. Morrell's *Handbook of Properties of Technical and Engineering Ceramics Part 2 Data Reviews* of which only Section I – *High-alumina Ceramics* – has yet been published, there has been no authoritative property compilation for very many years.

Such classic compilations as P. T. B. Shaffer, *High Temperature*

Materials – Materials Index (Plenum Press, 1964); Goldsmith, *Handbook of Thermophysical Properties of Solid Materials* (Armour Research Foundation); and Battelle Memorial Institute, *Engineering Properties of Selected Ceramic Materials* (American Ceramic Society, 1966), are out-of-date, though useful as a rough guide for the rarer materials. Only Samsonov's *The Oxide Handbook* (IFI/Plenum, 2nd edition, 1982) was revised.

The reason for great caution is threefold: first, property data for ceramics is generally statistical in nature anyway. Then, there are a great many ceramic materials and their properties are very dependent on small compositional changes. This applies especially to electrical properties. Finally, for the same material (in the sense of the same chemical composition) the method of processing may have a profound effect on the microstructure, on the porosity and texture, and consequently on the properties, particularly on the mechanical strength and on such thermo-mechanical properties as the resistance to thermal shock. Great advances have been made in recent years in strengthening ceramics by altering their microstructure or by the addition of small amounts of sintering aids or crystal structure modifiers, to produce denser or reinforced ceramics. Thus, for accurate property data, it is still necessary to refer to the original literature for results and detailed descriptions of the materials.

Often manufacturers' trade literature provides copious (or sparse) property data, but this is often given in the form of ranges, or as a general guide only. Efforts are being made to provide computer data-banks, wherein more precise property data for ranges of materials may be stored and retrieved, but these have yet to bear fruit. *The Handbook of Glass Data*, by O. V. Mazurin *et al.* (Elsevier, 1983) provides a wide-ranging review of thermal, mechanical, optical and chemical properties of glasses.

While not strictly property data, the monumental 5-volume *Phase Diagrams for Ceramists* (American Ceramics Society) is an indispensable work of reference in this field, which is the subject of an ongoing programme of updating by that Society. Well over 6,000 phase diagrams of inorganic systems have now been included.

Business statistics and company information

For the most part, the generally known compilations of business statistics must be used, with the problem that there is frequently an insufficiently detailed breakdown to make them really useful for an analysis of particular parts of the ceramics industry. Ceramics journals (particularly *Sprechsaal* and *Industrial Minerals*), as well as others from time to time, contain more detailed information. Memorial lectures are a useful source, as the invited lecturers often try to provide some previously unpublished detailed snippet from their own knowledge to add point to a prestigious occasion.

Two items which might well have been included under books and monographs are worth noting:

P. W. Gay and R. L. Smyth, *The British Pottery Industry* (London, Butterworths, 1974).
R. L. Smyth and R. S. Weightman, *The International Ceramic Tableware Industry* (London, Croom Helm, 1984).

Particularly relevant British Government statistics include:

Business Monitor Quarterly Statistics, HMSO, London;
PQ. 2.410. *Structural Clay Products*.
PQ. 2.460. *Abrasives*.
PQ. 2.489. *Ceramic Goods*.
PQ. 2.481. *Refractory Goods*.
PQ. 2.516. *Dyestuffs and Pigments*.
PQ. 2.551. *Paints, Varnishes and Painters' Fillings* (for glazes).
PQ. 4.126. *Animal By-products* (for bone for bone china!).
PQ. 2.471. *Flat Glass*.
PQ. 2.478. *Glass Containers*.
PQ. 2.479. *Miscellaneous Glass Products*.
PQ. 2.420. *Cement*.
PQ. 2.511. *Inorganic Chemicals except Industrial Gases*.
Lofty, G. J. *et al.*, *World Mineral Statistics 1979–1983* (HMSO, United Kingdom Mineral Statistics (annual)).
Iron and Steel Industry Annual Statistics, produced by the UK Iron and Steel Statistics Bureau, Croydon; contains information on refractories.
The US Dept of Commerce, Bureau of Mines, and the Refractories Institute, Pittsburgh publish useful regular statistical series.

In Britain, relevant company information is to be found in the publications of Inter-Company Comparisons Ltd., London:

ICC Financial Surveys
ICC Business Ratio Reports

for both Ceramics Manufacturers and Brick and Tile Manufacturers, an online service is also now available. Also useful is *Accounting in the Pottery Industry* (Institute of Cost and Management Accountants, London, 1985), as it provides an economic insight into the various pottery-making processes.

Directories and trade literature

Useful directories specific to the ceramics industry include:

American Ceramic Society Bulletin. January Issue – Company Directory (Westerville, Ohio; American Ceramic Society, Inc.).
Ceramic Industries Journal Directory Issue (London, Turret Press. Annual.).
Ceramic Industry: Databook and Buyers Guide (Des Plaines, Ill.; Cahners Publ. Annual.).
Ceramic Source '87. Vol. 2 (American Ceramic Society, Inc.).
Clarke, G. M. (ed.), *Industrial Minerals Directory* (Worcester Park, Surrey; Metal Bulletin Books, 1986).
Sprechsaal Databook für Keramik, Glas, Raustoff Production (Coburg; Sprechsaal Verlag, 1985).
European Ceramic Directory (Coburg; Sprechsaal Verlag).

Trade literature on both ceramic products and on machinery, materials and equipment for their production, is quite prolific, especially that for refractory goods, where property data is also to be found. Probably the largest collection, which includes much foreign as well as British material, and which is backed by product and trade name indexes, is in the library of British Ceramic Research Ltd., Stoke-on-Trent.

Market research reports

The burgeoning of interest in advanced ceramics, with large (actual) markets in electronic products and (potential) automotive components, has led to a minor flood of expensive market research reports. There are two dangers in spending large sums on such reports – the actual data is limited and to some extent, the reports quote each other, and other cheaper sources are readily available (to those who use abstracts bulletins) – while considerable parts of many of the reports describe actual and potential applications in the way of review articles except that their technical accuracy sometimes leaves something to be desired.

In the following list, the Japanese YANO reports have been often quoted, while the Mitchell Market Reports are an ongoing series not as costly as some. The British Ceramic Research Ltd. report of Advanced Ceramics stemmed from an EEC contract, so its price reflects publication costs only, and is considerably less than that of other comparable studies.

Industrial Aids Ltd	*Advanced Materials Engineering in Europe* Vol. 1. Advanced ceramics Vol. 2. Termoplastic composites	12/85
Stanford Research Inst. (SRI International)	*Report of Multi-Client Study* (US, W. European and Japanese market advanced ceramics)	/85
Business Communications Co. Inc. (BCC)	*Advanced Ceramics-Technologies Economics and Market Opportunities*	10/85

World Business Publications Ltd	*Advanced Engineering Ceramics*	6/85
Technical Insights Ltd	*Annual Report on High-tech Materials 1984: The Year that Was. . . . 1985: The Year to Come*	nd
Frost & Sullivan Ltd	*US Technical Ceramics Market*	Spring/85
Frost & Sullivan Ltd	*Industrial Ceramics Market in Europe*	Autumn/84
Frost & Sullivan Ltd	*Metallurgical Industry Consumers Market in Europe*	Autumn/84
Frost & Sullivan Ltd	*Europe Advanced Ceramics Market*	in preparation
Find/Sup	*High Technology Industrial Ceramics*	6/83
Find/Sup	*World Market for High-Technology Industrial Ceramics*	7/85
Gorham International	*US Market for Engineering Ceramics and Ceramic Matrix Composites*	4/86
Technomic Consultants	*Analysis of the Commercial Markets and Technologies for Engineering Ceramics: 1984–94*	6/83
Long-Term Credit Bank of Japan Ltd.	*Fine Ceramics Industry in Japan*	12/84
Charles River Associates	*Technological and Economic Assessment of Advanced Ceramic Materials*	1985
US Dept of Commerce	*A Competitive Assessment of the US Advanced Ceramics Industry*	1984
Yano Research Institute (Japan)	*New Ceramics Market in Japan*	1982
Yano Research Institute (Japan)	*Japan's Innovative Industrial Materials*	1984
Mitchell Market Reports	*Zirconia*	1985
Mitchell Market Reports	*Silicon Nitride and the Sialons*	1986
Mitchell Market Reports	*Alumina*	?
Mitchell Market Reports	*Boron Nitride*	?
British Ceramic Research Ltd and North Staffordshire Polytechnic	*Marketing Study: An Analysis of the Domestic Tableware Market in the UK* (Special Publication No. 120)	1986
British Ceramic Research Ltd and North Staffordshire Polytechnic	*An Analysis of the Market for Advanced Ceramic Products in EEC Countries* (Special Publication No. 122)	1987

Abstracts and other secondary sources

The two longest-established specialized abstracting services are

Ceramic Abstracts (American Ceramic Society, Columbus, Ohio).
British Ceramic Abstracts (British Ceramic Research Ltd., Stoke on Trent, UK).

The former covers traditional and advanced ceramics, including phosphorous and electronic ceramics; glass; cement and concrete) plaster and lime products. It publishes some 8,000 abstracts each year, and includes coverage of US Patents, appearing bimonthly.

The latter has more detailed coverage of traditional and advanced ceramics, including detailed coverage of conference proceedings, but much more sparse coverage of the other areas. It publishes some 4,000 abstracts each year, and includes British Patents, appearing monthly. It is backed by a document supply service from the library of British Ceramic Research Ltd.

In Britain, some 2,500 abstracts on glass technology are published in the Society of Glass Technology's bimonthly journals *Glass Technology* and *Physics and Chemistry of Glasses*. Other more recently introduced services (apart from the large general abstracting services) are: *Key Abstracts – Advanced Materials* (INSPEC, Hitchin) which draws some 250 abstracts per month from the INSPEC database to produce a bulletin, of which about half relate to ceramics, refractories and glasses; and *Polymer/Ceramics/Composites Alert* (Materials Information, London). Initially polymers and composites outweighed ceramics, but the proportion is growing.

These two last services draw their material on ceramics chiefly from journals and reports already scanned to serve the longer-standing interests of the bodies concerned. Thus, for the moment at least, the services complement each other and the established ceramic abstracting services, rather than being in direct competition.

Other less formalized secondary sources have concentrated on Japanese developments in ceramics, for Japan has emerged as the world's leading supplier and developer of advanced ceramics, especially in ceramics for electronic applications. Useful are: *Ceramic Industry Japan* (Industrial Research Center of Japan, Tokyo); *Japanese Economic Journal*; *Advanced Ceramics Report* (Elsevier, Amsterdam); and *Technical Ceramics Bulletin* (N. C. Dellow, Watford). All contain economic, business and technical news items on developments in ceramics and concerning ceramics manufacturers worldwide.

Miscellaneous – who's who?

Lacking any clear place to be mentioned, but certainly not to be ignored, is the valuable American Ceramic Society Roster of members, usually the October issue of their bulletin, which lists individuals active in all aspects of ceramics. The ceramics and materials sciences entries in *Scientific Research in British Universities and Colleges* (British Library) besides providing some picture of current academic research, also gives names of the researchers. Many journals feature new appointments, honours and retirements to keep readers abreast of who's who and who does what. Useful in this way are the brief author's profiles in the American Ceramic Society's bulletin and journal.

Online services

It might be thought from the foregoing that information on ceramics was confined to paper and that the electronic information revolution has passed by ceramics. This is not so, but nevertheless ceramics as a specialization has yet to be well-served by online databases.

Much information can be gained from the well-established large databases – *Chemical Abstracts*, INSPEC, COMPENDEX. However, the problem is that because ceramic materials range over such a wide area of application, from traditional to modern, no one of the large services provides all the required information. On the other hand, there is considerable overlap between them. This leads to expensive searching and time-consuming weeding out of duplicates.

Ceramics Abstracts from the American Ceramic Society has recently been mounted as an online database by Pergamon-Orbit Infoline, and might have provided the answer – unfortunately its indexing is extremely rudimentary at times.

The glass industry is served by Glassfile, from the International Commission on Glass, a database of some 40,000 references from 1970 to date, updated with about 500 references every two months, indexed according to the ICG's thesaurus and bearing English titles, although the database itself is in French. The various Materials Information services from the Institute of Metals, now include an online service. *Engineered Materials Abstracts* includes ceramics alongside polymers and composites. Published searches are a feature of this database and that of the American Ceramic Society. Other online services are being developed. *British Ceramic Abstracts* is at an advanced stage of negotiations to mount it online, and British Ceramics Research Ltd. is developing a ceramics companies database under an EEC contract. The EEC Joint Research Centre at Pelten in the Netherlands is working on a computerized property databank for ceramics.

CHAPTER THIRTEEN

Corrosion, corrosion-resistance and protection

D. R. GABE

Introduction

Corrosion resistance is generally regarded as an extrinsic property, unlike many properties of metals, because it involves the surface of a metal in interaction with its environment. Consequently the study of corrosion and protection is an interdisciplinary science and technology.

Older definitions of corrosion involved the principle of electrochemical dissolution of metals in their (liquid) environments, but a more modern definition ensures that corrosion implies degradation or deterioration as well. Corrosion is the interaction of a material with its environment, in an engineering context, leading ultimately to the failure of that engineering structure. Thus the whole spectrum of materials in a range of chemical environments may be involved in any one of several engineering assemblies. Clearly basic information will be obtained from the primary disciplines and interdisciplinary information only will be considered here.

Terminology is also of importance. Corrosion protection or prevention is an ideal state where corrosion is eliminated completely but in practice corrosion control is adequate to give reasonable service lives by reducing the corrosion rate to an optimum level. Diagnosis of corrosion is the first step in a two-stage process culminating in a means of minimizing corrosion by improving corrosion resistance. This is achieved in practice in many ways including environmental control (use of inhibitors in solution), improved engineering selection and design, use of surface coatings, application of electric current to reverse or stabilize corrosion. Other types of deterioration exist – wear is embraced in the subject Tribology – but will be disregarded in this chapter.

No sourcebook has been prepared specifically for information scientists in this field but two have been published for the corrosion fraternity including in them other information such as consulting services, university research and teaching syllabuses for schemes of corrosion education etc. These sourcebooks may be regarded as first sources for further information searching.

European Federation of Corrosion, *Corrosion Education Manual* (Stockholm, 1972).

Dept. of Industry, *Corrosion Prevention Directory* (HMSO, London, 1975).

Primary discipline information will be found in *Use of Engineering Literature* by A. J. Anthony (Butterworths, London, 1986) and other similar volumes.

Learned societies

In Western countries learned societies or institutions exist among other reasons to disseminate information by publishing results of recent research and critical reviews of important fields. They also may provide information banks on personnel and companies as well as hold important collections of books, periodicals and trade literature; they will normally provide information in a commercially disinterested manner. Those having corrosion and protection as a main object and who publish in a substantial way include:

Institution of Corrosion Science and Technology (I. Corr. S. T.), Milton Keynes, UK.
Institute of Metal Finishing (IMF), Birmingham, UK.
American Electroplaters Society (AES), Orlando, Fl., USA.
National Association of Corrosion Engineers (NACE), Houston, Texas, USA.
Metal Finishing Society of Japan.

A substantial number include corrosion and protection among many interests and publish some literature in this field.

Institute of Metals (IM), London, UK.
Institute of Marine Engineers (I. Mar. E.), London, UK.
Royal Society of Chemistry (RSC), London, UK.
Oil and Colour Chemists Association (OCCA), Wembley, UK.
Society for Chemical Industry (SCI), London, UK.
Electrochemical Society of America (ES), Manchester, NH, USA.
American Society for Metals (ASM), Metals Park, Oh. USA.

One further society may be noted which is primarily concerned with standards but also publishes a substantial number of monographs, often with multiple authorship, concerned with quality control and testing.

American Society for Testing Materials (ASTM), Philadelphia, Pa. USA.

In the UK the National Corrosion Advisory Service, Teddington, London is a government-sponsored agency charged with promoting industrial co-operation in corrosion and protection. It has accumulated a considerable wealth of information on industrial and commercial activity in the field.

Dictionaries and glossaries

No dictionary specifically devoted to corrosion has been published and readers are referred to primary discipline dictionaries as recorded by Anthony, W. F. (loc. cit).

Several glossaries of terms have been published partly to establish terminology but also to clarify equivalence in several languages.

NACE, 'Glossary of Corrosion Terms (English)', *Mat. Prot.*, **7** (10) (1968), pp. 68–71.

Swedish Centre of Technical Terminology, *Korrosionsordlista*, TNC40 (Stockholm, 1968) (English, German, Scandinavian languages).

DECHEMA, *Glossarium der Korrosionsfachsprache* (German, English, French, Russian, Spanish) (Frankfurt).

DIN, *Korrosion der Metalle* (German), DIN 50900 (Berlin, 1960).

A glossary will also be found in *Corrosion* by L. L. Shreir (Newnes-Butterworths, London, 1976).

Standards organizations

Although most countries have a standards organization, relatively few operate an extensive system. In general standards must be objective rather than subjective and are usually related to marketable commodities rather than phenomena. Thus standards related to corrosion are relatively few – methods of assessing corrosion being one important area – while those related to protection processes, methods and products are extensive. The British Standards Institution publishes three main types of document:

British Standard Specifications (BS).
Codes of Practice (CP).
Drafts for Development (DD).

The British Ministry of Defence operates a parallel system of Defence Standards (DEF STAN) and Defence Guides (DG) as do other governments. The most important organizations for specifications in the field of corrosion and protection are:

British Standards Institution (BSI), London.
American Society for Testing Materials (ASTM), Philadelphia, USA.
Deutsche Institut für Normung (DIN), Berlin, FDR.

Fuller information will be found in Anthony, W. F. (loc. cit).

Standard specifications are regularly revised and if active are maintained in print. BSI publish individual specifications as booklets while ASTM issue the whole range annually, bound in over 40 volumes, irrespective of whether recent revision has occurred or not. Standards and bodies dealing with these have been considered at length in the introductory chapter and under individual metals and applications.

Data sources

The collation of data on corrosion resistance is such a formidable task involving such a multiplicity of metals, alloys, environments, temperatures etc. that no comprehensive assembly is ever likely to exist. Private computerized data banks are beginning to be formed by commercially oriented bodies, generally on a highly selective basis. Textbooks present equally selective illustrative data and are generally irrelevant in the wider context, although some handbooks are attempting to collate simple data in tabulated or graphical form. Four types of data source may be identified.

Handbooks

ASM *Metals Handbook*, 9th edn, Vols 1–3, *Properties and Selection of Metals and Alloys*; Vol. 5, *Surface Cleaning, Finishing and Coating* (Metals Park, Oh, USA).
Dechema Institut (Dechema Werkstoff-Tabellen, Frankfurt, W. Germany).
Pourbaix, M. *Atlas of Electrochemical Equilibria in Aqueous Solutions* (Pergamon Press, Oxford, 1966) or (in French) (Gauthier-Villars, Paris, 1963).
Rabald, E. *Corrosion Guide* (2nd edn) (Elsevier, 1968).
Schweitzer, P. A. *Corrosion and Protection Handbook* (Marcel Dekker, New York, 1983). Also by the same author: *Corrosion Resistance Tables, Metals, Plastics, Nonmetallics and Rubbers* (Dekker, 1986).
Smithell's Metal Reference Book (ed. E. A. Brandes) 6th edn (Butterworths, London, 1983).
Shreir, L. L. *Corrosion* (Newnes-Butterworths, London, 1975).
Treseder, R. S. (ed.), *NACE Corrosion Engineer's Reference Book* (NACE, Houston, Tx, 1980).
Uhlig, H. H. *Corrosion Handbook* (Wiley, New York, 1955).

Databanks

At least two databanks are being developed by semi-independent research institutes on a commercial subscription basis, and cover wide ranges of properties.

Dechema Institut, Frankfurt am Main (loc. cit.), Dechema Werkstoffe-Tabellen
Fulmer Research Institute, Stoke Poges, Bucks, UK. Materials Selector

Marketing companies

This category covers major industrial companies marketing particular materials having corrosion resistant properties and who for prestigious and other reasons research, collate and publish data. Clearly these only deal with particular (sometimes proprietary) products.

British Steel Corporation	London
US Steel Corporation	Pittsburgh, Pa.
International Nickel Corpn. (INCO)	Toronto, Ont. or London, UK
Climax Molybdenum (part of AMAX)	New York
American Metal Climax Inc. (AMAX)	New York
Dupont de Nemours and Co.	Wilmington, Del.

Note that INCO and Climax/Amax make special efforts to promote the use of their metals as alloying elements and consequently offer data over a wide field of alloy steels.

Trade associations

As with marketing companies the trade associations represent very specific interests but on a co-operative basis and occasionally worldwide.

Aluminium Federation, London
Cadmium Development Association (see ZDA), London
Cobalt Information Centre, Brussels
Copper Development Association (CDA), Potters Bar, Herts, UK
Lead Development Association (see ZDA), London
International Tin Research Institute (ITRI), Uxbridge, Middx., UK.
Rubber and Plastics Research Association (RAPRA), Shrewsbury, UK
Zinc Development Association (ZDA), London

Books

The number of books relating to corrosion and protection is large and most attempt some type of bibliographical reference system. However, two outstanding bibliographies should be referred to as primary sources.

Corrosion Prevention Directory (HMSO, London, 1975).
E Jackson, *A Bibliography of Books on Corrosion and Protection of Metals* (I. Corr. S. T., Milton Keynes, 1980).

Among the textbooks three are of treatise standard. Evans is an older book describing classical work and is exceptionally well-referenced. Shreir is a multi-author review compilation. Tomashov is a valuable guide to Russian literature and thought.

U. R. Evans, *The Corrosion and Oxidation of Metals* (Arnold, London 1960); 1st supplementary volume 1968; 2nd supplementary volume 1976.
L. L. Shreir, *Corrosion* (2 vols) (2nd edn, Newnes-Butterworths, 1976; 3rd edn, in preparation).
N. D. Tomashov, *Theory of Corrosion and Protection of Metals*, transl. (Macmillan, New York, 1966).

The following classified list of textbooks is given on a selective basis as a means of introduction to various aspects of the subject.

Corrosion and electrochemistry

J. O. M. Bockris and A. K. N. Reddy, *Modern Electrochemistry* (Plenum Press, 1970, 2 vols).
M. G. Fontana and N. D. Green, *Corrosion Engineering* (McGraw-Hill, 1967, 2nd edn 1978).
P. J. Gellings, *Introduction to Corrosion Prevention and Control for Engineers* (Delft University Press, 1976).
F. L. LaQue and H. R. Copson, *Corrosion Resistance of Metals and Alloys* (Reinhold, 1963).
J. C. Scully, *Fundamental of Corrosion* (Pergamon, 1st edn, 1965; 2nd edn 1975).
H. H. Uhlig, *Corrosion and Corrosion Control* (Wiley, 1st edn, 1963; 3rd edn 1984).
J. M. West, *Electrodeposition and Corrosion Processes* (Van Nostrand, 2nd edn, 1971).
J. M. West, *Basic Corrosion and Oxidation* (Ellis Horwood/Wiley, 1981).
G. Wranglen, *An Introduction to Corrosion and Protection of Metals* (Stockholm, 1972) distrib. Chapman and Hall).

High temperature oxidation

N. Birks and G. H. Meier, *Introduction to High Temperature Oxidation of Metals* (Arnold, London, 1983).
J. Benard, *L'Oxydation des métaux* (Gauthier-Villars, Paris, 1964).
K. Hauffe, *Oxidation of Metals* (Plenum Press, New York, 1965).
O. Kubaschewski and B. E. Hopkins, *Oxidation of Metals and Alloys* (Butterworths, London, 1962).

Protection methods

W. H. Ailor, *Handbook on Corrosion Testing and Evaluation* (Wiley, 1971).
L. M. Applegate, *Cathodic Protection* (McGraw-Hill, 1960).
ASTM, *Testing of Metallic and Inorganic Coatings: A Symposium Sponsored by ASTM Committee B-8* (Chicago, 1986).
R. M. Burns and W. W. Bradley, *Protective Coatings for Metals* (Reinhold, 1955; 3rd edn 1967).
P. D. Donovan, *Protection of Metals from Corrosion in Storage and Transit* (Ellis Horwood, 1986).
L. S. Evans, *Selecting Engineering Materials for Chemical and Process Plant* (Business Book, 1974).
D. R. Gabe, *Coatings for Protection* (I. Prod. E., London, 1983).
D. R. Gabe, *Principles of Metal Surface Treatment and Protection* (Pergamon, 1978).
A. Kuhn (ed.), *Modern Techniques in Electrochemistry Corrosion and Metal Finishing* (with contributions).
J. H. Morgan, *Cathodic Protection* (Hill, 1959).
J. A. Murphy, *Surface Preparation and Finishes for Metals* (McGraw-Hill, 1971).
I. N. Putilova *et al.*, *Metallic Corrosion Inhibitors* (Pergamon, 1960).
O. L. Riggs and C. E. Locke, *Anodic Protection* (Plenum Press, 1981).
H. Silman, G. Isserlis and A. F. Averill, *Protective and Decorative Coatings for Metals* (Finishing/Public, 1978).

Electrodeposition

A. Brenner, *Electrodeposition of Alloys*, 2 vols (Academic Press, 1963).
W. Canning, *Canning Handbook on Electroplating* (1982).
J. K. Dennis and T. E. Such, *Nickel and Chromium Plating* (Newnes-Butterworths, 1972).
W. Goldie, *Metallic Coating of Plastics*, 2 vols (Electrochemical Publications, 1968).
A. K. Graham, *Electroplating Engineering Handbook*, 3rd edn (Van Nostrand/Reinhold, 1971).
F. A. Lowenheim, *Modern Electroplating* (Wiley 2nd edn 1963; 3rd edn 1974).
E. Raub and K. Muller, *Fundamentals of Metal Deposition* (Elsevier, 1967).
M. Rubinstein, *Electrochemical Metallizing* (Van Nostrand Reinhold, 1986).
A. T. Vagramyan and Z. A. Soloveva, *Technology of Electrodeposition* (Portcullis Press, 1981).

Diffusion coatings

N. S. Gorbunov, Diffuse Coatings on Iron and Steel (trans.) (Israel Prog., 1960).
H. Schaefer, *Chemical Transport Reactions* (Academic Press, 1964).
P. G. Shewmon, *Diffusion in Solids* (McGraw-Hill, 1963).

High temperature coatings

A. I. Andrews, *Porcelain Enamels*, 2nd edn (Garrard Press, 1961).
H. Bablick, *Galvanising*, 3rd edn (Spon, 1951).
W. E. Ballard, *Metal Spraying and Sprayed Metal* (Griffin, 1948).
W. F. Hoare *et al.*, *The Technology of Tinplate* (Arnold, 1965).
J. Huminik, *High Temperature Inorganic Coatings* (Reinhold, 1963).
C. F. Powell *et al.*, *Vapour Plating* (Wiley, 1955).

Conversion coatings

A. W. Brace, *The Technology of Anodizing Aluminium*, 2nd edn (Draper, 1979).
D. B. Freeman, *Phosphating and Metal Pre-treatment: A Guide to Modern Processes and Practices* (Industrial Press, 1986).
G. Lorin, *Phosphating of Metals* (Finishing/Public., 1974).
S. Wernick and R. Pinner, *Surface Treatment and Finishing of Aluminium Alloys*, 2 vols, 4th edn, (Draper, 1972).
W. Wiederholt, *The Chemical Surface Treatment of Metals* (Draper, 1965).

Non-metallic coatings

J. Boxall and J. A. von Fraunhofer, *Concise Paint Technology* (Elek Science, 1977).
V. Evans, *Plastics as Corrosion Resistant Materials* (Pergamon, 1966).
F. Fancutt *et al.*, *Protective Painting of Structural Steel* (Chapman and Hall, 1968).
A. A. B. Harvey, *Paint Finishing in Industry* (Draper, 1967).
W. M. Morgans, *Outlines of Paint Technology* (Griffin, 1969; 2nd edn (2 vols) 1982, 1984).
P. Nylen and E. Sunderland, *Modern Surface Coatings* (Wiley, 1965).
Paint Technology Manuals (Chapman and Hall).
D. H. Parker, *Principles of Surface Coating Technology* (Intersci./Wiley, 1965).

Current research

Abstracts

Four abstract services are devoted exclusively to corrosion and protection literature, each having particular fields – both geographical and topical – of speciality. It is not clear how interdependent they are.

Corrosion Abstracts, Houston, Texas (NACE).
Corrosion Abstracts, Stockholm.
Corrosion Control Abstracts (Referativny Zhurnal), transl. London.
Metal Finishing Abstracts, London.

A further information service surveys the above and issues topic profiles.

Corrosion Profile, (UKCIS), University of Nottingham, UK.

A substantial number of abstract services include corrosion topics, usually as a series of specialized sections headed corrosion properties, applied electrochemistry, metal finishing, coatings etc. Those issued by trade associations deal with single metals and may be published in several languages: these associations often offer a fee-paying copy service.

Aluminium Abstracts, Paris, France
Cadmium, Lead, Zinc Abstracts, Zinc Development Association, London.
Cobalt Abstracts, Brussels, Belgium.
Copper Abstracts, Birmingham, UK.
Rubber and Plastics Abstracts, Shrewsbury, UK.

Those issued by learned societies as part of a publishing programme are usually in one language only and are often but not exclusively derived from *Chemical Abstracts* and *Engineering Index* on a selective basis.

Chemical Abstracts, Easton, Pa., USA.
Engineering Index, New York, USA.
Chemisches Zentralblatt, Weinheim, FDR.
Materials Today, (ASM) Metals Park, Oh, USA.
Metal Abstracts, Materials Information (IM/ASM), London.
World Surface Coating Abstracts, London.
J. Appl. Chem., (SCI), London,
J. Paint Tech., Easton, Pa., USA.
Werkstoffe und Korrosion, Weinheim, FDR.

Journals

The following is a list of those journals devoted almost exclusively to corrosion and protection.

Anti-Corrosion Methods and Materials, London.
Australasian Corrosion Engineering, Sydney.
British Corrosion Journal (IM), London.
Corrosion (NACE), Houston, Texas, USA.
Corrosion Engineering, Tokyo, Japan.
Corrosion, Prevention and Control, London.
Corrosion Science (I. Corr. S.T.), Oxford, UK.
Corrosion, traitement, protection, finition, Paris, France.
Galvanotechnik, Saulgan, FDR.
Industrial Corrosion (I. Corr. S.T.), Birmingham, UK.
Korrosion, Rossendorf bei Dresden, DDR.
Korrosion, Weinheim, FDR.
Materials Performance (formerly *Materials Protection*) (NACE), Houston, Texas, USA.
Metaux, corrosion, industrie, St Germain en Laye, France.
Metalloberflache, Munich, FDR.
Oberflache/Surface, Zurich, Switz.
Oxidation of Metals, Plenum Press, New York, USA.
Plating and Surface Finishing (AES), East Orange, NJ, USA.

Protection of Metals (Zaschita Metallov), transl. New York, USA.
Surface Technology, Elsevier SA, Lausanne, Switz.
Transactions of the Institute of Metal Finishing (IMF), Birmingham, UK.
Werkstoffe und Korrosion, Weinheim, FDR.

Papers on corrosion and protection may appear in a substantial number of other journals, particularly those sponsored by the learned societies. Among them the following are most significant.

Archiv für Eisenhuttenwesen, Düsseldorf, FDR.
Electrochimica Acta, London.
Journal of the Electrochemical Society of America (ES), Manchester, NH, USA.
Journal of Metals (ASM), Metals Park, Oh, USA.
Journal of the Oil and Colour Chemists Assoc. (OCCA), London.
Metal Science Journal (IM), London.
Metals Technology (IM), London.
Soviet Electrochemistry (Electrochimija), transl. New York, USA.
Zeitschrift für Metallkunde, Stuttgart, FDR.

Reviews

An increasing number of review journals and book serials can be found dealing with state-of-the-art papers by established authors. Quality varies widely and authority of authors cannot be guaranteed; nevertheless, such papers provide important bases for bibliographical searches.

International Metallurgical Reviews (formerly Metallurgical Reviews) is a joint IM/ASM publication and includes papers from the whole metallurgical spectrum. Some journals feature occasional reviews. Specialized periodicals include:

Advances in Corrosion Science and Technology (Plenum Press, New York, USA).
Advances in Electrochemistry and Electrochemical Engineering (Wiley, New York, USA).
Corrosion Reviews (formerly *Reviews on Coatings and Corrosion*) (Freund Pub. Tel Aviv, Israel).
Reviews on High Temperature Materials (Freund Pub., Tel Aviv, Israel).
Oxides and Oxide Films (Marcel Dekker, New York, USA).

Congresses/conferences

Conferences which choose to publish their proceedings on a 'one-off' basis are generally indexed and catalogued as books and are included in book bibliographies mentioned earlier. Those sponsored by learned societies are published as separate volumes under their auspices notably ES, ASTM and NACE, or in their technical journals notably AES and IMF.

World congresses are organized on an international co-operative basis with proceedings published by either a commercial publishing house or by the most learned society. Such proceedings usually consist of a large number of short papers and give an excellent indication of current research interests worldwide. The two most important series are:

International Congress on Metallic Corrosion (three year intervals since 1961), most recently Mainz, 1981, Toronto, 1984, and Madrid, 1987.

World Congress on Metal Finishing (four year intervals since 1934), most recently Kyoto, 1980, Jerusalem, 1984, and Paris, 1988.

CHAPTER FOURTEEN

Design

R. F. FLINT

Many people think that design is only concerned with improving the appearance of objects. While this is an important part of design, it is only one of the aspects which have to be taken into account, because design is the complete concept of making an object fit a given purpose.

The interdependence of design and materials can now be understood. Without suitable materials available, the best designer in the world cannot design a satisfactory product. Could Leonardo da Vinci have brought more of his ideas to fruition if suitable materials had been available in his era? Even today when we are faced with a wide choice of materials, materials scientists are still pressed to produce even more exotic materials by the designers of advanced technological products.

The changes which have occurred during the last two hundred years are tremendous. The eighteenth century was a period of considerable political and scientific change with the American and French Revolutions, the Napoleonic Wars and change from alchemy to modern chemistry and physics which all resulted from the modern ideas propounded by the philosophers of the times. At the end of the eighteenth century, only a small range of materials existed, the main construction and engineering materials being stone and wood together with limited use of a few metals. For instance, the total annual world production of copper (see Figure 14.1) was then only some 10,000 tons with three-quarters of this being produced in the British Isles. Today output has reached some 8.3 million tonnes. The Industrial Revolution of the nineteenth century gave a boost to the production of materials and the second half of the century saw the invention of modern methods of steelmaking which were to revolutionize engineering. Aluminium is another important metal which was first made in the same era but this had to wait for the simultaneous invention by Hall and Heroult of the electrolytic process of aluminium manufacture for it to play an important role in engineering. The twentieth century saw the appearance of the wide range of plastics now available as well as the superalloys and other exotic alloys now used in engineering and electronics. Yet, as can be seen in Figure 14.1 stone and wood still top the production tables and, with steel, are the most important materials today from the point of view of quantity. In the eighteenth century the designer often lacked the materials suitable for his purpose. For instance, the Montgolfier was the balloon used by the first men to fly, because the Montgolfier Brothers were able to use paper which was available to make the envelope to contain the hot air, Charles had to develop new materials in order to contain hydrogen gas. The modern designer is faced by a wide range of materials to fill his purpose.

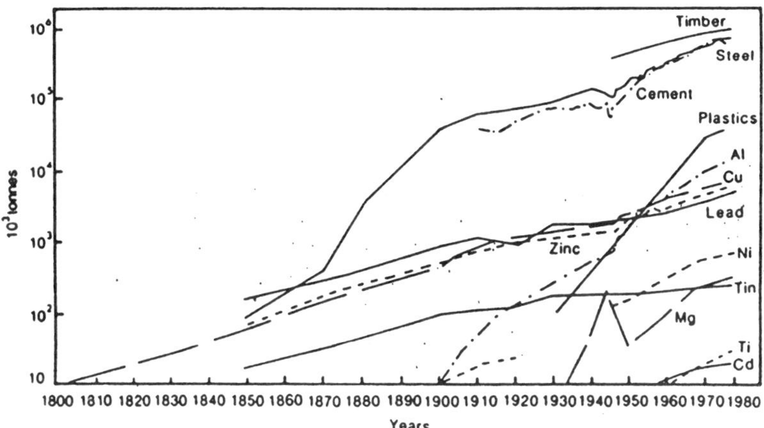

Figure 14.1. Production of materials by mass

Source: W. E. Duckworth, *Materials Handling 1983–84*, vol. 4, p. 924.

There are two facets to this problem: one is optimization of the choice available to the designer and the second is conformity of the products and materials, or standardization. When one moves away from individually made items and changes over to mass production, however primitive, one is often faced with the need for interchangeability of parts, and thus the need for standardization.

To achieve standardization a standard method of measurement is required. In its latest revision, the Système International or SI system has been chosen to replace both the Imperial and US systems where although the names were the same, the units sometimes had different values.

Standardization is necessary to answer the demands both of mass production and of international trade for materials and parts that are consistent with regard to size and quality. Many industrial companies and government departments issue their own standards, but in most cases, apart from some well known series, these are restricted to use within their own organizations or by subcontractors.

In the United Kingdom, European countries, Japan and the Soviet Union, standards for general use are issued by a single independent organization, the more important of these organizations being BSI in the UK, DIN in Germany, AFNOR in France, UNI in Italy, GOST in the Soviet Union and JIS in Japan. Many of these countries produce standards which are widely accepted elsewhere and their organisations act as agents for each others' standards. DIN standards are of particular importance and usefulness, one of their features being the use of werkstoffe (materials) numbers to distinguish individual alloys.

In the United States the position is quite different. The national body, the American National Standards Institute, only prepares a limited number of standards, but gives its own designation to some of the standards issued by its sponsoring members, often professional or scientific societies, who issue standards which are accepted in their own industrial fields. These standards are issued in answer to industrial and public needs for national consensus standards. For example, the 1987 catalogue of the American National Standards Institute (ANSI) lists nearly 2000 standards dealing with aerospace materials – both metallic and nonmetallic. European agents for these are American Technical Publishers. In view of the large number of organizations in the United States issuing industrial standards only a few of the more important ones dealing with materials will be mentioned. The Aluminium Association Inc. (AA): their numbering system for aluminium alloys has gained wide acceptance; The American Petroleum Institute; The American Society of Mechanical Engineers; The American Welding Society; The Society of Automotive Engineers Inc.

Perhaps the most important standards organization in the United States and dealing with materials is the American Society for Testing and Materials. ASTM, as it is known worldwide, was founded at the turn of the century and is an important scientific society with a worldwide reputation. Besides standards, ASTM issue books of high technical quality in a numbered series called *Special Technical Publications* covering all the subjects of interest to their members and of which over 800 have now appeared.

Their main journal is *Standardization News* which is issued monthly to all their members and is also available on subscription and gives news of standards and committees's actions. ASTM also publishes five journals which contain original articles:

Journal of Testing and Evaluation (bimonthly);
Cement, Concretes and Aggregates (monthly);
Geotechnical Testing Journal (quarterly);
Composites Technology Review (quarterly); and
Journal of Forensic Sciences (quarterly).

ASTM standards are numbered numerically as they appear in seven sections:

A. Ferrous Metals
B. Nonferrous Metals
C. Cementitious, Ceramic, Concrete and Masonry Materials
D. Miscellaneous Materials
E. Miscellaneous Subjects
F. Materials for Specific Applications
G. Corrosion, Deterioration and Degradation of Materials

Revisions are indicated by a change of date and all the standards are collected into volumes dealing with particular subjects.

Until 1961, all the Books of Standards were issued at three year intervals with annual supplements in 15 numbered parts plus an index. The number of parts was then increased to 33, with a comprehensive index and again to 48 parts in 1974. The Complete Book of Standards has since been published annually, so that it is comparatively easy to ensure that one always has an up-to-date set of the standards of interest to one. In 1981, ASTM changed the arrangement of the Book of Standards and the parts are not numbered strictly numerically but are divided into 16 sections but still published annually:

1. *Iron and Steel Products* – 6 parts
2. *Nonferrous Metal Products* – 5 parts
3. *Metals Test Methods and Analytical Procedures* – 6 parts
4. *Construction* – 9 parts
5. *Petroleum Products, Lubricants, and Fossil Fuels* – 5 parts
6. *Paints, Related Coatings and Aromatics* – 3 parts
7. *Textiles* – 2 parts
8. *Plastics* – 4 parts
9. *Rubber* – 2 parts
10. *Electrical Insulation and Electronics* – 5 parts

11. *Water and Environmental Technology* – 4 parts
12. *Nuclear, Solar and Geothermal Energy* – 2 parts
13. *Medical Devices* – 1 part
14. *General Methods and Instrumentation* – 2 parts
15. *General Products, Chemical Specialities, and End Use Products* – 9 parts

00. The Index

That is 66 volumes and more than 7000 standards.

Standards issued by the US Government are also used widely, especially the Federal and the Military specifications. In the case of metals and alloys a unified numbering system (UNS) has been established as a mean of correlating the many numbering systems used nationally which are currently administered by societies, trade associations and individual users and producers of metals and alloys and therefore avoiding the confusion caused by the use of more than one identification number for the same material or the opposite situation of having the same number assigned to two different materials. This publication, now in its fourth edition, is published by SAE and ASTM (1986) under the title *Metals and Alloys in the Unified Numbering System* (ASTM DS 56B) and covers over 6,500 metals and alloys. Although online systems are discussed elsewhere, it is useful to mention here the existence of a database, Standards and Specifications (1950 to the present), prepared by the National Standards Association, Inc., Bethesda, MD and which is available on DIALOG and ESA-IRS.

A new PC-based software package for searching metals and alloys data has beend eveloped jointly by SAE and ASTM. UNSearch allows one to search by chemical composition which the printed edition cannot provide. Search can also be made by UNS number, by society or federal specification number, by military specification, metal or alloy. This database contains over 102,000 records and provides access to all (US) government and industry standards, specifications and related documents which specify terminology, performance testing, safety materials, products or other requirements or characteristics of interest to a particular technology or industry.

Metal Bulletin's *Brands and Alloys*, compiled by E. Langford (1975) is still a useful source. A prolific publisher of standards information is the Society of Automotive Engineers (SAE). The 1988 *SAE Handbook* is published in four volumes and Volume 1 covers ferrous, nonferrous and nonmetallic materials. The volumes can be purchased separately. Other annual indexes from SAE include *Aerospace Materials Specification Index* and the *Ground Vehicle Standards Index*.

A good catalogue of engineering and materials design software has been produced by Comline Engineering Software, 1 Wallace Way, Hitchin, Herts, UK in association with American Technical Publishers.

ISO (the International Organization for Standardization) is the specialized agency for standardization at present comprising the national standards institutes of 84 countries. The object of ISO is to promote the development of standards in the world with a view to facilitating international exchange of goods and services and to developing mutual cooperation in the sphere of intellectual, scientific, technological and economic activity. The results of ISO technical work are published as International Standards.

International Standards are the result of an agreement between the member bodies of ISO and may be used as such or may be implemented through incorporation in national standards of different countries.

The scope of ISO is not limited to any particular branch and it covers both technical and non technical standards. ISO is however not active in the electrotechnical field which is the responsibility of the International Electrotechnical Commission (IEC).

Their catalogue contains besides a classified list of their standards a list of the addresses of the national standards bodies in the various countries which are members of ISO.

Design Council

The importance of good design to industry was recognized by the British Government with the formation of the Design Council in 1944. The Council is an official research centre affiliated to the Department of Trade and Industry. The object of the Council is to promote the improvement of design in products of British industry. It provides an advisory service to industry; evaluates consumer products and maintains records of products accepted for the Design Centre, the permanent exhibition centre located at the same address; promotes schemes to assist companies with selected products; develops improved methods of design education in secondary, further and higher levels; and administers Department of Industry research funds for curriculum development in design education and the management of design. Besides its annual reports, the Council publishes three periodicals *Designing* (3 a year), *Design* and *Engineering* (both monthly). *Design* gives news and descriptions of some of the best current British and foreign designs while *Engineering*, before becoming one of the official journals published by the Council, was one of the two independent (practical) journals published in the UK which had a worldwide reputation. The other one *The Engineer* is now one of the better controlled circulation journals and is published by Morgan-Grampian.

The Council publishes a wide range of books on design and related subjects and the Design Centre Bookshop provides a retail service of Council books as well as those of commercial publishers.

The UK Department of Trade and Industry has recently appointed the Design Council as national manager of its new Materials Information Centre, based in the Council's London office. This is a consultancy service on new materials complementing the Department's initiative PROMAT (Profit Through Material Technology) which is aimed at promoting new materials to senior industrialists and designers.

A useful series of booklets published by Oxford University Press on behalf of the Design Council are the *Engineering Design Guides* covering such subjects as *Brazing*, *Soldering*, *Welding*, *Seals*, *Helical Springs*, *Plastics Moulding*, *Selection of Materials*, etc. More than 30 of these guides have now been published.

The choice of the right materials is an important factor in successful design and a decision has to be made when a new product is on the drawing board or a change in material is required for economic or manufacturing considerations. Both for the new product or for material substitution, the factors which have to be taken in account are: ability of the materials to meet the properties required, availability and cost. Today, there is a wide range of materials available and except for very specialized applications a number of materials may prove to be suitable so the choice can often be difficult. An information system specially designed to help designers choose the right materials is the Fulmer Materials Optimizer whose aim is the selection, specification and costing of engineering materials. Published by the Fulmer Research Institute Ltd., it was first established in 1974. The system consists of four volumes contained in eight convenient loose-leaf binders of assimilated information on the performance and current costs of commercially available production engineering materials and related component manufacturing processes. A recommended method for selecting the optimum material is also provided with many examples and case studies.

The system is designed to enable a rapid comparison of materials, which after processing, can compete for the same application. The data is presented in tabular and graphical form, arranged in a progression of increasing detail. The user can review candidate materials and quickly proceed to relevant information without wasting time on unsuitable materials and manufacturing methods. The Optimizer is updated regularly and the whole set or separate volumes can be purchased.

Costs are an important factor in the choice of materials and it must be remembered when making a selection that costs are normally given by weight units and that the volume costs which are more relevant can be quite different. The section on costs includes a useful monogram for the conversion of costs per unit weight to costs per unit volume.

The Engineering Sciences Data Unit (ESDU) compiles, evaluates and issues authoritative data for application in engineering in the form of memoranda and data sheets. There are more than 750 data items available for application in a wide range of engineering design projects and a good detailed index is issued at regular intervals. To simplify the selection, ESDU have grouped data items into sub-series covering the following topics: acoustic failure; aerodynamics; computer software; dynamics of aircraft; fatigue; fluid mechanics; heat transfer; industrial aerodynamics; materials properties; mechanisms (cams, gears, linkages); noise; performance of aircraft; physical data; strength analysis; transonic aerodynamics; tribology; vibrations and wind loadings.

This choice of subjects reflects the fact that ESDU is sponsored by the Royal Aeronautical Society, the Institution of Mechanical Engineers, the Institution of Structural Engineers and the Institution of Chemical Engineers who have specified working design and physical data needed by engineers working in these fields. Each sub-series can be purchased separately, with its own updating service, on subscription.

There are a number of authoritative handbooks which should prove useful to the designer and among those of note are: *Smithells Metals Reference Book*, now in its 6th edition and edited by Eric Brandes (Butterworths, 1983). While some of the 36 sections are only of interest to the metallurgist, the majority will contain data that the designer working mainly with metals should find useful.

The American Metals Society, one of the more important American scientific societies, is the publisher of reference books as well as scientific periodicals in the materials field. It also publishes jointly with the Institute of Metals *Metals Abstracts* – an important secondary publication. ASM publishes one of the most important handbooks for metallurgists. *Metals Handbook*; the 8th edition in 11 volumes is still useful although the 1st volumes of the 9th edition appeared in 1978 and the 12th volume dealing with fractography appeared in 1987.

Other useful books published by ASM are the source books series which include *Tool and Die Failures*; *Industrial Alloy and*

Engineering Data; *Materials for Elevated Temperature Applications*; *Materials Selection* (2 vols); *Wear Control Technology*; *Innovative Welding Processes* and *Failure Analysis*, to name but a few. Another useful work is *ASM Metals Reference Book* (not to be confused with the larger work by Butterworths). Now in its 2nd edition, it consolidates data from a wide variety of ASM publications and is designed as a desk handbook. Its 555 pages cover the elements, compositions, properties, testing and inspection, heat treating and fabrication.

ASM was one of the first organizations to study the possibilities of computerized methods of information retrieval and *Metals Abstracts* is available online as Metadex as well as *World Aluminium Abstracts* and *Alloy Index*, the other secondary publications published by ASM. The information from *Metals Handbook* and a new *Metals Handbook* Desk edition (a single-volume concise version of *Metals Handbook*) is now available as a computer software system called MetSel 2 – a materials database management program designed by ASM and Physical Science Inc., for use on IBM PCs and compatibles. It should prove useful to sort through the data available in *Metals Handbook* and the designer can update the database to meet his own requirements. It comes complete with a starter database featuring 100 alloys in hundreds of conditions.

Selection of Materials for Component Design: Source Book, edited by H. E. Boyer was published by the ASM (1986) and is a collection of 45 key articles from the literature. It is unique in that it is organized by component rather than material and includes case histories dealing with criteria selection from size and shape to wear resistance and recycling feasibility. *Selection of Materials for Service Environments*, edited by H. E. Boyer (ASM, 1987) is a source book of key contributions to the literature of topics such as alloys for use in subzero or cryogenic temperatures and materials for magnetically soft and permanent magnet applications. *Fatigue in Mechanically Fastened Composite and Metallic Joints*, edited by J. M. Potter (ASTM STP 927, 1986) covers basic and applied research covering bridge structures, airframes, adhesive bonding, welding and metallic and composite materials.

Computer Simulation in Materials Science, edited by R. J. Arsenault *et al.* was published in 1987 as the Proceeding of ASM's Materials Science Seminar held the previous year and, using forging as a case study, covers topics such as atomistic models and grain-boundary modelling. *Application of Fracture Mechanics for Selection of Metallic Structural Materials*, by J. E. Campbell *et al.* (ASM, 1982) provides detailed information on fracture mechanics properties of alloy steels, stainless steels, aluminium alloys, titanium alloys and superalloys. An invaluable source of information is *Useful Books of Reference for Designers (1926–1983)* held by the Science Reference Library, The British Library (1984). This selection by M. M. Palyzo is in two parts with metallurgy receiving extensive coverage in Part 1 along with metrication and civil engineering. The detailed subject index is particularly useful. *Advanced Composites* is an ASM Conference volume published in cooperation with the Engineering Society of Detroit, 1986. Design, manufacturing and materials are reviewed in the automotive, aerospace and defence industries.

In the field of plastics, a useful publication is the *Modern Plastics Encyclopedia* which is published annually in October as a supplement to the American journal *Modern Plastics* (McGraw-Hill). The Encyclopedia is divided into four broad sections: Section 1 deals with materials and manufacturing processes including tooling and testing. The information is given in the form of short articles and the various plastics groups, the composites, fillers and other additives are covered. Section 2 – A design guide is a short section dealing with the selection of plastics materials and designing with plastics. Section 3 – Engineering Data Bank is in tabular form and covers the various properties of plastics including flammability, chemicals and additives charts and brief specification of plastics machinery. The last section is a directory of suppliers. The Encyclopedia can be purchased separately from the journal.

The American journal, *Materials Engineering*, contains useful articles on the application of materials and issues a yearly supplement, the *Materials Selector*. There are two main sections, the first is a buyers guide of some 200 pages and the other gives in tabular form the properties and applications of iron and steel and nonferrous metals; other materials; composite materials; parts and forms; and joining and sealing. Some useful tables are provided in which the materials are compared for given properties. In spite of its American bias, *Materials Engineering* and *Materials Selector* are most useful sources. The designer is faced not only by a wide choice of materials but also by a varied nomenclature with the same material often known under different names, because of the wide variety of standards and the use of trade names. This is especially so for metallic materials and guides to these names used either as trade names or standard nomenclature are useful to the designer.

Half way between handbook and guide to nomenclature is *Metallic Materials Specification Handbook* by Robert B. Ross now in its 3rd edition, which is described elsewhere. Properties, heat treatment, welding, testing and applications are described, followed by a list of metals and alloys covered by that section or subsections, giving for each name some composition and manufacturer names. There is an index of alloys and a list of manufacturers' addresses at the back of the volume.

A useful guide to trade names is *Woldsman's Engineering Alloys* now in its 6th edition and edited by Robert C. Gibbons (American Society for Metals, 1979). Its 1,832 pages list in alphabetical order more than 40,000 alloys of which 10,000 are new or have been modified since the last edition.

Also published by ASM are the *Worldwide Guide to Equivalent Iron and Steels*, 2nd edition published in 1987 and the companion volume *Worldwide Guide to Equivalent Nonferrous Metals and Alloys*, 2nd edition also published in 1987. Both cover specifications and designations for the United States, Germany, France, the United Kingdom, Canada, Japan, Italy, Sweden, Mexico etc. Similar alloys are grouped together and chemical compositions, mechanical properties and available product forms are given for each designation. Both volumes are indexed alphanumerically and by the UNS system.

In the case of aluminium there is an excellent publication, the Aluminium Zentrale *Key to Aluminium Alloys*, compiled by W. Hufnagel and available from Aluminium-Verlag GmbH. This book facilitates the comparison of the various standards for Austria, Australia, Canada, Switzerland, Czechoslovakia, Federal Republic of Germany, German Democratic Republic, Spain, France, Great Britain, Hungary, Italy, Japan, Norway, Poland, Sweden, USSR, Turkey, USA, Yugoslavia, the Republic of South Africa as well as the ISO standards. The book lists the full chemical composition and there is a section on temper designation as well as a list of trade names. Particularly helpful is the fact that there is an English language version.

Because of the large number of steel grades available the position can be particularly confusing and, besides the ASM Guides already mentioned, the following works can be useful: *Stahlschlüssel* (The Key to Steels) compiled by C. W. Wegst is now in its 13th edition and is published by Verlag Stahlschlüssel and is discussed in Chapter 3 on iron and steel.

The International Technical Information Institute, Toanomon-Tachikawa Bldg., 6–5, 1-Chome, Nishi-Shimbashi, Minato-ku, Tokyo, Japan, publishes a handbook of comparative World Steel Standards covering Japanese, French, British, USSR, German (DIN and Verein Deutscher Eisenhuttenleute) and American (ASTM and American Petroleum Institute) standards. Vol. 6, *Steel, World Standards Mutual Speedy Finder* is now in its 4th edition and provides a convenient way of comparing standards from these countries with each other, although it has a Japanese bias it can be used by people from other countries as the headings are in English and Japanese. The other volumes of *World Standards Mutual Speedy Finder* are: Vol. 1, *Chemicals*; Vol. 2, *Electrical and Electronics*; Vol. 3, *Machinery*; Vol. 4, *Materials* and Vol. 5, *Safety*.

A small publication issued by Market Promotion Department of the British Steel Corporation, Sheffield summarizes all the steel-related British standards including DTD and SAE, AISI, ASTM, API, ISO and Euronorms. It is pocket-sized making it convenient to carry but is at present out of print and there are doubts whether it will be reprinted.

The following journals are also available:

The Engineer (Morgan-Grampian).
Engineering Materials and Design (IPC Industrial Press Ltd).
Engineer Digest (Eyre and Spottiswoode Publications Ltd).
Eureka, a controlled circulation journal (Findlay Publications Ltd).
Materials and Design (Scientific & Technical Press Ltd).
Technology, a controlled circulation journal (Engineering Today Ltd).
Design Engineering (US, Morgan-Grampian Publishing Co.).

The number of scientific and technical journals is so large that for many years it has been impossible for the scientist or the technologist to read all the journals published in his own field and apart from reading a very limited number of core journals the only method of keeping up to date with the literature has been to rely on secondary publications or indexes. Abstracting journals are not new, for instance *Chemisches Zentralblatt* first appeared in 1830 under the title of *Pharmaceutisches Zentralblatt* and the Iron and Steel Institute, now incorporated in the Institute of Metals and which was founded in 1869 included references to other publications from early issues of its journal. As a result of research by many organizations after the Second World War, online information systems began to appear in the 1960s. Online systems run on large capacity computers called Hosts usually containing a number of databases or databanks covering a number of subjects. The databanks cover numerical data whilst the databases cover bibliographic data usually equating to abstracts published in printed form although the complete abstracts are not always given in full. Online systems are growing rapidly and it is quite possible that in the foreseeable future abstracts journals and indexes will only appear in the machine readable form. Therefore abstracts journals and online systems will be dealt with together.

Among the more important hosts the following can be included: DIALOG Information Services; Orbit Search Services; ESA-IRS

Esrin; Pergamon Financial Data Services; INKA Fachinformationzentrum Energie. Physik. Mathematik GmbH; Télésytèmes-Questel; and Data-Star, Radio Suisse Ltd. These Hosts have offices or agents in other countries. They all offer a large number of databases and besides these, the following will be found to be of use to the designer.

Compendex is the machine readable form of *Engineering Index* and covers the world's significant literature in the field of engineering and technology since 1970 although the printed version was well established long before that date. The publishers are Engineering Information Inc., and it is available on the first three hosts listed above.

Ismec is an international source in mechanical engineering which is available on DIALOG and IRS and began in 1973. It is supplied by Cambridge Scientific Abstracts. NTIS is the machine readable version of the semi-monthly *Government Reports Announcements* which covers business and economic data as well as scientific and technical reports literature released to the public. NTIS is an agency of the Department of Commerce and the files which start in 1962 are available on a large number of hosts. NTIS also sells directly or through their agents copies of the reports listed on their files. Their address will be found in the Appendix.

Pascal is the computerized version of *Bulletin Signalétique*, which covers literature of all types; and has appeared since 1973 although the bulletin itself first appeared shortly after the end of the Second World War. The supplier is Informascience, Centre National de la Recherche Scientifique, and it is available on ESA-IRS.

Cetim provides information useful to the mechanical industry since 1973. The printed version is *Bulletin de la construction mécanique* and is supplied by Centre Technique des Industries Mécaniques. It is available on ESA-IRS. There are a large number of more specific data bases and details can be obtained from the Hosts or the Online Information Centre at Aslib.

Suppliers' brochures and catalogues can be a useful source of information because many manufacturers give data on their materials which cannot always be easily found elsewhere. Many journals issue buyers' guides as supplements to their normal issues though independent directories do also exist.

Besides those mentioned earlier as adjuncts to encyclopaedias two useful specialized directories are available for metals: *Ryland's Directory* published biannually by Fuel and Metallurgical Journals Ltd; and the *European Plastics Buyer's Guide* (IPC Industrial Press Ltd). General directories can also be useful to trace suppliers and specially recommended are the various editions of Kompass which are available for most West European countries and Australia, Indonesia, Morocco and Singapore. Some of their data has been computerized and is now available online.

Advice on individual materials can be obtained from the development associations. These are organizations funded by industry to promote the use of the material and they issue publications as well as advice and sometimes carry out research into applications of materials. Some also run training courses for people who are using these materials in their work. Among the more important are:

Aluminium Federation Ltd
British Non Ferrous Metals Federation
The British Plastics Federation
The Cement and Concrete Association
Copper Development Association
International Tin Research Institute
The Lead Development Association
The Zinc Development Association
Timber Research and Development Association

Overseas offices of the above or equivalent associations can be found in some other industrialized countries. The United Kingdom Atomic Energy Authority National NDT Centre has launched an information and advisory service on NDT named QUALTIS which will provide members with free access to the centre's database and provide regular research reports on areas of interest.

This brief resumé of sources of information available to the designer can be supplemented by reference to other chapters of this book.

CHAPTER FIFTEEN

Welding, brazing and soldering

LINDA DUMPER and JOHN LOADER

Consider for a few moments the variety of components and structures which rely on welding and related joining methods for their strength and integrity. They range from bridges, offshore structures and earthmoving equipment to medical prostheses, semiconductor devices and jewellery. In addition, all stages of the production of a workpiece must be carefully monitored, from the choice of suitable parent materials and consumables to the testing of the completed assembly to assure the quality of the joints. The wide variety of subjects which impinge on welding, brazing and soldering can necessitate the use of information throughout the materials science and engineering disciplines and this broadens the scope of relevant fields of interests which must be considered.

Not only does this mean that the range of literature necessary to encompass the required information is large and varied, but also that many societies, research associations, government bodies etc. must be used as further sources to allow an adequate coverage of a specific area within the overall subject.

Information sources – general

Many organizations exist which can provide assistance to those engaged in welding and its allied processes. These can offer general services covering a wide range of subjects, or more specifically concerned with metallurgy, engineering, the machine tool industry, the offshore industry, health and safety, etc. The Institutions of Civil, Mechanical, Structural, Electrical and Marine Engineers and the Institute of Petroleum, all provide library and information services to members. The Institute of Quality Assurance has now set up the National Quality Information Centre (NQIC) which will provide a referral service to do literature searching and to direct enquirers to companies with experience in specific aspects of quality assurance. The British Institute of Non-destructive Testing, although it fills the role of learned society for its members, does not have information facilities.

In the United States, the American Petroleum Institute, the Institute of Electrical and Electronic Engineers (IEEE), the American Society for Metals, the American Society of Mechanical Engineers and the American Institute of Metallurgical Engineers, fill similar roles to the British learned societies mentioned above.

The Bundesanstalt für Materialprufung, Berlin, is West Germany's major metallurgical institution, producing an abstracting journal and offering information services.

In Japan, the National Research Institute for Metals, and the Iron and Steel Institute of Japan also offer library services, and publish their transactions in English.

Research and trade associations

The relevant associations cover a wider subject range than the learned institutions. The major ones offer, as well as library and information services, research facilities into their area of interest, now mainly sponsored by industry. The associations involved in various aspects of metallurgy are the Steel Castings Research and Trade Association (SCRATA), the Steel Construction Institute, the British Cast Iron Research Association (BCIRA), the British Independent Steel Producers' Association (BISPA) funded by the independent steel companies; the BNF Metals Technology Centre (formerly British Non-Ferrous Metals Research Association), the Copper Development Association (CDA), the Aluminium Federation and the Zinc, Lead and Cadmium Development Associations.

Numerous engineering associations exist which specialize mainly in various types of instrumentation. These range from the Electrical Research Association (ERA) to the Advanced Manufacturing Technology Research Institute (AMTRI), formerly MTIRA, the Production Engineering Research Association (PERA), which has a particular interest in mechanization and the use of robots in industry, SIRA (formerly the Scientific Instrument Research Association), and the British Robot Association (BRA). These associations exist to undertake research work and provide information for their members. PERA also administers the Department of Trade and Industry's Manufacturing Advisory

Service, whereby companies request assistance in planning relevant mechanized systems for their particular applications, and may either take a contract themselves or refer the company to other relevant research associations to do a feasibility study.

Other existing associations whose area of interest overlaps with that of welding include the Motor Industry Research Association (MIRA), British Maritime Technology Ltd (formerly BSRA), and the Stainless Steel Fabricators' Association (SSFA).

In the United States the National Association of Corrosion Engineers (NACE), the Society of Automotive Engineers (SAE), the Resistance Welder Manufacturers' Association (RWMA), the American Gas Association (AGA), the International Copper Research Association (INCRA), and the International Lead-Zinc Research Organisation (ILZRO), are bodies undertaking research into welding and related subjects.

Government sponsored information sources

A wide variety of sources can be grouped under this heading. The most obvious being the British Library Document Supply Division which, with the Science Reference Information Service (formerly the Patent Office Library) form the major back-up to the majority of small and medium-sized libraries. (Patents are still available to consult in the SRIS, but can only be purchased from the Patent Office.)

The British Standards Institution (BSI) is an example of a government-backed centre, though an attempt is being made to achieve profitability. The Milton Keynes complex now houses the enquiry service, the British and foreign standards libraries and Technical Help to Exporters services.

Health and Safety forms a major area of concern in the workplace, and the Health and Safety Executive offers a number of libraries and information points. The Red Hill Library, Sheffield, specializes in information on fume, and also compiles occupational fume exposure limits (OELs).

One major government funded body is the National Physical Laboratory. From this centre are run both the National Corrosion Service and the British Calibration Service. The latter centre is essential to British industry as a centre for measurement standardization.

For advice on non-destructive testing it is possible to become a member of the National Non-destructive Testing Centre at the Atomic Energy Research Establishment, Harwell, part of the United Kingdom Atomic Energy Authority.

In the field of transport, the Transport and Road Research

Laboratory, funded by the Department of Transport, undertakes research into structural safety, while the Department of Energy Petroleum Engineering Division, Underwater Engineering Group (UEG), and the Marine Technology Support Group (MATSU) are important sources of information for the offshore industry.

Similar bodies overseas include a number in the United States of America which are recognized federally. Relevant examples are the American National Standards Institute, the National Bureau of Standards, which undertake similar work to that of BSI and the British Calibration Service. The American Conference of Government Industrial Hygienists (ACGIH) publishes the threshold limit values (TLVs), similar to OELs, for exposure to fume. In addition the American Association of State Highway and Transportation Officials (AASHTO) co-ordinates research into road and bridge safety, though many local regulations exist for individual states. The Battelle Memorial Institute, Metals and Ceramics Information Center, is an invaluable source of detailed information on many materials. It also undertakes research into surfacing and non-destructive testing as well as welding processes.

Educational centres

Various institutions offer research or information facilities in areas of interest to the welding community. For example, in the United Kingdom, the University of Aston in Birmingham has set up the Wolfson Heat Treatment Centre as a specialist unit, whilst Herriot Watt University in Edinburgh has the Institute of Offshore Engineering, which undertakes research programmes as well as providing a large collection of information on all aspects of the offshore oil and gas industry.

The World Metals Index at Sheffield City Libraries maintains a wide collection of materials relating to UK standards and their foreign equivalents as well as information on trade names. The staff there have built up substantial links over the years with Sheffield University and its Department of Metallurgy and directs enquiries and requests for consultancy and training to both the University and companies in the region.

In the United States, the Ohio State University at Columbus has a strong welding research department, though this has now been subsumed along with the Battelle Memorial Institute, by the Edison Welding Institute. The Massachusetts Institute of Technology undertakes a wide range of research into materials as do the Battelle Memorial Institute, whose Metals and Ceramics Information Center publishes both detailed information on materials and comprehensive listings of alloys, and the Carnegie-Mellon University, Pittsburgh Centre for joining of materials.

In West Germany the Universities of Stuttgart, Aachen and Essen, undertake a wide range of research into welding, while in France the Ecole Nationale Superieure des Arts undertakes postgraduate research into welding.

General sources

Some private centres exist which offer information to industry. Examples are the Steel Advisory Centre run by the British Steel Corporation, the Royal Society for the Prevention of Accidents (RoSPA) and the International Translations Centre in Holland, which is a useful source for checking the possible existence of a translation, thus saving companies both time and money.

In addition, the Engineering Sciences Data Unit, funded by a number of associations, has been set up to provide evaluated working designs and physical data published in the form of Data Sheets.

Information sources – welding

Institutions and societies

A worldwide distribution exists of the member societies of the International Institute of Welding (IIW). This loose association of institutions has been developed for the exchange of information on new research in welding and allied processes, and for the publication of some guidelines recommending good practice for various subject areas, such as pipeline and pressure vessel welding, and also radiography. The IIW with headquarters in London and Paris, publishes a journal *Welding in the World*, and organizes 16 Commissions, with members drawn from the affiliated societies, which are involved in all aspects of welding research. These Commissions and the Annual Assembly of the IIW meet each year in different countries. The member societies offer a variety of services in their home country, of which the following are the most significant.

United Kingdom The UK nominal member for organization of activities for the IIW is the Welding Institute (WI), which has been affiliated since the creation of the IIW. The WI publishes the journal *Joining and Materials*, and translates selected papers from a number of welding journals as *Welding International*. It also holds conferences and publishes the proceedings, produces books and audio-visual aids on a variety of subjects connected with welding, and runs training and certification courses in welding and non-destructive testing. These facilities are available to all, but the research capability of the Institute is available only to members. The Library and Information services are also available to members, with the computerized information retrieval system Weldasearch available online from an international database host.

USA The American Welding Society (AWS) and the Welding Research Council (WRC) are both members of the IIW. AWS holds conferences and seminars, with the papers now often being published, produces the *Welding Journal* and publishes numerous monographs and, above all, standards. WRC, which co-ordinates extensive research into welding in the private and educational sectors, publishes conference proceedings, a bulletin, and reports of the progress of current projects. The roles of two new institutions in the United States have not yet been fully defined. The American Welding Institute recently set up in Tennessee, is intended to be the national centre of welding research. A similar role is envisaged for the Edison Welding Institute, established in Columbus, Ohio, which collaborates closely with the Welding Institute (UK).

France The Institut de Soudure (IS) offers research facilities, training and examination schemes, publications, journals and audio-visual aids, and some library and information facilities to members.

West Germany Deutscher Verlag für Schweisstechnik (DVS) provides library services for its members, and also publishes the journal *Schweissen und Schneiden*, and contributes abstracts to the Bundesanstalt für Materialprufung (BAM) database, from which an abstracting journal is produced.

Japan The Japan Welding Society (JWS) and Japan Welding Research Institute (JWRI) both hold conferences and publish proceedings and journals. Some papers from their journals are translated into English and appear as the Transactions of each Institute.

Australia The Australian Welding Institute (AWI) and the Australian Welding Research Association (AWRA) have now combined to form the Welding Technology Institute of Australia. Both have library services, and produce journals, conference proceedings and other publications.

Canada The Welding Institute of Canada (WIC) is beginning to play a more active role in holding conferences and producing proceedings. The Institute publishes a journal, and holds training and certification courses, while the Canadian Welding Bureau offers qualification for welders and undertakes research into many aspects of welding.

Denmark Svejsecentralen runs an information service and has published many reports, and is especially noted for its work on the effects of fume on the health of welders.

USSR The Paton Institute of Electric welding of the Ukrainian Academy of Sciences in Kiev is the largest welding institute in the Communist bloc, and is known worldwide for the volume of research work undertaken. The majority of articles from the Russian journals translated by the Welding Institute are written by members of the Paton Institute's staff.

Trade associations

Other than the member societies of the IIW, specialist societies and associations exist, often linked with manufacturers and factors. In the United Kingdom, for example, the Welding Manufacturers' Association (WMA) comprises British companies involved in manufacture of welding equipment and consumables, and also the Association of Welding Distributors (AWD) made up of factors and distributors, have both evolved through the need for groups of similarly occupied companies to pool their knowledge and experience to their mutual benefit. Similarly in the United States the Resistance Welder Manufacturers' Association among other activities, publishes standards. The National Welding Supply Association has a similar role to that of the equivalent UK association.

Information sources – brazing and soldering

While the Welding Institute, the American Welding Society and Deutscher Verlag dur Schweisstechnik carry out some research and hold occasional conferences on brazing and soldering, other institutions specialize in these subjects. The BNF (British Non Ferrous) Metals Technology Centre, formerly included the British Association for Brazing and Soldering (now at the Welding Institute), carries out research into non-ferrous metals, their application and joining techniques, publishes a journal and conference proceedings. Unfortunately the online information service previously provided, BNF Abstracts, has now been removed from its online host, and the files of abstracts are not being kept up to date owing to lack of resources.

The Copper Development Association publishes a number of technical notes on various aspects of the use of copper, brasses and aluminium bronzes, including joining methods, and also publishes *Copper Abstracts*.

International institutions concerned with brazing and soldering include the International Tin Research Institute and the Centre d'Information des Métaux Non Ferreux, which both hold conferences and conduct meetings concerned with joining tin and non-ferrous materials.

Manufacturers often provide detailed information of the brazing and soldering alloys they produce, and the methods of using them. For example, Johnson and Matthey, International Nickel (INCO), and Handy and Harman all provide technical publications.

Types of literature

Bibliographic services

Although few bibliographic services are devoted specifically to welding, a wide variety of abstracting journals and databases hold information related to the welding industry. Few abstracting services do not offer an online service now, but as previously noted, BNF Abstracts is no longer online, as it is not being updated. This is a loss particularly in the fields of brazing and soldering.

Databases

Databases can be accessed via the networks and hosts listed in Chapter 1, using a terminal connected to a modem installed to give telephone links to the networks via the Packet Switching systems. Those of particular relevance to welding are:

Chemical Abstracts (CA Search)
Engineering Index (COMPENDEX)
European Patent Office (INPADOC)
Health and Safety Abstracts (HSE LINE)
International Information Services to the Physics and Engineering Communities (INSPEC)
International Nuclear Information System (INIS)
International Occupational Safety and Health Abstracts (CISDOC)
International Service in Mechanical Engineering (ISMEC)
Metals Datafile (MDF)
Metals Abstracts (METADEX)
Rubber and Plastics Research Association Abstracts (RAPRA)

US Government Reports (NTIS)
US Patents (CLAIMS)
Weldasearch
World Aluminium Abstracts
World Patent Index
World Surface Coatings Abstracts (WSCA)

Current availability of databases should be checked with the producers. The majority of the databases mentioned also have a hard-copy abstracting journal produced. In addition, the following abstracting journals can be of use in welding:

BCIRA Journal (British Cast Iron Research Association)
BMT Abstracts (British Maritime Technology)
BAM Dokumentation Schweisstechnik (Schweissen und verwandte verfahren)
Cadmium Abstracts
China Science and Technology Abstracts, Series III Industrial Technology
Copper Abstracts
Leadscan
MIRA (Motor Industry Research Association) *Automobile Abstracts*
NDT Information (in *NDT International Journal*)
Production Engineering Abstracts
Robotics Technology Abstracts
Referativnyi Zhurnal-Svarka
Scientific and Technical Aerospace Reports (STAR)
Steel Castings Abstracts
Welding Abstracts
Zincscan

In addition, a number of abstracting services which are no longer published can be used to trace papers:

Bibliographic Bulletin of Welding (BBW) ceased publication 1971
BNF Abstracts ceased publication 1983
Journal of the Iron and Steel Institute (JISI) ceased publication 1974
Nickel Bulletin ceased publication 1968

Journals

There are approximately 50 welding journals at present in publication. The range of subject matter included in the journals varies widely according to their origin and purpose. For example, journals from welding societies and institutes such as *Welding* Journal from the American Welding Society, *Joining and Materials* from the Welding Institute and *Journal* (in English – *Transactions*) *of the Japan Welding Society*, contain a number of research papers designed for the welding engineer and metallurgist. There are, however, a number of commercial and trade journals such as *Welding and Metal Fabrication* and *Welding Design and Fabrication* which have a larger proportion of space devoted to more practical papers and short descriptions of process or equipment developments, tailored more to the production welder and supply industry. A third category of journals includes those produced by manufacturing companies such as *Svetsen*, published by ESAB, and *Oerlikon Schweissmitteilungen*, which offer papers largely with relevance to their own products and developments within their company. The major welding journals worldwide are listed in Table 15.2.

Table 15.2. Welding journals

	Country	*Frequency (p.a.)*
Australian Welding Journal	Austraia	6
Automatic Welding (now *Welding International*) (Translation of *Avtomatischeskaya Svarka*)	UK	12
Avtomatischeskaya Svarka)	USSR	12
BEFA Mitteilungen	FRG	12
Canadian Welder and Fabricator	Canada	12
Founding, Welding Production Engineering (FWP) Journal	South Africa	12
Hanjie Xuebao (Transactions of the China Welding Institution)	People's Republic of China	4
Hitsaustekniikka Svetsteknik	Finland	6
Indian Welding Journal	India	4
Journal of the Japan Welding Society	Japan	12
Quarterly Journal of the Japan Welding Society	Japan	4
Journal of Light Metal Welding and Construction	Japan	6
Lastechniek	Netherlands	12
Metal Construction (now *Joining and Materials*)	UK	12
Norgas Norweld	Norway	c. 2
Oerlikon Schweissmitteilungen	Switzerland	3
Philips Welding Reporter	Netherlands	c. 3
Prace Instytutu Spawalnictwa	Poland	4
Praktiker (Der)	FRG	12
Przeglad Spawalnictwa	Poland	12
Revista de Soldadura	Spain	4
Revue de la Soudure/Lastijdschrift	Belgium	4

Table 15.2. Welding journals

	Country	*Frequency (p.a.)*
Rivista Italiana della Saldatura	Italy	6
Schweissen und Schneiden	FRG	12
Schweisstechnik (Berlin)	GDR	12
Schweisstechnik (Wien)	Austria	12
Soudage et Techniques Connexes	France	6
Souder	France	6
Svarochnoe Proizvodstvo	USSR	12
Sveiseteknikk	Norway	6
Svejsning	Denmark	6
Svetsaren	Sweden	c. 2
Svetsen	Sweden	6
Technica (incorporating: *Zeitschrift fur Schweisstechnik – Journal de la Soudure)*	Switzerland	26
Transactions of JWRI (Japan Welding Research Institute)	Japan	2
Transactions of the Japan Welding Society	Japan	2
Varilna Tehnika	Yugoslavia	4
Welding Design and Fabrication	USA	12
Welding International	various	12
Welding Journal	USA	12
Welding and Metal Fabrication	UK	10
Welding News	Australia	4
Welding Production (now *Welding International*) (Translation of *Svarochnoe Proizvodstvo*)	UK	12
Welding in the World	UK	6
Welding Research Abroad	USA	10
Welding Review	UK	4
Welding Technique	Japan	12
WRC Bulletin	USA	c.10
Zavarivac	Yugoslavia	4
Zavarivanje	Yugoslavia	c. 4
ZIS Mitteilungen	GDR	12
Zvaracske Spravy/Welding News	Czechoslovakia	4
Zvaranie	Czechoslovakia	12

Standards

Standards on all aspects of the welding industry are produced by a variety of bodies and cover welding equipment, consumables, health and safety and safety equipment, welder qualification, procedure qualification, inspection and testing, inspector qualification, quality assurance and design and also many materials.

The most obvious sources of standards in each country are the respective national standards institutions, for example, British Standards Institution (BSI), American National Standards Institute (ANSI), Deutsches Institut für Normung (DIN), Canadian Standards Association (CSA), Japanese Standards Association (JIS), Association Française de Normalisation (AFNOR) etc. These institutions issue standards and act as a clearing house for obtaining national standards of other countries.

In the UK, BSI sets up committees of experts on the subject in question comprising BSI staff, users and suppliers from industry, universities, learned societies and research associations. In some countries, organizations produce their own standards, which are then adopted by the national standards institution, as for example American Welding Society standards which are adopted by ANSI.

In other cases, standards are produced which are not adopted by the national standards institution but are nevertheless still used in industry. In the UK, the Compressed Gases Association, the Institute of Petroleum and the Heating and Ventilating Contractors' Association produce standards as do many companies for their staff and contractors, for example Shell, British Gas, British Coal, CEGB and UKAEA. Some Government departments such as the Ministry of Defence, the Health and Safety Executive, and the Department of Energy issue specifications, but these may be restricted to bona fide customers and contractors who can demonstrate a 'need to know' before the standards are released.

In the United States a seemingly endless number of bodies produce standards. Of specific interest to the welding industry are those by the American Welding Society, American Society of Mechanical Engineers, American Society for Testing and Materials, Society of Automotive Engineers, American Petroleum Institute, Resistance Welder Manufacturers' Association, National Electrical Manufacturers' Association, Underwriters Laboratories, American Conference of Government Industrial Hygienists and American Association of State Transportation Officials. In addition to this selection of societies, many states have their own standards, far too numerous to mention here, and again the US Defense department publishes its own standards with similar restrictions in obtaining them. Most other countries have a proliferation of standards producing societies such as det Norske Veritas for Norwegian offshore regulations, although on a smaller scale than the United States.

There are also standards produced by more than one country. The main ones are European and International Standards. The European standards are of three types, the European Standards

(EN), the European iron and steel standards (EURONORM) and European electrical standards (CENELEC). The International standards are International Standards Organization (ISO) standards and International Electrotechnical Commission (IEC) standards. ISO frequently adopts standards issued by national bodies with the intention of gaining international acceptance.

Reports

The rapid release of information to other interested researchers is of great importance in all branches of science and technology. The results of research are often published as journal articles or books, or presented as conference papers, but the initial release of the information can be more quickly completed by the internal production of a report. Such reports are normally in booklet form, but some companies, government departments and research associations release their reports as microfiches. Once the original microfiche has been produced either from a paper original or as direct computer output, the reproduction of a microfiche is much less expensive than the production of paper copies.

Most research centres impose various classifications of restriction on their reports, depending on the degree of confidentiality of the information. It may not always be easy to obtain copies of reports, even if they are referred to in the published literature.

There are four main categories of report producer. Examples of the major ones for each type of report relevant to the welding industry will be given, although it will not be an exhaustive list.

The first type of publisher is the government department with the major UK ones being the Departments of Energy, Environment, Industry, Transport, and Defence with their sub-divisions. Similar foreign government departments are, for example, the Australian Defence Scientific Service, the US Department of Defense, again with sub-divisions for the different services, and the US Department of Commerce, Maritime Administration.

Most countries have government sponsored laboratories for such work as power generation, for example the UKAEA, CEGB and British Gas in the UK, the Danish Atomic Energy Commission, and the US Nuclear Regulatory Commission (NUREG). Others cover subjects such as aerospace, for example, the National Aerospace Laboratory in the Netherlands, NASA in the US; transport, for example the Transport and Road Research Laboratory (TRRL) in the UK, the National Transportation Safety Board, and the Ship Structure Committee in the US. Health and Safety research is often government sponsored, with the Health and Safety Executive (HSE) in the UK and the

American Conference of Government Industrial Hygienists (ACGIH) and the National Institute of Occupational Safety and Health (NIOSH) as examples.

The second type of report producer is the research and trade organization and the learned society, whose main aim is to further research and publish the results of that research. Examples for the UK are the Compressed Gases Association (CGA), Construction Industry Research and Information Association (CIRIA), Copper Development Association (CDA), British Cast Iron Research Association (BCIRA), Steel Castings Research and Trade Association (SCRATA), Electrical Research Association (ERA), BNF Metals Technology Centre, Zinc, Lead and Cadmium Development Associations, Aluminium Federation, and finally the Welding Institute. In the US the main associations are the American Gas Association (AGA), Aluminium Association, Resistance Welder Manufacturers' Association (RWMA) and International Copper Research Association (INCRA). The Welding Technology Institute of Australia, the West German Aluminium Zentrale, the French Centre de Recherches Métallurgiques, and the Italian Centro Sperimentale Metallurgico are also research bodies which produce reports.

Institutions or learned societies include in the UK the Institutions of Chemical, Electrical, Mechanical, Production, Aeronautical and Structural Engineers. Many such societies exist in the US, for example, the American Welding Society (AWS), Society of Automotive Engineers (SAE), Society of Manufacturing Engineers (SME), Welding Research Council (WRC), the Institution of Electrical and Electronics Engineers (IEEE), the American Institute of Steel Construction (AISC) and the American Institute of Non-destructive Testing (ASNT), while further north is the Welding Institute of Canada (WIC).

In Europe, most countries have national welding institutes many of which produce reports. Examples of these are Deutscher Verlag für Schweisstechnik (DVS) in West Germany, Institut de Soudure (IS) France, Svejsecentralen in Denmark, Svetskommissionen in Sweden, and the Instituto Italiano della Saldatura in Italy. There are also numerous metallurgical and construction institutions in European countries which produce reports, such as the Swedish Institute of Steel Construction, the Royal Institute of Technology Sweden, Technical Research Centre of Finland, the Institut de Recherches de la Siderurgie Française (IRSID), Centre Recherches Scientifiques et Techniques de l'Industrie des Fabrications Métalliques (CRIF) Belgium, and National Research Institute for Metals (NRIM) Japan.

A major source of research work, which can be written up either as doctoral theses or postgraduate research reports, is the university. There are several which specialize in research in welding. In the UK these are Cambridge; Oxford; Edinburgh; Birmingham; Aston in Birmingham; Loughborough; Bradford; Imperial College, London; Liverpool; Queen's University, Belfast; University College, Dublin; and Cranfield Institute of Technology. Overseas universities with similar interests include Queensland and Sydney, Australia; Alberta, Canada; Osaka and Tokyo, Japan; Harbin and Peking, China; Liege, Belgium; Delft, The Netherlands; Aachen, Stuttgart and Bochum, West Germany; and in the United States those of Texas, Virginia, Washington, Illinois, Iowa, Monash, Ohio State, Lehigh, Massachusetts Institute of Technology, and the Battelle Memorial Institute.

Finally, a number of private companies with research facilities produce reports. The best known of these in the welding field include the British Steel Corporation, British Oxygen Company, INCO, Vickers, Yorkshire Imperial Metals, Bethlehem Steel, Climax Molybdenum, General Electric Company, Lincoln Electric, Rockwell, US Steel, Westinghouse, ESAB, Estel Hoesch, Japan Steel Works, Kobe Steel, Nippon Kokan, Nippon Steel, and Yawata Steel.

Subject literature

Welding processes

The intention of this section is not to give an exhaustive list of publications on the various processes, but rather to indicate some of the more useful texts for information on process applications. Common abbreviations for welding processes and charts of welding, brazing and cutting processes, reproduced from BS 499 are shown at the end of this section.

General

The general textbooks covering many processes are The American Society for Metals, *Metals Handbook*, 9th edn, Vol. 6, *Welding, Brazing and Soldering* (Metals Park, Ohio, ASM, 1983), also the 5 volumes of the American Welding Society *Welding Handbook*, 7th edn (Miami, AWS, 1976–84).

Introductory texts on all processes are P. T. Houldcroft, *Welding Processes* (Engineering Design Guides 06, London, Oxford University Press, 1975); A. C. Davies, *The Science and Practice of Welding*, 8th edn, 2 vols (Cambridge, Cambridge University Press, 1984); and W. Kenyon, *Welding and Fabrication*

Technology (London, Pitman, 1982). There are also several texts of a slightly more advanced level, for example, P. T. Houldcroft, *Welding Process Technology* (Cambridge, Cambridge University Press, 1977), and two compilations by T. B. Jefferson, *The Welding Encyclopaedia*, 17th edn (Morton Grove, Ill., Monticello Books, 1974) and also *Welding Engineer Data Sheets*, 9th edn (Morton Grove, Ill., Monticello Books, 1985). For newer welding processes consult H. B. Cary, *Modern Welding Technology* (Englewood Cliffs, Prentice-Hall, 1979) and M. M. Schwartz, *Metals Joining Manual* (New York, McGraw-Hill, 1979).

A readily available text containing a good basic introduction to welding processes is *Kempe's Engineer's Yearbook* (annual), J. P. Quayle (ed.) (London, Morgan Grampian, 1987), in which Ch. D5, by E. N. Gregory, is entitled 'Welding and cutting'.

In addition to these general publications the main sources of information for the different welding processes are as follows:

Arc welding

A number of processes are included under this general heading. Some publications cover all arc welding processes, for example, the Lincoln Electric Company, *The Procedure Handbook of Arc Welding*, 12th edn (Cleveland, Oh., Lincoln Electric Company, 1973); J. W. Giachino, *Arc Welding* (Chicago, American Technical Society, 1977); BOC Murex, *Manual Arc Welding, Application Guide* (Waltham Cross, BOC Murex, 1979 (no longer available); and I. H. Griffin, *Basic Arc Welding*, 3rd edn (New York, Delmar, 1977). At a more advanced level, the American Welding Society, *Welding Power Handbook*, by A. F. Manz (New York, Union Carbide, 1973) concentrates on power sources for arc welding, while specific information on timing and electrode consumption, to achieve the most economic welding procedure possible is given in the Welding Institute, *Standard Data for Arc Welding* (Abington, the Welding Institute, 1975, 2 vols).

A more detailed selection of publications for each type of arc welding includes: 'Manual metal arc welding – Murex', *Manual Arc Welding: Application Guide* (Waltham Cross, Murex, 1979); Hobart Brothers, *Technical Guide for Gas Metal Arc Welding* EW-473 (Troy, Oh., Hobart, 1980), and American Welding Society, *Recommended Practices for Gas Metal Arc Welding* (AWS C5.6 – 79) (Miami, Fl., AWS, 1979). Submerged arc welding – the two main texts on this subject are Union Carbide, *Submerged-arc Welding Handbook* (New York, Union Carbide, 1974), and the Welding Institute, *Submerged-arc Welding* (Abington, the Welding Institute, 1978). Metal inert gas welding (MIG or

GMA) – the texts available include Union Carbide, *MIG Welding Handbook* (New York, Union Carbide, 1974). The Welding Institute, *Exploiting MIG Welding Developments* (Abington, the Welding Institute, 1983), and also I. H. Griffin and E. M. Roden, *Basic TIG and MIG Welding* (Albany, NY, Delmar Publishers, 1971), and *I. H. Griffin et al.*, *Basic Welding Techniques: Three Books in One – Arc, Oxyacetylene and TIG and MIG* (New York, Van Nostrand Reinhold, 1977).

Tungsten Inert Gas Welding (TIG or GTA) – the final two publications given for MIG welding also cover the subject of TIG welding. To these should be added two more: the Welding Institute *TIG and Plasma Welding* (Abington, the Welding Institute, 1978), and American Welding Society, *Recommended Practices for Gas Tungsten Arc Welding* (AWS C5.5 – 80) (Miami, AWS, 1980).

CO_2 welding

The basic textbook for this subject is still A. A. Smith, *CO_2 Welding of Steel*, 3rd edn (Abington, the Welding Institute, 1970). A more recent publication on the process is ESAB, *The CO_2 Welding Process* by N. E. Andersen, 2nd edn (Sweden, ESAB, 1977). The most recent is Distillers Company, *CO_2 Welding Questions and Answers* (Reigate, Distillers Company, 1981). Finally, a more specific text is by J. W. Jones, *CO_2 Welding for Pipeline Construction* (London, Temple Press Books, 1963).

Gas (oxyacetylene) welding

This process relies on a mixture of oxygen and acetylene in a torch or blowpipe, burnt to form a high temperature flame which can be used either for welding or cutting. There are many practical manuals for gas welding, with the following being some of the most important: R. J. Baird, *Oxyacetylene Welding: Basic Fundamentals*, 2nd edn (South Holland, Ill., USA, Goodheart-Willcox, 1980); P. H. M. Bourbousson (rev. by K. Leake), *Questions and Answers: Gas Welding and Cutting*, 2nd edn (Sevenoaks, Butterworths-Newnes, 1982); British Oxygen Company, *Handbook of Operating Instructions for Gas Welding and Cutting* (Skelmersdale, BOC, 1977); J. W. Giachino, *Gas Welding* (Chicago, American Technical Society, 1977); Messer Griesheim, *The Art of Gas Welding* by W. Schwenzer (London, Messer Greisheim, 197–?); N. Parkin and C. R. Flood, *Welding Craft Practice. Vol. 1: Oxy-acetylene Gas Welding and Related Studies*, 2nd edn (Oxford, Pergamon Press, 1979); F. R. Schell and B. Matlock, *Industrial Welding Procedures: Oxyacetylene, Electric*

Arc, MIG and TIG, Design and Special Processes (New York, Van Nostrand Reinhold, 1979); and finally Union Carbide Corporation, *Oxy-acetylene Welding, Brazing and Cutting for the Beginner* (New York, Union Carbide, 1973).

Resistance welding

The textbooks available specifically on resistance welding were mostly written some years ago, but are still indicative of current practice. These are American Welding Society, *Resistance Welding – Theory and Use* (Miami, AWS, 1981); E. J. Del Vecchio (ed.), *Resistance Welding Manual* Vols. 1 and 2, 3rd edn (Philadelphia, Resistance Welder Manufacturers' Association, 1956 and 1961), and W. A. Stanley, *Resistance Welding* (New York, McGraw-Hill, 1950); while the Welding Institute has published the papers of a seminar *Resistance Welding: Control and Monitoring* (Abington, the Welding Institute, 1977).

Electroslag welding

This process is used mainly for heavy civil engineering, shipbuilding, and thick-walled pressure vessels. The main reference is the book by B. E. Paton, *Electroslag Welding* (translated from Russian), 2nd edn (New York, American Welding Society, 1962) while Hobart Brothers have published *Technical Guide for Electroslag Welding*, EW 493 (Troy, Ohio, Hobart, 1980).

Electron beam welding

There are three different types of book available on this subject. The simplest introduction to the subject is by M. J. Fletcher, *Electron Beam Welding* (M & B. Monograph ME/2, London, Mills & Boon, 1971). Secondly a booklet published by a company which produces electron beam welding machinery is by Sciaky, *Electron Beam Welding*, 2nd edn (Vitry-sur-Seine, Sciaky, 197–?) (in English). Thirdly, there are three general textbooks wholly or partly on electron beam welding. The standard text for some years has been A. H. Meleka (ed.), *Electron Beam Welding: Principles and Practice* (London, McGraw-Hill for the Welding Institute, 1971); while R. Bakish and S. S. White, *Handbook on Electron Beam Welding* (New York, John Wiley, 1964) is still a basic reference work. A more recent publication is the American Society for Metals, *Source Book on Electron Beam and Laser Welding*, compiled by M. M. Schwartz (Metals Park, Ohio, ASM, 1981).

Laser welding

In addition to the American Society for Metals publications mentioned above on electron beam and laser welding, one textbook by M. M. Schwartz, *Metals Joining Manual* (New York, McGraw-Hill, 1979) can be recommended, and one by N. Rykalin *et al.*, *Laser Machining and Welding*, transl. from Russian by O. Gleloov (Oxford, Pergamon Press, 1978). The Welding Institute has published *Laser Welding, Cutting and Surface Treatment* (Abington, The Welding Institute, 1984).

Solid phase welding

Five welding processes are covered by this overall heading, which denotes surfaces brought together by mechanical deformation or atomic diffusion. These processes are cold pressure welding, for which no known text books exist; friction welding, explosive welding, ultrasonic welding, and diffusion bonding. Two books cover the whole range of these processes: R. F. Tylecote, *The Solid Phase Welding of Metals* (London, Edward Arnold, 1968), and M. M. Schwartz, *Metals Joining Manual*, referenced above. The American Society for Metals, *Source Book on Innovative Welding Processes* (Metals Park, Oh., ASM, 1981) covers friction, explosive and diffusion welding. For the four processes which have specific textbooks, these references are as follows:

Friction welding – an introductory book, from the same series as the one mentioned for electron beam welding is M. J. Fletcher, *Friction Welding* (London, Mills & Boon, 1972). Two publications, both translated from Russian are firstly A. F. Vavilov and V. P. Voinov, *Friction Welding*, J. H. Dixon, ed. BWRA (Boston Spa, National Lending Library for Science & Technology, 1968) and secondly, I. V. Vill, *Friction Welding of Metals* trans. from Russian (New York, American Welding Society, 1962). The Welding Institute has published seminar papers on the subject *Exploiting Friction Welding in Production* (Abington, the Welding Institute, 1979).

Diffusion bonding – the only substantial publication on this subject is again a collection of seminar papers by the Welding Institute, *Diffusion Bonding as a Production Process* (Abington, the Welding Institute, 1979), although there is also a chapter on the subject in the *ASM Source Book on Innovative Welding Processes* mentioned previously, and Schwartz, *Metals Joining Manual*.

Explosive welding – firstly a set of conference papers has been published: T. Z. Blazynski (ed.), *Explosive Welding, Forming and Compaction* (London, Applied Science, 1983). The second publication is B. Crossland, *Explosive Welding of Metals and Its Application* (Oxford, Clarendon Press, 1982). An earlier seminar is the Welding Institute, *Explosive Welding* (Abington, the Welding Institute, 1976).

Ultrasonic welding – a chapter of Schwartz, *Metals Joining Manual* referenced above deals with ultrasonic welding and soldering.

Hardfacing

Of growing importance in the welding industry is the hardfacing of materials for increased wear resistance to increase the life of components, such as earthmoving machinery. A simple introduction to the process is BOC Murex Ltd, *Application to Cobalarc Hardfacing* (Waltham Cross, BOC Murex, 1980) while the standard text on this subject is M. Riddihough, *Hardfacing by Welding*, 4th edn (Solihull, Deloro Stellite, 1975). Further to these publications the following may be consulted, the Welding Institute, *Weld Surfacing and Hardfacing* (Abington, the Welding Institute, 1980), and Welding Research Council, *Hardfacing Structures* (New York, WRC, 1980).

Brazing

There are three major types of book on brazing, those produced by the companies which manufacture brazing materials, those published as handbooks by learned institutions, and several published textbooks. To give some indication of the level of complexity of the books cited, they will be divided into three types.

An introductory level of publications includes: P. M. Roberts, *Brazing* (Engineering Design Guides 10, London, Oxford University Press, 1975); H. R. Brooker and E. V. Beatson, *Industrial Brazing*, 2nd edn (London, Newnes-Butterworths, 1975); Handy and Harman, *The Brazing Book* (New York, USA, Handy and Harman, 1977); Union Carbide Corporation, *Oxy-acetylene Welding, Brazing and Cutting for the Beginner* (New York, USA, Union Carbide, 1973). At a more advanced level the following will be of assistance: American Society for Metals, *Source Book on Brazing and Brazing Technology* (Metals Park, Ohio, ASM, 1980); R. J. C. Dawson, *Fusion Welding and Brazing of Copper and Copper Alloys* (London, Newnes-Butterworths, 1973); and Southern Cross Steel Company, *The Welding, Brazing and Soldering of Stainless Steels* (Sandton, USA, Southern Cross Steel, 197–?).

The most complex and comprehensive texts on brazing are

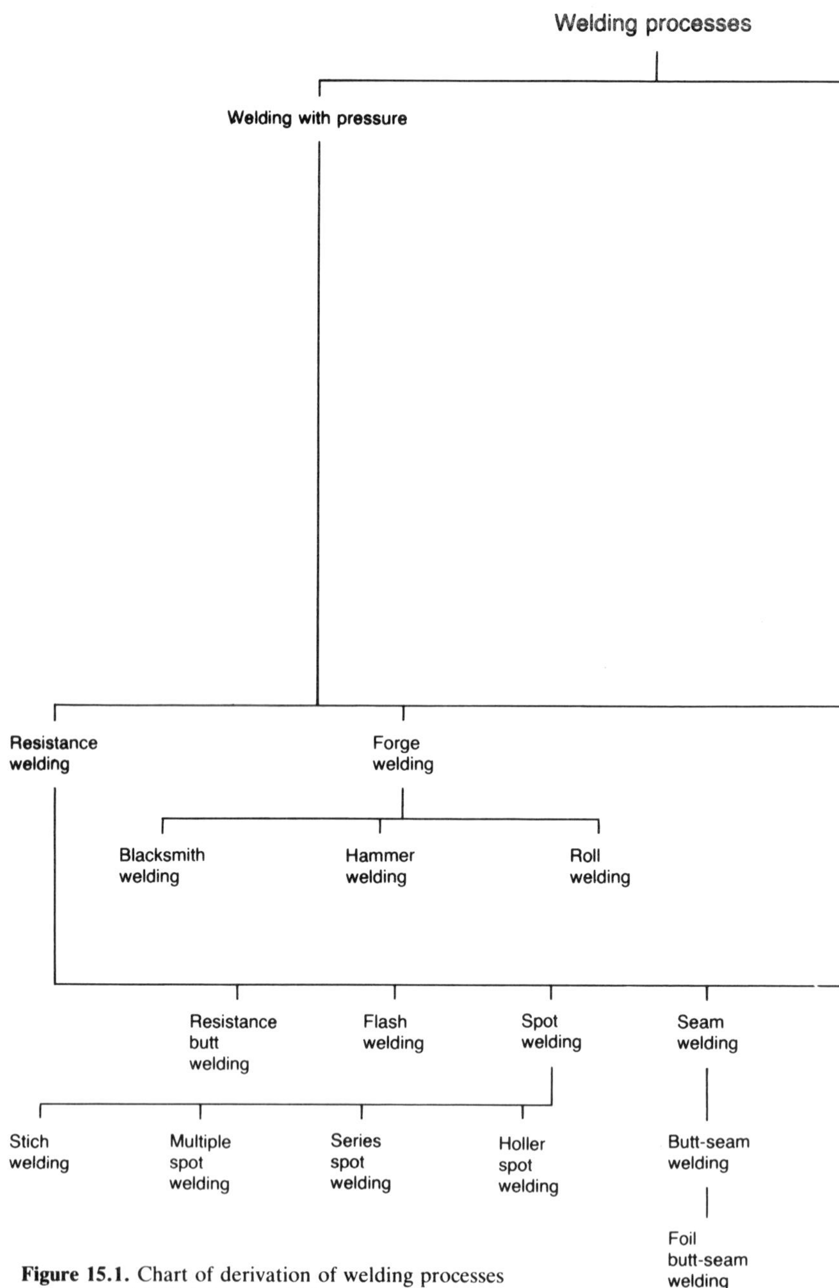

Figure 15.1. Chart of derivation of welding processes (Courtesy of British Standards' Institution)

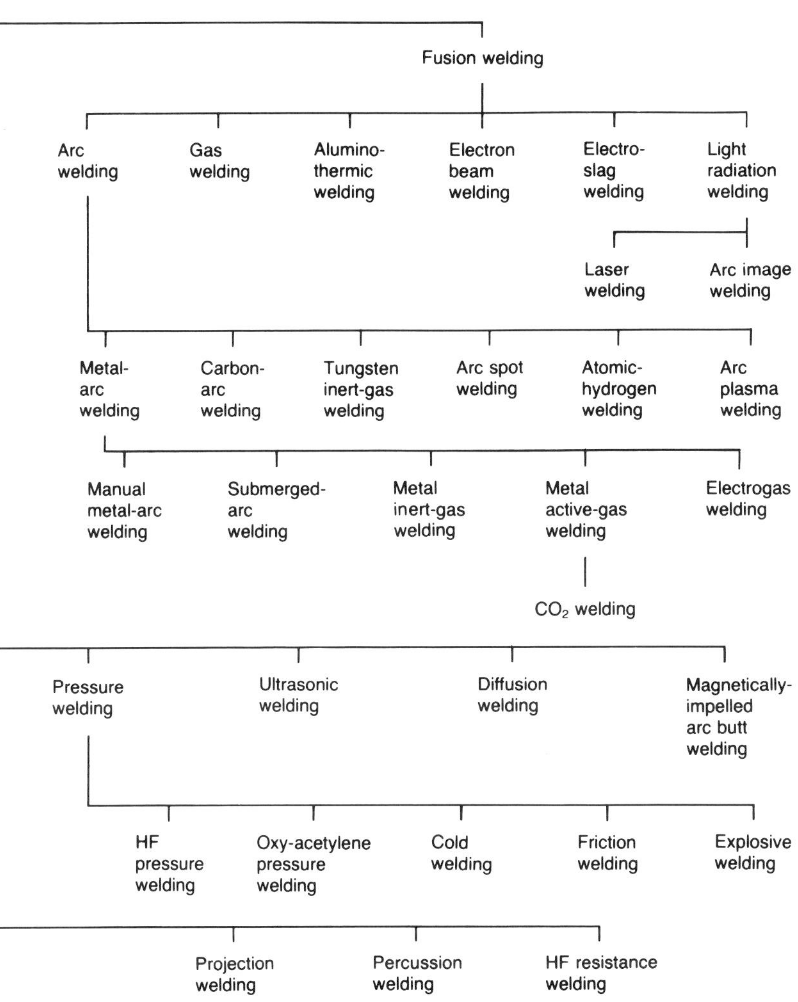
Fusion welding
Arc welding
Gas welding
Alumino-thermic welding
Electron beam welding
Electro-slag welding
Light radiation welding
Laser welding
Arc image welding
Metal-arc welding
Carbon-arc welding
Tungsten inert-gas welding
Arc spot welding
Atomic-hydrogen welding
Arc plasma welding
Manual metal-arc welding
Submerged-arc welding
Metal inert-gas welding
Metal active-gas welding
Electrogas welding
CO_2 welding
Pressure welding
Ultrasonic welding
Diffusion welding
Magnetically-impelled arc butt welding
HF pressure welding
Oxy-acetylene pressure welding
Cold welding
Friction welding
Explosive welding
Projection welding
Percussion welding
HF resistance welding

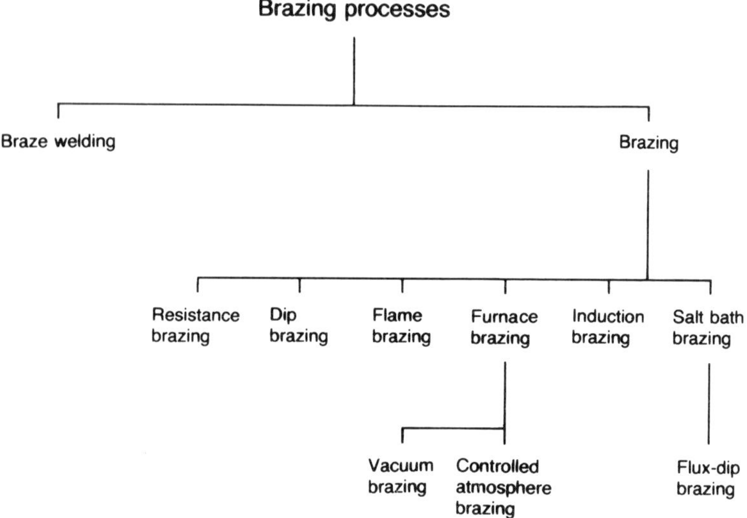

Figure 15.2. Chart of derivation of brazing processes

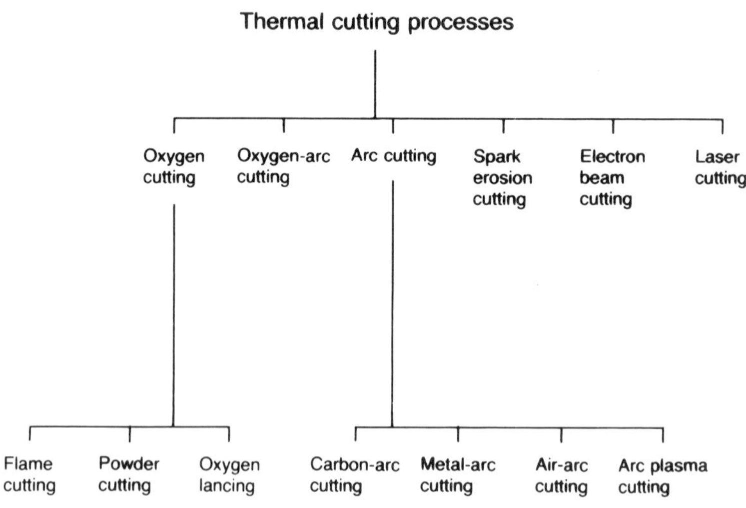

Figure 15.3. Chart of derivation of thermal cutting processes

American Society for Metals, *Metals Handbook*, 9th edn, Vol. 6, *Welding, Brazing and Soldering* (Metals Park, Ohio, ASM, 1983); American Welding Society, *Brazing Manual*, 3rd rev. edn (Miami, AWS, 1976).

Soldering

Simple introductory texts are: Aluminium Association, *Aluminium Soldering Handbook*, 3rd edn (Washington, Aluminium Association, 1976), and C. J. Thwaites and B. T. K. Barry, *Soldering* (Engineering Design Guides No. 7 (London, Oxford University Press, 1975). More detail is given in the Southern Cross Steel book on *Welding, Brazing and Soldering of Stainless Steels* referred to in the section on brazing books. Again the most advanced texts include: Vol. 6 of the ASM *Metals Handbook* referred to in the brazing section, and, in addition, the following two publications: American Welding Society, *Soldering Manual*, 2nd edn (Miami, AWS, 1978), and H. H. Manko, *Solders and Soldering*, 2nd edn (New York, USA, McGraw-Hill, 1979).

Table 15.3. Welding processes: common abbreviations

MMA	Manual metal arc welding. In USA known as SMAW
SMAW	Shielded metal arc welding (USA). Same as MMA
MIG	Metal inert gas welding. Also GMA or GMAW
GMA, GMAW	Gas metal arc welding. Same as MIG
MAG	Metal active gas welding
TIG	Tungsten inert gas welding. Also GTA, GTAW, TAGS
GTA, GTAW	Gas tungsten arc welding. Same as TIG
TAGS	Tungsten arc gas shielded welding. Same as TIG
SAW	Submerged arc welding
EB, EBW	Electron beam welding
MIAB, MBL	Magnetically impelled arc butt welding
MIAF	Magnetically impelled arc forge welding
ESW	Electroslag welding
FCA	Flux cored arc welding
CO_2	Carbon dioxide welding
PAW	Plasma arc welding

Welding metallurgy

The effects of welding on the structure and properties of both the weld metal and the parent material must be studied carefully to ensure the correct choice of material, consumable and welding conditions.

Two general texts refer to the use of ferrous and non-ferrous material for each welding process, ASM, *Metals Handbook*, 9th edn, Vol. 6: *Welding, Brazing and Soldering* and AWS, *Welding*

Handbook, Vol. 4: *Metals and Their Weldability*, 7th edn, both mentioned previously.

The best known books on welding metallurgy are the following: D. Seferian, *The Metallurgy of Welding*, trans E. E. Bishop (London, Chapman & Hall, 1962); T. B. Jefferson and G. Woods, *Metals and How to Weld Them* (Cleveland, Oh., James F. Lincoln Arc Welding Foundation, 1962); M. D. Jackson, *Welding Methods and Metallurgy* (London, Griffin, 1967); R. D. Stout and W. D'Orville Doty, *Weldability of Steels*, 4th edn (New York, Welding Research Council, 1987); J. F. Lancaster, *Metallurgy of Welding*, 4th edn (London, Allen & Unwin, 1987); K. Easterling, *Introduction to the Physical Metallurgy of Welding* (London, Butterworths, 1983). To these should be added two publications by the Welding Institute and its predecessor the British Welding Research Association. The first, one of a series of Design Memoranda is K. G. Richards, *Weldability of Steels* (Abington, British Welding Research Association, 1967), while a more specialized book is F. R. Coe, *Welding Steels without Hydrogen Cracking* (Abington, the Welding Institute, 1973).

Two volumes on this subject appeared in 1987 namely Sindo Kou, *Welding Metallurgy* (Wiley) and J. F. Lancaster, *Metallurgy of Welding* (Allen & Unwin), both of which contain useful bibliographies. Publications can also be cited which cover a specific type of material. For carbon and carbon manganese steels two books may be consulted. The first is a classic reference work in welding G. E. Linnert, *Welding Metallurgy, Carbon and Alloy Steels*, 3rd edn, Vol. 1: *Fundamentals*; *Vol. 2: Technology* (New York, American Welding Society, 1967). The second publication, which is available in both English and French is International Institute of Welding, *Guide to the Welding and Weldability of C-Mn Steels and C-Mn Microalloyed Steels* (IIS/IIW 382, ex doc. IX–646–69, Stockholm, Svetskommissionen for IIW, 1969).

Several references can be given for the metallurgy of stainless steels. The most widely used textbook is R. Castro and J. J. de Cadenet, *Welding Metallurgy of Stainless and Heat Resisting Steels*, trans. from the French edn (Dunod, Paris, 1967; London, Cambridge University Press, 1974). Useful publications from American societies are the American Society for Metals, *Source Book on Stainless Steels* (Metals Park, Oh., American Society for Metals, 1976), Section VI – 'Welding', pp. 226–312; and the American Iron and Steel Institute, Committee of Stainless Steel Producers, *Welding of Stainless Steels and Other Joining Methods* (Washington, AISI, 1979). Two publications by stainless steel manufacturing companies can also be consulted, Stoody Company, *Stainless Welding Engineering* (Whittier, Ca., Stoody Company,

1977), and also Smitweld *Welding of Stainless Steels* (Nijmegen, Netherlands, Smitweld, 198–?).

There are fewer texts concerning the welding metallurgy of non-ferrous materials. A general series of data sheets which can be used to find grades, properties and details of similar materials and standards, is *Metal Construction Data Sheets* Series 2. Welding Properties of Nonferrous Metals (Abington, the Welding Institute, September 1980 onwards).

For aluminium, two main texts contain relevant information Kaiser Aluminum and Chemical Sales Inc., *Welding Kaiser Aluminum* (Oakland, Ca., Kaiser Aluminum, 1967), and American Society for Metals, *Source Book on Selection and Fabrication of Aluminum Alloys* (Metals Park, Oh., ASM, 1978), Section X, *Welding*; Section XI, *Brazing*; Section XII. *Soldering*. For copper, information can be found in R. J. C. Dawson, *Fusion Welding and Brazing of Copper and Copper Alloys* (London, Newnes-Butterworths, 1973). In addition two publications on copper nickel alloys by the International Nickel Company, under different names are INCO, *Guide to the Welding of Copper Nickel Alloys* (Birmingham, INCO, 1973), and Huntingdon Alloys, *Joining Huntingdon Alloys*, 24th edn (Huntingdon, West Virginia, Huntingdon Alloys (an INCO company), 1981). For other nickel alloys, those for use at high temperature known as nimonic, the following publication has a section on welding, W. Betteridge and J. Heslop, *The Nimonic Alloys* (London, Edward Arnold, 1974). Finally for information on titanium one text is available, American Society for Metals, *Source Book on Titanium and Titanium Alloys* (Metals Park, Oh., ASM, 1982), pp. 301–22.

Market surveys and statistics

By its very nature market intelligence poses problems for the information searcher: anyone having, or obtaining, market information is unlikely to want to share it and thereby lose a competitive edge. There is a similar unwillingness to divulge where information can be obtained, and so difficulties are acute in the search for both sources and data. Many companies employ or contract professional market researchers who are familiar with the standard works of reference. Individual manufacturers in the UK and European welding industry are still reluctant to disclose details of their production and sales, which makes market surveying more complex than it would otherwise be. The Welding Manufacturers' Association and Association of Welding Distributors maintain welding industry statistics, but these are not generally available to non-members. Market surveys have been published by Frost & Sullivan for the UK, Europe and USA (Frost & Sullivan, 104 Marylebone Lane, London, W1M 5FU); Marketing Strategies for Industry (UK) Ltd for UK, Europe and France (MSI, 32 Mill Green Road, Mitcham, Surrey, CR4 4HY); Techonomics for Europe (Technomics Ltd, Manor House, Moreton, Dorchester, Dorset); Sanpo for Japan, *Japan Welding Industry Fact Book* (Tokyo, 1983); and Jefferson for USA (annual review in July issue of *Welding Design and Fabrication*).

Some welding journals carry occasional articles or commentary on their national markets, although usually in no great detail: *Schweissen und Schneiden* (West Germany), *Souder* (France), *FWP Journal* (South Africa), *Welding and Metal Fabrication* (UK), *Welding Design and Fabrication* (USA), *Welding Journal* (USA), *Canadian Welder and Fabricator* and *Australian Welding Journal*.

Buyers guides are issued with the journals *Welding and Metal Fabrication* (UK), *Welding Design and Fabrication* (USA), *FWP Journal* (South Africa), *Canadian Welder and Fabricator*, *Souder* (France). *Welding and Fabricating Data Book 1984/5* (Cleveland, USA, Penton/IPC, 1984) is published separately, as is *Directory of Welding Manufacturers and Suppliers* (Abington, Welding Institute, 1988).

Welding consumables

Consumables are the rods, electrodes, filler wires, fluxes, wire gas combinations, fuel and shielding gases required by various arc welding processes to complete satisfactory welds. There are probably some ten thousand welding consumables that are now in use or have been used in the past. Finding details of these can be a most time-consuming and difficult task. Consumable manufacturers go to considerable lengths to help potential customers by issuing catalogues of their products. These vary from straight listings to detailed compilations arranged by consumable type, application or material. Some include guides to selection, safe use, good practice, and practical hints to obtain maximum benefit from the product. Many list product name, composition, national specifications, approvals, and even, in some cases, listings of competitors' similar products.

Product catalogues can be freely obtained from manufacturers provided they can be identified, and commercial or competitive reasons do not prevent a direct approach. Many queries about consumables are related to trade names with no reference to manufacturer or country of origin. The American Welding Society, *Filler Metal Comparison Charts* (AWS A5.0–86) contains an index by company and product name for American consumables conforming to the 30 standards issued by AWS. Many of the products listed are also available in other countries so its usefulness is not limited only to North America. Three directories listing trade names and manufacturers but without technical details are *Fabguide: Buyers Directory for (UK) Welding and Fabrication Engineers* (Haywards Heath, Business Press, Annual, issued with *Welding and Metal Fabrication* journal); *Directory of Welding Consumables Tradenames* (Abington, Welding Institute, Annual) and *Welding and Fabricating Data Book 1988/89* (Cleveland, USA, Penton/IPC, 1988).

Consumables are usually classified according to the national standards of their country of origin. This often causes confusion in identifying standard specifications, gradings within them, and their issuing body. The International Standards Organization (ISO) devised a universal designation system for electrical codings which was published as ISO Standard 2560 in the hope that it would be adopted by individual countries. This was the basis in the UK for BS 639: 1976 which combined two earlier standards BS 639 and BS 1719 although the latter are still referred to.

The major designation systems for covered electrodes for mild and low alloy steels are given in ISO 2560, BS 639 (UK), AWS A5.1 and A5.5 (USA), DIN 1913 (West Germany), UNI 5132 (Italy), AFNOR NF A81–309 (France), JIS Z 3211 (Japan), and GOST 9466, 9467 (USSR). Submerged arc welding wires are designated in BS 4165, AWS A5.17 and A5.23, DIN 8557, JIS Z 3311, GOST 2246; and gas shielded wires in BS 2901, AWS A5.18, DIN 8559 and JIS Z 3312. The AWS consumables standards (A5 series) have all been approved and issued by ANSI as US national standards and most are also printed identically in the *ASME Boiler and Pressure Vessel Code Part 2c. Welding Rods, Electrodes and Filler Metals* (New York, ASME, published every three years). AWS-IFS-87 attempts to match similar grades from major national standard specifications.

Manufacturers wishing to supply consumables for construction of vessels to be classed by a classification society must obtain special approval for them, normally based upon an annual test conducted on each consumable product. The annual lists issued by these societies therefore provide a useful guide to consumable availability. Their arrangement is under country, by manufacturer and trade name, and includes brief details of type. The commonly available listings are those of Lloyds Register of Shipping (UK), Det Norske Veritas (Norway), American Bureau of Shipping (USA), Bureau Veritas (France), and Registro Italiano Navale (Italy).

The Canadian Welding Bureau certifies electrodes for use in Canada and their listing contains manufacturers' and product name classed by type and grade of consumable, *List of Electrodes Certified to CSA W. 48 Series of Standards and Applicable AWS A5 Specifications* (Toronto, Canadian Welding Bureau). V/O Promsyrioimport exports Soviet electrodes which comply fully with GOST standards and their catalogue lists each product with detailed descriptions of applications and properties, *Electrodes for Arc Welding and Depositing* (V/O Promsyrioimport, 13 Tchaihovsky St, Moscow G-99, 121834, USSR). *The Guide to British Arc Welding Electrodes and Consumables* (formerly *BEAMA Guide*) (London, Welding Manufacturers' Association, 1979) is a British listing by consumable type and manufacturer of products with brief details of each. No attempt at comparison is made and only products of WMA member companies are included. In 1988 the Welding Institute launched a series of *Guides to Arc Welding Consumables* with brief details of each. No attempt at comparison is made but coverage is worldwide.

Health and safety

Health and Safety in Welding and Allied Processes (Abington, Welding Institute, 1983), now in its third edition, has become the standard work relating to welding and contains many references and contact addresses for further information. Less exhaustive texts have been issued in the UK – *Welding Safety* (London, Heating and Ventilating Joint Safety Committee, 1978) and Ross, D. S., 'Welders' health' series of articles in *Metal Construction* 1977; July pp. 303–4, October pp. 475–9, December pp. 568–71; and 1978: March pp. 119–21, May pp. 204–8, and US Dept Health, Education and Welfare: *Safety and Health in Arc Welding and Gas Welding and Cutting* (Washington, USGPO, 1978).

Medical aspects of the effects of welding on health have been the subject of many studies and are briefly reviewed by Ross (above). A clear and comprehensive study was issued by the Australian Welding Research Association, *Health and Safety in Welding* (Sydney, AWRA Technical Note 7, July 1982) and articles appear frequently in a number of journals including *Welding Journal* (USA), *Australian Welding Journal*, *Welding Design and Fabrication* (USA), *Welding and Metal Fabrication* (UK) and *Joining and Materials* (UK).

Personnel protection and safe practices are the subject of many AWS standards and codes, publications of the British Compressed Gases Association and Health and Safety Executive Guidance Notes series. The *Reference Book for Protective Equipment*, 6th edn (London, Industrial Safety (Protective Equipment) Manufacturers' Association, 1982) is a buyers guide, and suppliers and manufacturers in the UK can also be identified from *Fabguide* (Haywards Heath, Business Press International). To protect the health of workers HSE gives guidance on the control of levels of exposure to toxic substances. Previously based on the American Conference of Government Industrial Hygienists (ACGIH) lists of the Threshold Limit Values (TLVs), the HSE list is now independently compiled as HSE Guidance Note EH40 Occupational Exposure Limits (OELs). In the US the AWS has issued many reports on fume, radiation, noise and other medical factors.

The Welding Manufacturers' Association produces leaflets *Fumes from Arc Welding Processes*; *The Arc Welder at Work* and a series on resistance welding machine safety, while the ISPEMA issues a buyers guide and source book, *Reference Book for Protective Equipment*, mentioned above.

Nondestructive testing

Nondestructive testing societies have been established in the major industrialized countries to co-ordinate activities and develop training and research into all aspects of nondestructive inspection. In the UK, the British Institute of Nondestructive Testing (BINDT), Institute of Quality Assurance (IQA) and National Nondestructive Testing Centre at Harwell all play a significant role. New bodies have also been formed. For example, Lloyds Register of Shipping now have a special Quality Assurance Certification Association (LRQA) while the Associated Offices Technical Committee (AOTC) has a newly formed section, Plant Safety Ltd. The longer established Pressure Vessel Quality Assurance Board (PVQAB) run by the Institution of Mechanical Engineers also deals with inspection, and the National Quality Information Centre was inaugurated in 1984 by the IQA to give advice and act as a clearing house for information on all aspects of quality and testing.

In Europe, the French Institute de Soudure inspects through the Comité Française d'Etude des Essais Non Déstructifs (COFREND), while the Scandinavian countries have co-operated to form NORDTEST, a scheme which covers inspection and qualification for Norway, Denmark and Sweden. The USA has four main inspection bodies: American Society for Nondestructive Testing (ASNT), American Welding Society (AWS), American Society of Mechanical Engineers (ASME), and National Board of Boiler and Pressure Vessel Inspectors.

Training courses for inspectors in the UK are run by the

Welding Institute's School of Applied Non-Destructive Testing (SANDT), the Scottish School of NDT and elsewhere, and the generally recognized qualification scheme is the Certification Scheme for Weldment Inspection Personnel (CSWIP) and PCN. In France COFREND offers qualification schemes, as does NORDTEST for Scandinavia. In the USA, AWS and ASNT have qualification schemes and the Welding Institute of Canada offers courses recommended by the Canadian Standards Association (see below, Welding inspector training establishments).

Inspection societies

ASME (American Society for Mechanical Engineers)
ASNT (American Society for Nondestructive Testing)
AWS (American Welding Society)
AOTC (Associated Offices Technical Committee)
CSWIP and PCN (Certification Scheme for Weldment Inspection Personnel)
COFREND (Comité Française d'Etude des Essais Non Déstructifs)
LRQA (Lloyds Register Quality Assurance Certification Association)
National Board of Boiler and Pressure Vessel Inspectors
NORDTEST (Norsk NDT-Forening and Svejsecentralen)
PVQAB (Pressure Vessel Quality Assurance Board)

Welding inspector training establishments

American Society for Nondestructive Testing
American Welding Society
Australian Institute for Non-Destructive Testing
Canadian Society of Non-Destructive Testing
Deutsche Gesellschaft für Zerstorungsfreie Prufung EV
General Dynamics, Convair Division
Institut de Soudure, Comité Français d'Etude des Essais Non Déstructifs (COFREND)
Lloyds British School of Welding
Nondestructive Testing Information Analysis Centre (NTIAC)
Scottish School of NDT
Welding Institute, School of Applied Non-Destructive Testing (SANDT)
Welding Institute of Canada

A number of training handbooks are available, all published in the USA. The most widely used textbooks are the ASNT series of recommended practices ASNT-SNT-TC-1A. *Personnel Quali-*

fication and Certification in Nondestructive Testing, 1984, and two series published by General Dynamics Convair Division *Classroom Training Handbooks* and *Programmed Instruction Handbooks*. The *ASNT Recommended Practices* can also be described as an inspection manual, as can G. P. Hayward, *Introduction to Nondestructive Testing* (American Society for Quality Control, 1978), the *Guide to Welding Inspector Qualification and Certification (WIQC) (Miami, AWS, 1984)*, and American Petroleum Institute, *Guide for Inspection of Oil Refinery Equipment*, 4th edn (API, 1982) which has an appendix on inspection of welding.

Simple introductions to NDT are the Engineering Design Guide *Nondestructive Testing* by D. Birchon (Oxford, Oxford University Press, 1975) and AWS, *Fundamentals of Welding Inspection* (Miami, AWS, 1980). One of the standard general texts is that by J. Blitz, King and Rogers, *Electrical, Magnetic and Visual Methods of Testing Materials* (London, Butterworths, 1969). A number of textbooks cover the whole range of techniques and provide useful first sources of information on any specific method. R. S. Sharpe (ed.), *Quality Technology Handbook*, 4th edn (London, Butterworths, 1984); *ASM Metals Handbook*, Vol. 11: *Nondestructive Inspection and Quality Control* (Metals Park, Ohio, ASM, 1976) and R. C. McMaster, *Nondestructive Testing Handbook* (New York, Ronald Press, 1959, 2 vols), have been standard texts for some time, the latter in a new edition with separate volumes for each method of testing (Vol. 1. *Leak Testing*; Vol. 2. *Liquid Penetrant Tests*; Vol. 3. *Radiography and Radiation Testing*; Vol. 4. *Electromagnetic Testing*) is now being published by ASM.

At a more advanced level is R. S. Sharpe's series *Research Techniques in Non-destructive Testing* (London, Academic Press, 1970–84), Vols 1–7, and *NTIAC Handbook* (San Antonio, Nondestructive Testing Information Analysis Center, 1979, NITAC–79–1).

NDT as applied to welding is the subject of AWS, *Welding Inspection* (2nd edn, Miami, AWS, 1980), and their standard AWS B1.0–1984 *Guide to Nondestructive Inspection of Welds*. A substantial volume edited by N. T. Burgess was *Quality Assurance of Welded Structures* (London, Applied Science, 1983). For information on materials and testing BINDT has reprinted a series of articles by J. L. Taylor entitled *Basic Metallurgy for NDT* (Northampton, BINDT, 1974) and ASNT have published *Materials and Processes for NDT Technology* (Columbus, ASNT, 1981).

Further to these general textbooks, specialist literature exists for each brand of the subject. A selection are: J. C. Seymour, *Acoustic Emission: Techniques and Applications* (Evanston, Intex

Pub. Co., 1974), the only significant textbook on this subject as most other volumes comprise conference papers; T. F. Drouillard, *Acoustic Emission: A Bibliography with Abstracts* (New York, IFI/ Plenum, 1979); *Eddy Current Testing* (Chalk River, Atomic Energy of Canada Ltd, 1981), Report AECL 7523; R. M. McMaster, *NDT Handbook*, 2nd edn, vol. 1: *Leak Testing* (ASM, 1982); C. E. Betz, *Principles of Magnetic Particle Testing* (Chicago, Magnaflux Corp., 1966); C. E. Betz, *Principles of Penetrants* (Chicago, Magnaflux Corp., 1963); International Institute of Welding (IIW), *Handbook of Radiographic Apparatus and Techniques* (Abington, Welding Institute for IIW, 1973); J. C. Drury, *Ultrasonic Flaw Detection for Technicians* (Swansea, Unit Inspection Co., 1978), and for more advanced work J. Krautkramer, *Ultrasonic Testing of Materials*, 3rd rev. edn (New York, Springer Verlag, 1983) is the standard text.

Three bodies have produced reference radiographs of typical weld defects to assist in the interpretation of radiographs taken during testing procedures. The IIW have two sets for steel and one for aluminium welds, but these are merely a comparison for reference purposes and should not be quoted as standards. The American Society for Testing and Materials have a number of standards with collections of radiographs, numbered ASTM E155, E186, E192, E242, E272, E286, E390 and E446. Deutscher Verlag für Schweisstechnik at Düsseldorf have produced *DVS Evaluation Catalogue DIN 8563 Part 3. Reference Code for Evaluation of Radiographic Exposures of Butt Welds* (1981).

Design for welding

Designers are often worried by the complexities of various welding processes but need to be able to specify materials, welding processes, joint design, and the consumables to be used for their product. All these factors will affect the fabrication but, discouraged by the knowledge that so many past failures have been associated with welds, designers are easily tempted to overdesign and err on the safe side. This approach can lead to increased fabrication costs and lowering of the chances of a particular design being accepted by the client. On the other hand, overdesign may be more acceptable than a failure, especially in bridge, nuclear, offshore, petroleum and chemical environments. K. G. Richards has written a series of design memoranda on *Joint Preparation* (1966), *Weldability of Steel* (1967), *Fatigue Strength of Welded Structures* (1969), and *Brittle Fracture* (1971), which have proved to be best sellers. The publishers of that series, the Welding Institute, have issued a number of volumes covering *Control of*

Distortion, 2nd edn (1968), *Improving Fatigue Performance of Welded Joints* (1983), T. R. Gurney, *Fatigue of Welded Structures*, 2nd edn (Cambridge, CUP, 1979), and W. S. Pellini, *Guidelines for Fracture-safe and Fatigue-reliable Design of Steel Structures* (1983). One of the major standard works is F. Koenigsberger, *Design for Welding in Mechanical Engineering* (London, Longmans Green & Co., 1948) and, in spite of its age, still gives good coverage of the principles and practice. More recent attempts have been made by J. G. Hicks, *Welded Joint Design*, 2nd edn (Oxford, BSP Professional Books, 1987); T. G. F. Gray and J. Spence, *Rational Welding Design*, 2nd edn (London, Butterworths, 1982); T. B. Jefferson, *Welding Engineer Data Sheets*, 9th edn (Morton Grove, USA, Monticello Books, 1987), pp. 128–49; R. H. Thornley, 'Aspects of welding design', Ch. 6 in D. R. Andrews, *Soldering, Brazing and Adhesives* (London, Institute of Production Engineers, 1978), pp. 128–65; and V. A. Lucas, *Designer's Basic Guide to Welded Structure Performance*, SAE Paper 770528.

O. W. Blodgett has made major contributions to design through publications of the James F. Lincoln Arc Welding Foundation. His massive *Design of Welded Structures* (Cleveland, 1966) includes load and stress analysis, column, girder and welded connection design, joint design, and a reference section on formulae. *Modern Welded Structures* (3 vols, Cleveland, 1963–70) is a series of reports reviewing designs as is *Design Ideas for Weldments* (3 vols, Cleveland, 1963–1979). Lincoln issued many publications including two excellent volumes, *Procedure Handbook for Arc Welding. Section 2.* 'Designing for arc welding', 12th edn (Cleveland, 1973) and *Principles of Industrial Welding*. Chapter 7. 'Joint design and welding instructions' (Cleveland, 1978).

A general introductory guide to the welding of structural steelwork including design, has been issued by Constrado, J. C. Pratt, *Introduction to the Welding of Structural Steelwork* (Croydon, Constrado, 1979), while specific recommendations are available from the International Institute of Welding, *Recommended Welded Connections for Pipework. Document IIS/IIW–146–64* (London, IIW, 1965), and *Recommended Welded Connections for Pressure Vessels. Document IIS/IIW–237–66* (London, IIW, 1967). The (US) Welding Research Council has many Bulletins relating to pressure vessel design, and others on tubular joints for offshore structures. Few texts deal with pipelines in much detail but that by J. B. Herbich, *Offshore Pipeline Design Elements* (New York, Marcel Dekker, 1981) devotes a whole volume to environmental considerations. Much of the work at Massachusetts Institute of Technology on residual stresses and distortion, brittle fracture and fatigue, especially for offshore, has been summarized in a book prepared by K. Masubuchi, *Analysis of Welded Structures: Residual Stresses, Distortion, and Their Consequences* (Oxford, Pergamon Press, 1980).

Many volumes have been published of studies relating to failures in bridges, buildings, ships, pipelines, offshore structures, pressure vessels and storage tanks. In addition to individual case studies (Kings Bridge, Alexander Kielland and Sea Gem oil rigs, John Thompson pressure vessel etc.), books reviewing aspects of such failures and their causes include G. M. Boyd, *Brittle Fracture in Steel Structures* (London, Butterworths, 1970); W. D. Biggs, *The Brittle Fracture of Steel* (London, Macdonald & Evans, 1960); J. W. Fisher, *Fatigue and Fracture in Steel Bridges. Case Studies* (New York, Wiley-Interscience, 1984); R. Hammond, *Engineering Structural Failures* (London, Odhams Press, 1956); C. F. Tipper, *The Brittle Fracture Story* (London, Cambridge University Press, 1962); and S. S. Ross, *Construction Disasters: Design Failures, Causes and Prevention* (New York, McGraw Hill, 1984).

The UK standards for bridges have developed considerably over the last decade largely as a result of failures in box girder bridges and the findings of the Merrison Committee of Enquiry. BSI has issued a completely new code in ten parts as BS 5400, which covers materials, design, workmanship, fatigue and construction. In the USA the AWS, *Structural Welding Code* (AWS D1.1) and American Association of State Highway Transportation Officials (AASHTO) specify nationally for bridges and buildings, and ASME for boilers and pressure vessels, while in the UK, BS 5500, the pressure vessel code, is following a similar three-year cycle of editions.

Shipbuilding and offshore structures are covered in the Classification Societies' rules (Lloyds, American Bureau of Shipping, Det Norske Veritas, Bureau Veritas, Germanische Lloyd, Registro Italiano Navale and Nippon Kaiji Kyokai). In the UK, the Department of Energy Guidance Notes, *Offshore Installations: Guidance on Design and Construction*, 3rd edn (London, HMSO, 1984), and BS 6235 are also relevant, and in the USA the Structural Welding Code has a section on tubular joints.

Education and training

Training establishments

All those involved in the welding industry must be trained to a level of skill which will permit acceptable work to be designed, carried out and tested.

Innovation and design in welding are largely the province of the graduate welding engineer. There are few courses specializing in welding as a first degree or postgraduate subject. These are at Brunel and Strathclyde Universities, and postgraduate MSc study can also be undertaken at Cranfield Institute. To these should be added several part-time courses at Teesside Polytechnic, Newcastle-upon-Tyne Polytechnic, and Gwent College of Higher Education, leading to a BTEC qualification and thence to C. Eng. grade.

In addition, several universities have a welding component to their engineering or metallurgy degrees. These are:

City of London Polytechnic	(with Metallurgy)
W Bromwich College H. Ed.	(with Metallurgy)
University of Surrey	(with Metallurgy)
University of Aston in Birmingham	(with Engineering)
Paisley College	(with Engineering)

Many colleges of higher or further education train welding technicians up to Higher BTEC level, either on a full-time or part-time basis.

It is largely from the ranks of welding engineers that the professional members of the Welding Institute are drawn, though some work their way into professional membership from the technician grades. All members have to satisfy the Institute in their skill and knowledge of the industry before being accepted into membership. Examinations are an essential part of this process. Similar qualifications exist in other countries, as for example, those offered by the Institut de Soudure in France, and the AWS in America.

In addition to the training received by welders, they may well wish to attend intensive welding technology courses for basic training and to widen their knowledge, enabling them to undertake more advanced work. Short term courses, aimed towards satisfying the requirements of a particular specification can be undertaken at courses on welding technology. A selection of those institutions offering such courses in the UK are listed below.

Anglian Welding School
Babcock & Wilcox Limited
Bowford Engineering Services Ltd, Welding Technology Centre (Murex Appointed Training Centre)
British Shipbuilders (Training, Education & Safety) Ltd, Scottish Regional Centre (Murex Appointed Training Centre)
British Shipbuilders (Training, Education & Safety) Ltd (Murex Appointed Training Centre)
EAGIT (Engineering) Ltd
Engineering Industry Training Board

Eutectic Co. Ltd
Foster Wheeler Automated Welding Ltd
ITM-Head Wrightson Teesdale Ltd
Lloyds British School of Welding
Murex Training Centre
National Engineering Training Association Ltd
North West Welder Training (Murex Appointed Training Centre)
Oilfab Group Ltd
The Scottish School of NDT
SITMS
Stubs Welding Ltd
The Welding Institute
The Welder Training Centre (Lancaster) Ltd

Literature sources

The literature on training of welders can be divided into two categories. Firstly, handbooks and textbooks for use by instructors, and secondly, those for use by the students. There are naturally some books which will be useful to both. These include A. C. Davies, *The Science and Practice of Welding*, 8th edn, 2 vols (Cambridge, Cambridge University Press, 1984); D. Geary, *The Welder's Bible* (Blue Ridge Summit, PA, Tab Books Inc., 1980); L. Koellhoffer, *Shielded Metal Arc Welding* (New York, Wiley, 1983); R. P. Schmidt, *Welding Skills and Techniques* (Reston, Virginia, Reston Publishing Co., 1982); J. R. Walker, *Arc Welding: Basic Fundamentals* (G-W Career Series, South Holland, Ill., Goodheart-Willcox Co. Inc., 1977); I. H. Griffin *et al.*, *Basic Arc Welding*, 3rd edn (New York, Delmar, 1977); I. H. Griffin *et al.*, *Welding Processes*, 2nd edn (New York, Van Nostrand Reinhold, 1977). Further to these publications, a number of publications by Hobart Brothers, Linde, Union Carbide and tne James F. Lincoln Arc Welding Foundation, previously referenced (p. 293) should be consulted.

Texts of particular interest to instructors are led by the numerous training elements and instruction manuals produced by the Engineering Industry Training Board, covering all manual welding processes. The main elements are listed in the Welding Institute's list, Books on Welding (d) Training, obtainable on application to the Welding Institute Library. The following are also used widely, The Aluminium Association, *Aluminium Welders' Training Manual and Exercises* (Washington DC, AA, 1978); Australian Welding Institute *et al.*, *Submerged Arc Welding: Basic Training Manual* (Milsons Point, AWI, 1981); Hobart Brothers, *Training in Shielded Metal-Arc Welding I and II* (EW–269, Troy, Oh., Hobart Welding School, 1969); F. R. Schell, *Welding Procedures: Electric Arc. Instructors Guide* (Albany, USA, Delmar 1977), and finally H. A. Sosnin, *Teachers'*

Manual. Arc Welding Instructions for the Beginner (Cleveland, Oh., James F. Lincoln Arc Welding Foundation, 1964).

As well as these manuals for the trainer there are two publications designed to give assistance in setting up and monitoring training schemes: F. Clark and J. Twining, *Standards of Welding Achievement during Training* (London, Heinemann Educational Books, 1966); Hobart Brothers Co., *Recommendations for Teaching Welding, Setting up Training Programs and Shops* (Troy, OH, Hobart Brothers Co., 1971).

For the trainees, the EITB produces a number of trainee booklets, while the Hobart Welding School and the James F. Lincoln Arc Welding Foundation publish work books and welding lessons. A number of UK publishers, such as Macmillan, Longman, Cassell, Newnes-Butterworths, Arnold, Hutchinson, Stanley Thornes Ltd, and Van Nostrand Reinhold, produce textbooks geared towards the various levels of the Technician Education Council (TEC) courses which UK trainees attend.

CHAPTER SIXTEEN

Materials for the aerospace industry

Dr D. K. THOMAS

General

The aerospace (aircraft) industry makes use of a very wide and seemingly ever increasing range of materials, and the prospect of systematizing and describing the sources of information on properties is an extremely daunting one. In fact, such is the volume of materials property data of likely relevance to aerospace applications that it is impossible to deal with the subject in anything approaching a comprehensive manner in an article of this length. Numerous corners will have to be cut and attention concentrated on what might be termed materials of primary importance, these being usually regarded as the ones used for the manufacture of engine and airframe components which are critical to flight safety. Outside this category there is a larger number of materials, many of which are non-metallic, whose properties and performance characteristics need to be known in some detail in order that material choices can be made from the very wide range of proprietary materials available. Although many of the components made from non-metallics will be used in 'non-structural' applications they can nevertheless have a considerable influence on the reliability, safety and cost of operation, and therefore need to be taken very seriously at the stage of materials selection.

The primary structural materials, and here one is really referring to metallic alloys, have, because of their importance in determining structural integrity and performance, been the most deeply studied and evaluated. Those alloy types of particular importance to aerospace designers have been identified during the course of long and successful usage in aircraft manufacture, and not surprisingly the large and specialized database on aerospace metals has over the years been separated from the enormous volume of general metals engineering data and published in a number of concise and valuable handbooks. In relation to these materials therefore the task undertaken in the present article has been much lightened and simplified, and can be viewed as a further digest of already thoroughly assessed and methodically assembled data. The position on non-metallics generally and on the newer structural composites in particular is very different.

For reasons of practicality the article is divided into three subject areas, these being metals, composites, and non-metallics. As implied already, information relevant to each category will differ greatly in respect of quantity and acknowledged quality, this being a consequence of the differing needs for close specification and depth of evaluation as between materials destined for structural and non-structural applications. Clearly the most stringent standards will be applied to aero-engine materials and to those used in primary load bearing parts of the airframe structure. By and large the non-structural materials require a lesser data base, but nevertheless because of the special functions which many of them perform they are still required to meet fairly stringent quality and consistency standards as laid down in approved aerospace materials specifications.

Airframe construction is still dominated by metallic materials, although advanced fibre/resin composites are beginning to make their presence felt. The much heralded carbon fibre/epoxy resin structural composites are slow to appear in production aircraft, and the usage of composites on commercial fixed wing aircraft and helicopters is more based on aramid/epoxy and glass/epoxy rather than carbon/epoxy at the present time. However in the longer term there seems little doubt that carbon fibre/organic resin composites will make significant inroads into what have been traditionally areas of application of light alloys, and substantial databases are being assembled by most major airframe constructors in order to support this change. Understandably much of this data for use in design is not being made generally available at this early stage, and there is therefore no immediate access to suitably assessed and critically appraised data as in the case of airframe metals. For other non-metallic materials such as elastomers, sealants, adhesives, optical and radar transparencies, dry bearings, thermal and electrical insulants, and interior furnishing materials generally, much of the property data available in the engineering materials literature may well be relevant to aircraft usage. However it will, in most cases, be necessary to assess the value of the data against the standards set by the aerospace specifications generated nationally and internationally to cover this wide range of materials.

The chief sources of information on aerospace materials published in English are those emanating from the UK and the United States. Each country produces major handbooks of aircraft materials data and also source books giving substantial advice on the many important aspects of design, fabrication and manufacturing processes, joining methods, protective schemes etc. The accumulated experience embodied in these publications is enormous and they are completely open publications. Likewise each country has generated and continues to generate specifications governing the quality of materials suitable for use in aircraft manufacture, and these represent in aggregate a further large source of direct guidance on the quality and property levels required in aircraft materials. There have also been international initiatives in attempting to assemble materials property data of the kind suitable for use in aircraft design. A particularly notable example of this being the activity of the Advisory Group for Aerospace Research and Development (AGARD) which brings together property data on aircraft materials from the leading NATO member countries.

In the UK perhaps the most significant publication on aircraft design matters is what is now known as Defence Standard 00–970 on *Design and Airworthiness Requirements for Service Aircraft*. This handbook replaces what was previously Aviation Publication 970 (AvP 970) – *Design Requirements for Service Aircraft*, produced primarily for the use of the Crown and its contractors in the execution of contracts for the Crown. The Standard is published in a number of volumes and is extremely broad in its scope, covering far more than materials property data. The property data is dealt with in the context of the basic requirements of design, strength and stiffness seen as necessary for the aircraft to meet its operational requirements with an acceptably low risk of structural failure. Design detail and strength of materials is dealt with in generalized fashion in Vol. 1 Part 4 Chapter 400. The arrangement of information and general guidance in this Defence Standard publication is extremely well done. All information on design requirements is printed on white paper whereas further information and recommended practices which amplify the requirements are printed on green paper. This allows for easy recognition in the relevant chapter of the publication.

An essential first step in the consideration of materials selection for aircraft applications is to identify the grade of part into which that material is going. A rational and meaningful grading of parts will serve to ensure that all the materials and processes used in the production of that part are of a suitable standard, and that the quality control and testing are appropriate to the design requirements and application envisaged for the part. In grading parts all, except standard parts, are designated Grade A or Grade B. according to the strength and stiffness requirements as promulgated in the Standard, quality requirements, maintainability and inspectability requirements, and also such factors as failure by leakage, malfunction, or other defect. The grading of parts is described in Vol. 1 Chapter 400 of the Standard with further amplification in leaflet 400/1. Design data for metallic materials are covered in the Chapter 401 reference page of Vol. 1 where attention is drawn to the sourcebook of physical and mechanical properties used as design values, this being the *Metallic Materials Data Handbook* (AvP 932) which is published by the Engineering Sciences Data Unit. A list of useful references to technical data published in report form is given in leaflet 401/0. The processing and working of materials is covered in Vol. 1 Part 4 Chapter 402, where jointing processes, except for mechanical fastening, are considered in terms of requirements together with extensive guidance on good practices in adhesive bonding, brazing and soldering. Chapter 403 deals with castings and their static strength approval and gives excellent guidance on acceptable levels of variability in static strength and sources of data on strength properties.

Helicopters are dealt with separately from fixed wing aircraft in the Standard and feature in Vol. 2 Book 1. Detail design and strength of materials aspects are found in Part 4 where the lay-out and scope is similar to that of Part 4 of Vol. 1 Book 1. The source of metallic materials property data is again the *Metallic Materials Data Handbook* (AvP 932), and Chs 402 and 403 deal with processing and working of materials and castings respectively, again after the pattern of Part 4 Vol. 1. Chs 405, 406 and 407 and their associated leaflets deal extensively with corrosion and its prevention. Information on the relative susceptibilities of various wrought aluminium alloys to exfoliation corrosion and stress corrosion, and the relative susceptibilities to stress corrosion of various cast aluminium alloys and steels is given in a series of tables in the leaflets to these chapters.

In many ways an American publication which parallels the UK *Metallic Materials Data Handbook* is the *Aerospace Structural Metals Handbook* produced in the USA by the Mechanical Properties Data Center. This Handbook was first published in 1966 and is produced under US Department of Defense sponsorship. The updated 1974 Handbook contains information on physical, chemical and mechanical properties of over two hundred metals and alloys of interest for high efficiency structural applications. The Handbook (reference AFML-TR-G8-115) is presented in four volumes, and in addition to the data referred to above there are also data source references, a general discussion of properties, a glossary of terms, a discussion of fracture properties. SI conversion tables and factors, and a cross-index of the alloys contained therein. New and revised chapters of the annual revision supplements are distributed on a quarterly basis to ensure currency of content.

The four volumes of the Handbook encompass ferrous, non-ferrous, light metal alloys, and non-ferrous heat resistant alloys. Property data is presented in both graphical and tabular form and useful general information on fabrication is also included. The information presented on this latter aspect is intended to convey a picture of the relative fabricability of the alloy and to pin-point areas in which material properties may be adversely affected by fabrication techniques. In the context of this Handbook fabrication covers all processes which may normally be employed in the manufacture of parts or components from materials as supplied by commercial producers. The processes include formability (forging, rolling, drawing, forming etc.), material removal (machining, grinding etc.), joining (welding, brazing etc.), and the corresponding post-operational treatment that may be required (heat treatment, surface treatment etc.). General information on formability relates primarily to the forming of sheet, strip, and plate in various conditions. A very limited amount of information on machining is presented, primarily to illustrate the performance of different alloy conditions in various machining operations. On welding the information given sets out to draw attention to areas where the mechanical or physical properties are influenced. The weldability of an alloy is clearly a matter of importance in terms of materials selection and it is therefore discussed whenever information of significance is available. The Handbook identifies all alloys with Aeronatical Materials Specifications (AMS) of the Society of Automotive Engineers, these being the most complete with regard to new alloys, and groups them according to particular features of composition or properties. Reference is also made to military specifications and occasionally Federal specifications, but there is no cross-reference to non-US specifications. Extensive lists of references to sources of technical information are included against each alloy type covered in the Handbook.

A further substantial handbook of aircraft materials data is that published by the Advisory Group for Aerospace Research and Development (AGARD). This handbook, also published by the Engineering Services Data Unit in the UK, is compiled through an international co-operative effort to make information on the properties of aircraft materials produced in various NATO countries accessible to all. The AGARD Panel on Structures and Materials is responsible for the handbook and they set out to give designers reliable information in a convenient form on the mechanical properties of materials produced in the NATO countries. A further objective was to give encouragement to international collaboration in problems related to research, design and production.

The Handbook is printed in a series of four volumes: Vol. I deals with aluminium alloys and was first published in 1959; Vol. II deals with steels and was first published in 1960; Vol. III covers the alloys of titanium, nickel and magnesium and was first published in 1963; and Vol. IV on heat resistant alloys was published in 1966. In each volume the information is assembled in materials sections, each section dealing with the alloys produced by one country. Each material section has, for each distinct alloy, a set of sheets giving data on the physical and mechanical properties at both room and elevated temperatures. Preceding these sheets there are introductory notes which give general information and amplify such points as the designations and specifications of the alloys, heat treatments, cladding thickness of plates, and testing procedures. On all sheets giving mechanical properties there is a statement on the statistical basis of the information, this being a matter of great importance in relation to the design of airframe components. Data are given the status of A, B or C values as appropriate, A being minimum guaranteed values, B being 90 per cent probability values, i.e. values which will be met or exceeded by 90 per cent of the material supplied, and C being typical values (the basis for typicality not being specified). In addition to the aforementioned Handbook the AGARD organization also publishes bound volumes containing papers presented through its own lecture series and conferences on specialist aeronautical subjects. Typical examples of publications dealing with aircraft materials and related processes and manufacturing technology are those on *Advanced Manufacturing Techniques in Joining Aerospace Materials* (lecture series number 91) (1977), *Bonded Joints and Preparations for Bonding* (lecture series number 102) (1979), and *Characterisation, Analysis and Significance of Defects in Composite Materials* (conference proceedings number 355) (1983). The papers included in these publications are invariably authoritative in nature and are compiled by specialists in their own fields. They generally contain a great deal of technical detail, including materials data, and are usually very well referenced to other sources of technical information in the same subject area. Initial distribution of AGARD publications is made to nominated centres within the member nations, i.e. the NATO countries, and for the UK, for example, the national distribution agency is the Defence Research Information Centre (Station Square House, St Mary Cray, Orpington, Kent, BR5 3RE). Copies can also be obtained from any one of three purchase agencies, e.g. the National Technical Information Service (NTIS) (5285 Port Royal Road, Springfield, Virginia 22161) in the USA. Full lists of AGARD publications should be available from the Secretariat which is based in Paris.

Reference has already been made to the work of the Engineering Sciences Data Unit (ESDU) in connection with the publication of AvP 932 the UK *Metallic Materials Data Handbook*. ESDU maintains a far-reaching activity in the fields of aeronautical, mechanical, chemical and civil engineering,and publishes a number of volumes of direct interest and value to the aircraft designer. The acknowledged purpose of these publications is to make authoritative information readily available to design engineers in a form suitable for direct application, and compilation and critical appraisal of the data is carried out by ESDU technical staff assisted by committees of specialists drawn from industrial companies, government research laboratories and universities. Of particular interest in the present context are publications in the *Structures Sub – Series* and *Fatigue Sub – Series* of the aeronautical engineering series. The former consists of twelve volumes, one of which – Vol. 12 on bonded joints – has a strong materials and processing content, and the latter containing 5 volumes each of which incorporates valuable materials property data. Vol. 5 of this series is of special importance in that it presents fatigue data on aluminium alloys, high-strength steels, and a range of titanium alloys.

Another important series of publications covering the broad spectrum of aerospace materials is the *Science of Advanced Materials and Process Engineering* Series produced by the Society of Aerospace Materials and Process Engineers (SAMPE) in the USA. These publications are volumes of technical papers presented at national SAMPE symposia or national SAMPE technical conferences, the distinction between the two being that the symposia give broad coverage to the subject while the conferences are more restricted in scope and deal with a narrower aspect of aerospace materials and processes. Examples of the former are SAMPE Vol. 15 – *Symposium on Materials and Processes* (April 1969); Vol. 16 – *Symposium on Materials* (April 1971); and Vol. 17 – *Symposium on Materials* (April 1972), and of the latter NSTC 2 –

Conference on Aerospace Adhesives and Elastomers (Oct. 1970), and NSTC 4 – *Non-Metallic Materials, Selection, Processing and Environmental Behaviour* (Oct. 1972).

Finally mention should be made of the sophisticated computer based information retrieval systems now in operation worldwide and which cover the materials science and technology literature in comprehensive fashion for both metals and non-metals. Although not aimed specifically at aerospace interests they can clearly be used for identifying and retrieving the materials data with an allegedly aerospace connection. In the metals field the abstracts compiled and edited by the British Non-Ferrous Metals Technology Centre are a good example. They report on the literature relevant to non-ferrous metals and prepare abstracts on papers appearing in over four hundred journals and periodicals published worldwide. Subjects covered include metallurgy, metal finishing, corrosion, environmental aspects, extraction, economics, analysis and testing, industrial processing, statistics metallography, company information and book reviews. All the abstracts published since 1961, which amount to some 100,000 items, have now been placed on computer, and additions are made at the rate of about 500 per month from the current literature. This huge database can be searched rapidly and effectively and the output routed to a computer terminal with great speed. Running costs are minimized by making use of inexpensive international data telephone lines, and two online services are now available through the European Space Agency's Information Retrieval Services and the Lockheed Corporation's DIALOG Service.

In the non-metals field a comparable activity is maintained by the Rubber and Plastics Research Association (RAPRA) at Shawbury, Shrewsbury, UK. The RAPRA abstracts, which are designed to meet the needs of rubbers and plastics producers, processors and users, give comprehensive coverage of the world's polymer literature and abstracts are presented from over 400 periodicals, conference proceedings, books, trade and technical literature, standards, patents and government publications. This database is also on computer and searches can be made by RAPRA on request or online using one's own terminal through the Pergamon Information Line. Subject areas covered by the database are raw materials and monomers, polymers and polymerization, compounding ingredients, intermediate and semi-finished products related to particular industries and fields of use, applications, processing and treatment, and properties and testing.

Metals

Aside from the speciality aerospace materials data handbooks referred to previously there are a number of outstandingly good books of metallic materials physical and mechanical property data which are aimed at the design engineering fraternity in the broad. A few of these should be mentioned here because they identify some alloys and property information as being of particular interest to the aerospace/aircraft industry.

The American Society for Metals publishes a *Metals Handbook* in seven volumes which give an excellent coverage of important engineering alloys with reference to favoured areas of application. In Vol. 1 on *Properties and Selection of Metals* the importance of alloys based on aluminium and titanium in the aerospace field is highlighted and individual alloys currently finding usage in aircraft construction are identified. Because of the large proportion of the present industrial output of titanium and its alloys which goes into aerospace applications a separate chapter is devoted to 'The selection of titanium alloys for high temperature aeronautical service'. In the chapter on 'Selection and application of aluminium alloys' aerospace applications are also treated separately in a short section dealing with corrosion resistance and fabrication. There is also a fairly interesting section on 'Examples of light metal parts for aeronautical construction' where the emphasis is on the presentation of cost information for a wide variety of metal parts. Castings and forgings comprise the bulk of the examples presented. As implied earlier the purpose of this *Metals Handbook* is to present metals data of good quality in an orderly fashion for industrial users generally; aerospace is just one of many outlets for the usage of light alloys and heat resisting alloys and, apart from the rather special case of titanium alloys, the information of direct relevance to the aerospace industry is consequently much less visible than in specialist aerospace materials texts.

A second extremely useful source book of metals property data is *Smithells Metals Reference Book* (Butterworths in association with the Fulmer Research Institute). The 6th edition was edited by E. A. Brandes (1983) and, as with the *Metals Handbook*, it addresses the engineering user at large. Chapter 22 on 'Mechanical properties of metals and alloys' summarizes the mechanical properties of the more important industrial metals and alloys; all the information is given in tabular form and when a material is regarded as being suitable for consideration in aircraft or aerospace applications a comment to that effect is made in the final 'remarks' column.

Separate sections are devoted to aluminium alloys, copper and its alloys, magnesium and its alloys, nickel and nickel alloys, titanium and its alloys, zinc and zinc alloys, zirconium and its alloys, steels, and bearing alloys. An extremely useful aspect of the information given in these sections is that the various materials specifications which have been developed in the USA, UK, France, Germany, and by international aviation groups such as Association Européenne des Constructeurs de Material Aerospatial (AECMA), are identified and associated with a nominal alloy.

Aluminium alloys

The prime sources of data on aerospace quality alloys are the *Metallic Materials Data Handbook* (AvP 932), the *Aerospace Structural Metals Handbook* (AFML-TR-G8-11) and *AGARD Handbook* Vol. 1, and these will be considered in turn. Reference should also be made to Chapter 5 on this metal.

Metallic Materials Data Handbook (AvP 932)

Aluminium and its alloys are dealt with in Section 6 and the scope is indicated on the guide card at the beginning of the section. The materials covered are pure aluminium and the following alloys: Al-Cu, Al-Mn, Al-Si, Al-Mg, Al-Mg-Si, and Al-Zn. The basic material conditions are designated in all cases and the forms in which the materials are used, i.e. whether as sheet, strip, plate, bar, tube, forging or casting, are also indicated. Where appropriate the alloys are given their International Alloy Designation (from the *International Alloy Designations and Chemical Composition Limits for Wrought Aluminium Alloys* as administered by the Aluminium Association Inc. of USA). This is followed by a checklist of specification numbers against which property data sheets have been issued and the date of issue and issue number of each sheet. The data sheets themselves carry a range of physical and mechanical property values for alloys in their favoured conditions and forms, the physical properties being usually thermal expansion characteristics, electrical and thermal conductivity, specific heat and density. The mechanical properties given are derived mainly from room temperature tests in tension, with the results having the status of A, B or S values, where A and B relate to statistical significance and confidence levels and S is the specified minimum of an appropriate material specification. Usually only tensile ultimate and proof stresses in a specified testing direction are available in sufficient numbers for A and B values to be derived from them. Each data sheet also carries a generalized statement on characteristic properties of the alloy such as weldability by a variety of techniques and resistance to corrosion, stress corrosion, and exfoliation corrosion.

Aerospace Structural Metals Handbook

Aluminium alloys are covered in Vol. 3 of the Handbook and are grouped according to whether they are casting alloys, wrought heat-treatable alloys, or wrought non-heat-treatable alloys. The information on individual materials is arranged in the order of general, physical and chemical properties, and mechanical properties, and the data is presented in both tabular and graphical form. The 'general' section deals with alloy composition, commercial designation, specification, and characteristics such as heat treatment, hardenability, and the forms and conditions in which the alloy is available. The basic specifications referred to in this section, and indeed throughout the Handbook are all of American origin, usually the AMS series but also US military and Federal specifications where appropriate. Physical properties included in the data sheets are typically density, specific heat, melting point or range, phase changes, thermal conductivity, thermal expansion coefficients, electrical resistivity, and magnetic characteristics. The emphasis in chemical properties is on corrosion and oxidation resistance. Mechanical properties data form the largest section of each data sheet and they cover room temperature and elevated temperature static strength properties, fatigue properties, creep and creep rupture properties, and elastic properties. The data presented is regarded in all cases as being representative of that obtainable from material in current commercial production. For each alloy there is advice on suitable forming procedures, machining, and welding, and at the end of each data sheet there is a list of references to open literature sources of additional and more detailed information.

AGARD Materials Properties Handbook

Aluminium and its alloys is covered fully in Vol. 1 of the Handbook. Information is presented on a National basis with a group of data sheets being allotted to each of the alloy producing countries. Alloys are identified against national specifications.

Information is presented under the headings of general, physical, static, and fatigue properties, and the mechanical properties are accorded the status of A, B. or C. values depending on their statistical significance. Physical properties given include density, specific heat, thermal conductivity and coefficients of thermal expansion, static properties cover room temperature strength, yield stress, ultimate shear stress, ultimate bearing stress, yield stress in bearing, modulus of elasticity in tension and compression, shear modulus, and ultimate elongation. Where appropriate there is also data on short-term properties at elevated temperatures, recovered properties after longer exposure to elevated temperatures and elevated temperature creep rates and creep rupture times. Under fatigue properties there are data from rotating bend tests, reversed flexure tests, repeated axial loading and alternating torsion tests. In total Vol. I contains data on 30 alloys of stated composition, the countries of origin being France – 3, Germany – 5, Italy – 4, Norway – 5, UK – 6, and USA – 7.

Titanium alloys

In addition to the three major materials data handbooks already mentioned there is an additional and substantial information source on titanium and its alloys, i.e. the *Titanium Alloys Handbook* prepared by the Metals and Ceramics Information Center (Battelle, Columbus Laboratories) under US Air Force sponsorship. This Handbook sets out to present a single comprehensive reference source on titanium alloys and is produced in a loose leaf form which permits updating as further information on existing and/or new alloys becomes available. Since much of the world's production of titanium alloys finds its way into airframe and aeroengine applications this Handbook can be viewed very much as a source book of materials data for the aerospace industry. The information presented is drawn from various US Government Agencies, the titanium producers, and airframe and engine companies, and in terms of quality it is regarded as being representative of that required of 'good engineering practice'. The majority of structural applications for titanium are currently met by about 10 alloy compositions, the most widely used being the Ti-6Al-4V material, and Section 1 of the Handbook addresses the metallurgy of these principal variants. For each alloy there is a separate section giving compositional details, information on microstructure and metallurgical nature, deformation practice and its effects on static and fatigue properties, heat treatment practices and effects, thermal and chemical stability and corrosion behaviour. Section 3 on machinability and formability is a substantial and very important section, and it covers in considerable detail such practical aspects as conventional machining and cutting, grinding and abrasive cutting, chemical milling and electrochemical machining, stretch forming, deep drawing, drop hammer forging, roll forming, tube forming, etc.

Section 4 of the Handbook is devoted to joining methods with a particular emphasis on welding (by inert gas, electron beam, and arc spot methods), resistance welding, brazing, adhesive bonding, and mechanical fastening. It is a generally informative section and it presents some design advice and a limited range of mechanical property data. The majority of the materials property data is located in Section 5 on 'Mechanical property data', and this covers pure titanium and 11 major alloy compositions. The property data is well assessed and is quoted on the basis of A, B and S values, where A represents a value above which at least 99 per cent of the population is expected to fall with a confidence of 95 per cent, B indicates a value above which at least 90 per cent of the population is expected to fall with a confidence of 95 per cent, and S is a value which meets the minimum quoted in the governing materials specification. The statistical significance of the S value is not known but it is considered to represent current production capability. For the well-established alloys the A and S values are often identical. This Section is a very good source of tabular and graphical information on mechanical property data for the various forms and products of relevance to aerospace application, and there is good coverage of fatigue properties and temperature effects.

Metallic Materials Data Handbook (AvP 932)

Property data on commercially pure titanium and 4 alloy compositions is covered in Section 12 of the Handbook. The alloy compositions have associated with them a number of products forms which are governed by UK and European (AECMA) specifications, and the Imperial Metal Industries (IMI) designation is also given. The product forms covered are sheet, strip, plate, bar, tube, forgings and castings, and in addition to room temperature physical and mechanical properties there is also reasonable coverage of the elevated temperature strength properties. Each data sheet carries general guidance on weldability and corrosion resistance.

Aerospace Structural Metals Handbook

The amount of information on titanium and its alloys is large and is to be found in Vol. 4 of the Handbook. The lay-out of data sheets is as for aluminium alloys, with general, physical and chemical, and mechanical property sections. For this group of materials there is a somewhat extended treatment of chemical properties covering resistance to acids, alkalis, organic and inorganic salts, stress corrosion, galvanic corrosion, hydrogen embrittlement, oxidation resistance, and sealing. Special mention is made of nuclear properties in the context of neutron capture cross-sections and the effects of irradiation on physical and mechanical properties. Fabrication aspects are, because of their importance, also given a full treatment in a section dealing with forming and casting, machining, welding, heating and heat treatment, and surface treatment.

AGARD Materials Properties Handbook

Data sheets on titanium alloys occupy part of Vol. III of the Handbook. The sheets are again grouped according to the national origin of the alloys and there are a total of 9 entries covering commercially pure titanium and its alloys, 2 from France, 3 from the UK and 4 from the USA. Property data are presented in tabular form and the statistical basis for test values is given.

Steels

The information on steels is arranged somewhat differently within the three major materials data handbooks; the UK handbook (AvP 932) sub-divides the materials into 'Corrosion resisting steels', which are dealt with in Section 10, and 'Non-corrosion resisting steels' which are covered by Section 11, whereas the AGARD Handbook treats steels as a class in Vol. II, but groups heat resisting alloys of iron along with cobalt and nickel base alloys in Vol. IV which is entitled 'Superalloys'. *The Aerospace Structural Metals Handbook* is in some ways more conveniently arranged in that it takes all steels together in Vols 1 and 2. There is of course a vast steels literature available, and the three handbooks referred to above are thus of particular value in giving direct access to materials data for alloys which have found successful application in the aerospace industry. Reference should also be made to Chapter 3 on steel.

Metallic Materials Data Handbook (AvP 932)

In the section on corrosion resisting steels (Section 10) five alloy compositions are covered and each is identified against a UK and European (AECMA) specification. Product forms encompass sheet, strip, bar, forgings, tube, and castings, and the data are given in tabular and graphical form. There is physical property data at normal and elevated temperatures and mechanical property data at room temperature on test bars and/or cut-up specimens as appropriate. The graphical information presents the variation with temperature of strength properties, elongation, and some physical properties such as coefficients of thermal expansion, thermal conductivity, electrical conductivity and specific heat.

Creep and creep rupture data are also given for the appropriate alloys, and test temperatures are usually in the range up to 100°C. The data sheets for all alloys carry broad comment on weldability and corrosion resistance.

Section 11 on non-corrosion resisting steels is a large one, encompassing 29 alloy compositions variously in the form of sheet, strip, bar, forging, tube, and casting. All the alloys are identified by approved UK specifications, and, where possible AECMA specifications. By and large the data are presented in tabular form but in a few instances there is graphical information indicating the continuous effect of increasing temperature on some mechanical and physical properties. All data sheets indicate the broad characteristics of the alloy in relation to weldability and corrosion resistance.

Aerospace Structural Materials Handbook

Vol. I deals with carbon and low alloy steels and ultra high strength steels, the former being a small section and the latter a particularly large one reflecting the relative importance of these alloys in aerospace applications. Vol. II covers Austenitic stainless steels, Martensitic stainless steels, age hardening steels, and nickel chromium steels.

The data sheets are arranged as for other non-ferrous alloys, with information on composition, general characteristics, physical and chemical properties, and mechanical properties. Additionally there is information and advice on fabrication, e.g. forming, casting, machining and welding, and the various aspects of heating, heat treatment and surface treatment are also addressed. For those alloys such as the nickel chromium steels, which are used primarily at elevated temperature, much of the property data is given as a function of temperature up to maximum use levels.

AGARD Materials Properties Handbook

Materials data on steels, other than those adjudged to be in the 'superalloy' class, are given in Vol. II. Sets of charts are presented for each distinct steel or group of steels, and information is given on the various forms and conditions of each alloy and on the chemical composition, with specified limits for each alloying element. Typically for the AGARD Handbook the alloys are grouped according to their nation of origin and against their national specifications. The volume covers the 42 alloys whose origins are as follows: France – 7, Germany – 9, Italy – 10, UK – 8, and USA – 8.

Each data sheet carried values for selected physical properties at normal and elevated temperatures, and static mechanical properties are quoted at room temperature for the various forms, heat treatment conditions and thicknesses of material. Fatigue data are presented in chart form and have been selected as being reliable data derived from tests under various loading conditions on test specimens which are described on each chart. Elevated temperature performance is covered by data on mechanical properties after various times of exposure to temperatures up to the maximum working temperature for the alloy in question.

Information on the heat resisting ferrous alloys is found in Vol. IV on 'Superalloys', and the data given are predictably biased towards the high temperature values for physical and mechanical properties with an emphasis on creep and creep/fatigue data. There is a considerable volume of data on stress range properties with charts giving properties under combined fatigue and creep loading conditions. There is also information on the maximum stress of a fatigue cycle required to cause tensile failure (or a specified creep deformation) in a specified time at various temperatures and for a specified frequency of loading for fatigue cycles of various stress ratios, all of which is presumably of considerable interest to the design engineer.

Magnesium alloys

The importance of magnesium alloys in aerospace engineering derives mainly from their use in the production of castings, and the data in the various handbooks is therefore derived in the main from cast products.

Metallic Materials Data Handbook (AvP 932)

The magnesium alloys are identified by UK specifications and also by AECMA and ASTM designations. The data sheets present information in tabular and graphical form, the tables covering a selection of physical properties at room and elevated temperatures and mechanical properties at room temperatures on bars and/or cut-up test specimens. The graphical information covers the influence of elevated temperatures on mechanical properties and long-term creep behaviour. Each data sheet carries compositional details for the alloy and gives broad guidance on weldability and resistance to corrosion. The product forms covered in all cases are castings, and the property data presented are derived from tests on sand cast and chill cast materials.

Aerospace Structural Materials Handbook

Information relating to magnesium alloys is found in Vol. 3 of the Handbook and it covers casting alloys, wrought heat-treatable alloys, and wrought non-heat-treatable alloys. There are brief statements regarding the melting and casting practices normally employed for the alloys, and the data sheets follow the usual format of a general section dealing with heat treatment, hardenability, forms and conditions available, and compositions. This is followed by sections on physical and chemical properties and mechanical properties. Corrosion resistance and paint protection attract special comment and there are references to other more detailed sources of information.

AGARD Materials Properties Handbook

Magnesium alloys are dealt with in Vol. III along with nickel and titanium alloys. The data sheets are again grouped into national sections, and the number of alloys covered is as follows: Canadian – 4, French – 8, German – 12, UK – 7 and USA – 10.

Property data are presented in tabular form and all data sheets giving mechanical test results carry a statement on the statistical basis of the information. A large proportion of the magnesium alloys are used in cast form and the types of tests included for castings necessitated the introduction of a classification of three types of test for use in generating static properties. These tests are distinguished by the letters a, b, and c on the charts, a indicating tests on special separately cast test specimens, b indicating tests to find the minimum properties of castings using specimens cut from various locations in production castings, and c indicating tests to find the average properties throughout castings using specimens cut from various locations in production castings.

Data sheets carry the usual details of alloy composition, physical properties such as density, specific heat, thermal conductivity, coefficient of thermal expansion, and emissivity.

Heat resisting alloys

The three major handbooks group the heat resisting alloys rather differently and only the AGARD publication deals with heat-resisting alloys of iron in the same section as the nickel and cobalt base alloys.

Metallic Materials Data Handbook (AvP 932)

Heat resisting alloys are dealt with in Section 8, and the materials covered are the nickel and cobalt base alloys in a variety of basic conditions. The majority of the information relates to nickel based alloys Nimonic 75, 80, 90 and C263. Each alloy is identified by an AECMA specification and also by any previous UK DTD (Military) specification, a heat resisting (HR) specification is also given for each alloy in its particular basic condition and form.

The data sheets cover the physical and mechanical properties of the alloys in the form of bars, extruded sections, forgings, sheet, plate and strip, and property data are presented in tabular and graphical form. Properties at elevated temperature are of particular importance in this section and the graphical information tends to show the continuous variations with temperature of properties such as coefficient of linear expansion, thermal conductivity, electrical conductivity, specific heat, strength, stiffness and elongation. Additionally there is a volume of graphical information on creep at elevated temperature and creep rupture properties. Temperatures covered are in the range up to 1,200°C as appropriate, and each data sheet advises on weldability and on corrosion resistance in broad terms.

Aerospace Structural Materials Handbook

Nickel base alloys containing less than 5 per cent cobalt are dealt with in Volume 4 of the Handbook while those alloys containing more than 5 per cent cobalt and the cobalt base alloys are included in Volume 5. The section on nickel-based alloys is extensive and it carries a large volume of data on static and fatigue properties with a special emphasis on the effects of exposure to high temperatures. Temperature effects feature most strongly on all data sheets and aspects of performance attracting particular attention are creep, creep-rupture, and fatigue. Under physical properties thermal effects are specially well covered. The section on cobalt base alloys is quite small and, because as with nickel based alloys they are regarded as high temperature alloys for aeroengine applications, there is the same emphasis on the effects of high temperature on physical and mechanical properties.

AGARD Materials Properties Handbook

Heat resisting alloys (superalloys) are dealt with in Vol. IV and into this category fall the heat resisting alloys of iron, nickel and cobalt. For each distinct alloy or group of alloys a set of charts is given which details for the various forms and conditions of the alloy information on chemical composition and physical and mechanical properties at normal and elevated temperatures. The origins of the alloys covered are as follows: French – 14, German – 7, UK – 9, and USA – 12. Each material section of the volume states the alloys covered and relates them to national specifications. The lay-out of the data charts takes account of the special emphasis on high temperature effects on physical and mechanical properties, and the performance of the alloys in creep and under combined creep and fatigue is given particular attention.

Other alloys

The UK and USA Handbooks deal with a few additional classes of alloy which find some application in the aerospace field. AvP 932 devotes a section (section 7) to copper alloys and gives limited physical and mechanical property data for pure copper, for aluminium bronze, nickel-silicon-copper alloy, phosphor bronze, nickel-silicon-aluminium brass, and 80/20 brass. The US Aerospace Structural Materials Handbook deals in Vol. 5 with some speciality metals such as beryllium, niobium, molybdenum and its alloys, tantalum and its alloys, commercially pure tungsten, commercially pure vanadium, and zirconium alloys. The largest section is that devoted to niobium (columbium) and for all these materials the emphasis tends to be on their special thermal, nuclear and chemical properties.

Composites

With the advent of carbon fibres of high strength and stiffness it became evident that composites in which such fibres were controllably and accurately aligned in suitable organic polymer matrices could be regarded as materials suitable for use in airframe construction. The first 15 years has witnessed a great deal of materials development, associated with both the fibres and the matrix polymers, and also design, fabrication and ground and flight testing of composites airframe components. Although the volume of carbon fibre/polymer matrix composite structure in production aircraft is still quite small nothing has arisen during the course of development and flight testing which seriously shakes the conviction that materials of this type will eventually make significant inroads into primary airframe structure. Up to the present it is carbon fibre/epoxy resin composites which have been most closely studied in research and assessed in greatest detail by aircraft design engineers. Their attractions and disadvantages as airframe materials have been fairly clearly identified and this has provided the incentive for materials suppliers to modify both fibre and matrix resin properties in order to give a more desirable overall balance of properties, performance, and processing characteristics.

The published literature on structural composites (based presently on carbon, glass, and aramid fibres) has built up over a number of years and is now very extensive. It is scattered through numerous learned journals, textbooks, and conference proceedings and there are already a few journals devoted exclusively to composites, e.g. *Journal of Reinforced Plastics and Composites*, *Journal of Composites*, and *Composites*. An accumulation of data suitable for use in design has clearly been taking place in most of the major airframe and aeroengine manufacturing companies in the Western world, but it is understandably too early to expect free and open access to such information. Property data published by the suppliers of carbon, aramid and glass fibre, and of matrix resins and resin impregnated fibres, much of which is aimed at the aircraft industry, have improved greatly in respect of scope and quality as the materials property requirements have become more clearly understood. Without doubt the materials suppliers at present provide, in aggregate, one of the richest open sources of physical and mechanical property data on materials of aircraft quality.

A major difficulty in attempting any sort of critical assessment of the published property and performance data on composite materials and trying to assemble that which will be of lasting value is that significant changes in the basic materials, i.e. fibres and matrix resins, are still taking place. Quality standards are also changing in response to an improved appreciation of how basic materials variations can influence composite properties and performance. Currently the effects of absorbed moisture when combined with elevated temperature (i.e. ›100°C) and the impact resistance and damage tolerance of carbon fibre/epoxy composites are matters of some considerable concern for the airframe designer. It is still far from certain what the finally preferred fibre/resin combination will be for future aerospace structural applications, and in the mean time it is unlikely that major materials design data handbooks of the kind available for airframe and aeroengine metals will be produced, simply because they would be of only transient value.

During the 1970s a few design-guide type publications were produced in the structural composites field, but they tended to be collections of specialist articles aimed at the engineering community at large rather than compilations of materials data in the traditional style of aircraft materials handbooks. Typical of this breed were *Designing with Fibre Reinforced Material* (Mechanical Engineering Publications Ltd of London and New York for the Institution of Mechanical Engineers in London, 1977) which covered both carbon fibre and glass fibre reinforced plastics in a series of specialist articles; *Composite Materials: Testing and Design* (an ASTM Special Technical Publication 617, published in 1976 by ASTM, 1916 Rose Street, Philadelphia, Pennsylvania 19103) which presented an excellent series of specialist papers which had been presented at an ASTM conference in May 1976; and *The Properties of Fibre Composites* (IPC Science and Technology Press Ltd) which brought together a series of specialist papers on the fundamentals of composites which had been presented at a conference held at the National Physical Laboratory, Teddington, England in 1971. There is, of course, a considerable amount of useful materials and property data in volumes such as these, and the ASTM publication 617 is noteworthy in this respect. It is sharply focused on testing and design and has an impressive breadth of coverage, with articles on fracture of carbon fibre/epoxy composites, crack growth in fibre reinforced composite materials, fatigue of fibre composite materials, impact performance and testing, laminate design, composite curing stresses, environmental effects including thermal cycling and moisture effects, and non-destructive testing of composites. This publication has been most recently updated by Specialist Technical Publications (STP) 787 printed in December 1982 and which presents the Proceedings of the 6th ASTM conference on composite materials testing and design. This conference was sponsored by the ASTM Committee D-30 on *High Modulus Fibres and Their Composites* and has sections dealing with test methods, materials characterization, fracture and failure analysis, fatigue, non-destructive testing, time dependent and dynamic response, and the testing of composite structures.

Overall the ASTM organization has contributed greatly to the publication of high quality information and materials data on advanced composites, and they have produced a steady stream of volumes through the late 1970s and early 1980s, e.g. *Composite Materials: Testing and Design* STP 674 (1979), *NDE and Flaw Criticality for Composite Materials* STP 696 (1979), *Commercial Opportunities for Advanced Composites* STP 704 (1980), *Test Methods and Design Allowables for Fibrous Composites* STP 734 (1981), *Joining of Composite Materials* STP 749 (1981), *Composites for Extreme Environments* STP 768 (1982), *Short Fibre Reinforced Composite Materials* STP 772 (1982), *Damage in Composite Materials* STP 775 (1982), *Composite Materials: Testing and Design* STP 787 (1982) and *Long Term Environmental Behaviour of Composites* STP 813 (1983). The materials data embodied in these publications are aimed at the engineering community at large, and hence although much of it will be relevant to aerospace applications the design engineer will have to exercise his judgement in sifting through it.

The AGARD organization has also taken a close interest in the development of structural composite materials and this has been reflected in its specialist meetings and the publication of conference proceedings. The material published through AGARD is of course of direct relevance to aerospace interests and it is usually the important practical and operational issues which are addressed. A good example of this is the 1983 publication on *Characterisation, Analysis and Significance of Defects in Composite Materials* (AGARD Conference Proceedings No. 355). The section on characterization and analysis of defects covers fractographic analysis of failures in CFRP, NDE techniques for composite laminates, monitoring of defect growth by acoustic emission, growth of delaminations under fatigue loading, experimental investigations of delaminations in carbon fibres composites, characterization of cumulative damage in composites during service, and an empirical appraisal of defects in composites. On the significance of defects the subjects covered are correlation between NDE results and the performance of carbon fibre/epoxy structural parts, the effects of damage on the tensile and compressive performance of carbon fibre laminates, the influence of fabrication defects on static and dynamic properties of carbon fibre/resin composite structures, advanced NDE techniques for composite primary structures, fracture mechanics of sub-laminate cracks in composite laminates, effects of defects on aircraft composite structures, etc. Publications such as this are good sources of references to work on composites directly appropriate to aerospace applications, and the data contained therein will have been generated from state-of-the-art aircraft quality materials.

Perhaps the nearest approach to a comprehensive materials data handbook on composites is that entitled *Handbook of Composites* (ed. George Lubin, sponsored by the Society of Plastics Engineers in the USA) (Van Nostrand Reinhold Company, New York, 1982). This Handbook contains information and substantial amounts of data on processes and materials, test methods, and design and analysis techniques, and its avowed aim is 'to communicate the state-of-the-art in composites technology to the designer'. Advanced composites using organic matrix materials are highlighted and there is a major emphasis on presenting well validated and useful data and simplifying the understanding of composites.

The Handbook is arranged in chapters on materials, processes, design and analysis, and typical applications including aerospace, and there are appendices giving materials data in tabular and graphical form. Materials data are concentrated in Chapter 1 on new materials, which cover both fibres and matrix resins, and in appendices A, B. and C. Appendix A gives typical physical, mechanical and thermal properties of fibre reinforced composites and covers fibre glass, carbon fibre, quartz, and boron fibres and a range of resin matrices including epoxy, polyester, polyimide, phenolic and silicones. Appendix B. concentrates on carbon fibre composites and gives typical properties for materials based on a wide range of commercially available woven and non-woven fibres. Appendix C. gives typical property data for commercially produced glass fibre/thermoplastic matrix composites. The treatment of the major fibre composites based on carbon, aramid, and boron is extremely good and overall the Handbook represents an exceptionally good starting point for a consideration of what advanced composites have to offer the design engineer. The Handbook is well endowed with references to other sources of composites information.

Specialist textbooks on structural composites abound, but none can be viewed as materials data handbooks, and, indeed, that is not their aim. These specialist volumes aim to transmit a knowledge and understanding of the mechanics and physics of composite materials and show how their properties and performance can be placed on a firm theoretical basis. Typical of such books is *Load Bearing Fibre Composites* by M. R. Piggott (Pergamon Press), which deals in depth with the mechanics of composites, reinforcement processes and failure processes. Another book of this type but which has a particularly strong engineering design bias is *Composite Structures*, ed. I. H. Marshall (Applied Science Publishers, London), which presents the proceedings of the 1st International Conference on Composite Structures held at Paisley College of Technology, Scotland in 1981. This volume deals with structural analysis; modelling techniques, structural evaluation techniques, design, fracture and failure analysis, finite element studies, physical and mechanical characteristics, and environmental effects.

Undoubtedly aerospace requirements are driving much of current research in advanced structural composites, and consequently much of the published output in specialist composite journals, in specialist textbooks, and in composites conference proceedings will be likely to contain information and materials data of direct interest and value to the aircraft design engineer. Eventually data handbooks on aerospace quality composite materials will emerge to match those already in existence for metallic materials, but this is unlikely to happen until the selection of component materials is better resolved and is supported by significant experience in manufacture and usage of advanced structural composites on production aircraft.

Two other recent titles of interest are *The Engineer's Guide to Composite Materials* and *Advanced Composites*. *The Engineer's Guide to Composite Materials*, ed. J. W. Weeton (ASM, 1986) looks specifically at metal, polymer, ceramic, glass, and carbon/graphite matrix materials. There is a useful bibliography of composites reference books as well as lists of consultants, laboratories, manufacturers, suppliers and trade associations. *Advanced Composites* is an ASM Conference Book published in co-operation with the Engineering Society of Detroit derived from the Proceedings of the 1985 Advanced Composites Conference held in Dearborn, Michigan in which materials in the automotive, aerospace and defence industries were reviewed. As well as describing new materials, the use of CAE for the design and development of composite laminate structures is discussed.

Non-metallics

Non-metallic materials are used in aerospace applications to fulfil a whole range of special functions which are beyond the scope of metals. These materials are primarily organic polymer based and fall into a number of broad categories, the major ones being elastomers, adhesives, and thermoplastics. The usage to which these materials are put in aircraft construction has much in common with their usage in other areas of engineering application, and for that reason no distinctly separate aerospace materials database has been developed. Where particularly stringent quality requirements arise and where the aircraft environment presents an especially difficult combination of operating conditions these will be covered by an appropriate national or international aerospace material specification. It follows that the nearest approach to the definition of an aerospace material database is that which will be found in the aerospace series specifications which are generated around the world. These specifications will state property requirements, minimum property levels, and variability in properties which are seen to be necessary in aerospace quality materials.

In the UK the authoritative national specifications are to be found in the British Standards Aerospace Series. These are drawn up by committees composed of representatives from industry, government establishments, and the Institution. A further series of specifications of importance produced in the UK is that issued by the Ministry of Defence through the Royal Aircraft Establishment, Farnborough, Hants. This is known as the DTD series and it covers materials and processes for which more than one supplier could tender and for which there is a reasonably constant demand for significant quantities. Generally the materials covered by this series are not so widely used as those covered by the BS Aerospace series, and once usage is established at an appropriate scale and for an appropriate length of time a conversion to the latter series will be made. An index of valid DTD specifications is published periodically and can be purchased from HMSO Government Bookshops in London, Edinburgh, Cardiff, Manchester, Bristol, Birmingham and Belfast. The HMSO also produce daily and monthly lists of government publications in which new issues in this series are listed. The most recent index is the 48th edn (1980), and it covers specifications for non-metallic materials in the following categories: fabrics and cordage, glass, lubricants and hydraulic fluids and ingredients, organic corrosion preventatives and ingredients, paints and dopes and ingredients, plastics, rubbers, and process specifications. Specifications marked as out-of-print in the index can be obtained from the Department of Materials and Structures of the Royal Aircraft Establishment, Farnborough, Hants.

European activity in the specification of non-metallic materials is presently focused in AECMA which is supported by the European consortium of aircraft constructors. AECMA is engaged in developing specifications for use on a European scale, and such specifications when issued will be called up in preference to national specifications. Non-metallics are dealt with through a committee (C7) which has separate expert sub-committees considering and agreeing requirements for elastomers, transparent and thermoplastic materials, structural adhesives, surface protection, and reinforced plastics. An important feature of the standards generated by AECMA is that they must define interchangeable European materials, i.e. materials which will be used by several European manufacturers or produced by at least one and preferably more than one European manufacturer. Furthermore the standards must be able to be used without seriously upsetting the certification rules or National Standards in use in aerospace construction. Harmonization of European and United States materials specifications (MIL specs) is a matter of particular importance in order to facilitate and simplify the maintenance of aircraft built in the United States and operated by European countries. Such activities are supported, for example, by the NATO Committee AC/82 (Group of Experts for the Conversion of Dimensions, Drawings and Materials Specifications for US aircraft) and its sub-comittees AC/82/1 and AC.82.2, the latter dealing with non-metallic materials.

As for composite materials the most substantial and detailed materials property database on non-metallics will be that produced by the materials suppliers for their specialist customers in the field of general engineering application. The properties of non-metallics are strongly influenced by ingredients other than the base polymer, which are introduced during formulation, and the major materials suppliers tend to cover the widest possible range of compounding options in presenting their most versatile and favourable range of products to the consumer market. Numerous examples of the scale and excellence of this materials suppliers database are to be found in the field of speciality synthetic elastomers and sealants. Speciality elastomers and sealants of prime interest to the aerospace industry include the nitrile rubbers, fluoroelastomers, silicon elastomers and epichlorohydrin rubbers, and each of these categories is covered admirably in terms of materials formulation, processing and property data by periodic publications emanating from the materials suppliers. The commercially most significant suppliers of the various categories of non-metallics can be identified by reference to directories or almanacks such as *Sells Directory* (which through its products section will allow identification of suppliers against materials categories) or the *Plastics Industry Directory* (a joint publication by Maclaren Publishers Ltd and the British Plastics Federation) which identifies companies against material type and product ranges. Once the relevant company has been identified an excellent source of addresses is that provided by *Kelly's Manufacturers and Merchants Directory*.

There is, of course, a huge published literature on the main classes of non-metallic materials, with innumerable specialist journals, textbooks, reviews and conference proceedings. There are a number of good publications seeking to give general guidance on materials selection and design to the engineering community, examples being the *Materials Selector and Design Guide* (Design Engineering, Morgan-Grampian (Publishers) Ltd, 1974), *Design Engineering Handbook* (Product Journals Ltd), and *Thermoplastics: Properties and Design*, which is a compilation by members of the ICI Plastics Division, edited by R. M. Ogorkiewicz. The first of these covers thermoplastics, thermosetting resins, composites, elastomers, adhesives, coatings and insulating materials and includes useful property data and guidance on limitations to performance. Much of the information has perforce to be of a fairly generalized nature and reference will have to be made to more specialist texts to get the detail required to make a final judgement on materials selection for a particular end use. The second publication (2nd edn 1968), deals with filled and unfilled thermoplastics and thermosetting resins and is a good initial guide to the selection and processing of plastics materials for engineering applications. The handbook is composed of contributions from industrial specialists in the main classes of materials, i.e. acrylics, epoxies, fluorocarbons, polyimides, polyamides, etc. Each contribution indicates the special characteristics, practical advantages and uses, important design considerations, and processing and fabrication features of the materials, and makes a tabular presentation of typical mechanical, thermal, electrical and chemical properties. The final publication mentioned gives good background on the properties and behaviour of thermoplastics relevant to their use as engineering materials, with a particularly informative treatment of short-term strength and impact behaviour, general mechanical properties, electrical, thermal and chemical properties, and design principles and processing methods.

The field of elastomers is particularly well served by a number of technical journals covering both science and technology, and the publication *Rubber Chemistry and Technology* (produced by the Division of Rubber Chemistry of the American Chemical Society Inc.) is worthy of special mention. This publication aims to make available 'in convenient form under one cover important and permanently valuable papers on fundamental research, technical developments, and chemical engineering problems relating to rubber and its allied substances'. There are five publications per annum and they include papers and informative review articles on a very selective basis. Representative of the excellent articles on elastomers of particular interest for aerospace application are those on 'Fluoroelastomers' and on 'Polysulphides'. The necessary background information on rubber types and their particular characteristics, both mechanical and chemical, is best obtained from the many textbooks written on the subject of elastomer science and technology.

For the design engineer books with a strong applications orientation are of particular value, and one such volume is *Rubber Technology and Manufacture*, ed. C. M. Blow (Butterworths for the Institution of the Rubber Industry). This book keeps the chemistry content to a minimum consistent with clear presentation, and is very application and end use oriented. It aims to provide an up-to-date guide for, among others, commercial users of rubber products. Chapter 4 on 'Raw polymeric materials' gives excellent coverage of natural and synthetic rubbers, the latter being grouped according to whether they are regarded as general purpose non-oil resisting materials, special purpose materials, e.g. neoprenes, nitriles, fluorocarbons, silicones or other speciality rubbers, e.g. polyurethanes, polysulphides, epichlorohydrins. An excellent feature of this book is that it identifies the main commercial sources of all the polymeric materials discussed, and this in turn, harking back to a point made earlier, will identify the material supplier who is likely to be the custodian of the most detailed database relating to his specialized products.

Finally, on elastomers, mention should be made of the volume *Use of Rubber in Engineering*, Allen, Findlay and Payne (eds) (Maclaren and Son Ltd, London, 1966). This contains an authoritative series of papers on those aspects of elastomeric behaviours which are of prime importance in design, i.e. dynamic properties, time and temperature effects, swelling in fluids, ozone attack and fatigue, seal behaviour, bearing behaviour, and sound and vibration insulation. The volume contains much valuable materials property data, and is as relevant today as when it was written almost twenty years ago.

Adhesives, and particularly structural adhesives, i.e. adhesives which can be used in the manufacture of primary load bearing structures, are of particular interest for aerospace applications because of the design, performance and manufacturing advantages which they appear to offer. The subject of adhesion and adhesives is sufficiently important generally to have spawned a few specialist learned journals of its own, e.g. *Adhesive Age*, and *Journal of Adhesives*, and also to have led to the publication of a limited number of specialist textbooks and review articles. The *Materials Selector and Design Guide* referred to earlier in this section devotes a chapter to the subject and gives good background on such matters as joint design, choice of adhesive for particular applications, methods of application, compatibility with adherends, chemical resistance, and service temperature ranges. All the main types of adhesive, i.e. elastomeric, thermoplastic and thermosetting, are covered.

Of more direct interest in respect of aerospace applications are the more specialized review articles which concentrate attention on systems suitable for aircraft usage; one such recent review (presented at a conference on The Joining of Metals: Practice and Performance at the University of Warwick in April 1981) is *The Use of Adhesives in Aircraft Structures* by P. Poole of the Royal Aircraft Establishment, Farnborough. This considers the extent to which adhesive bonding is used in airframe structures and the advantages which it offers over mechanical methods of joining. There is discussion of how the important bonding parameters, i.e. adherend surface preparation and adhesive type, influence bond strength and durability. There is little in the way of property data but a good deal of informed comment on in-service performance, durability and mechanisms of deterioration in adhesive bonds. The AGARD organization has also given some attention to adhesive bonding and its lecture series number 102 was devoted to *Bonded Joints and Preparation for Bonding*. These lectures were published in 1979 and they present an authoritative view of the whole subject with a good range of references to more detailed sources of information, the subject areas covered being operational experience with adhesive bonded joints, interfacial fracture mechanical aspects of adhesive bonded joints, behaviour of adhesive bonded joints under cyclic loading, failure in adhesively bonded structures, adhesion mechanisms and the influence of surface treatment on the behaviour of bonded joints, surface preparation – the key to bond durability, and non-destructive inspection of adhesive bonded joints.

The range of non-metallic materials of potential use in aerospace engineering is extremely large, and the lack of a discrete aerospace materials database raises the spectre of an enormously time-consuming search through the published literature in order to find materials data of the appropriate kind. The aerospace materials specifications series provide a useful basis for the search, and, although pointers have been given as to how further to pursue the search, it might eventually be more cost-effective to engage a professional information-gathering organization to carry out the data-retrieval task once the category of material for the application has been identified. One such organization operating in the non-metallic materials field is the Rubber and Plastics Research Association (RAPRA, Shawbury, Shrewsbury, England), and the scope of their database activity and the retrieval services offered have already been referred to in the first section of this chapter.

CHAPTER SEVENTEEN

Metals in construction

B. A. HICKS

Introduction

Before the nineteenth century, metals played a very small part in building structures except in such things as Greek and Roman equivalents of wall ties – bronze cramps to join blocks of stone or wrought-iron chains and rods used in Renaissance times to counter the thrust of masonry vaults. The earliest iron bridge was built over the Severn River in England in the eighteenth century and is still in use. Earlier iron-framed buildings were expensive to build in wrought iron and unpredictable when cast iron was used. Thanks to Henry Bessemer, the production of good cheap steel in the 1850s paved the way for the building of the first skyscrapers which began to dominate American town skylines by the 1880s. Since then the usage has grown apace – not only in steel but also in non-ferrous metals – particularly aluminium, copper, lead and zinc. A comprehensive volume which deals with materials at some length and serves as a good introduction to the subject is *Fundamentals of Building Construction: Materials and Methods* by Edward Allen (John Wiley, 1985).

With metals permeating the whole of construction, both as structural materials and in individual components, the extent of information is vast and it is a difficult task to separate the literature into that which is relevant to metallic materials and that which is not. Many worthwhile publications have, therefore, been omitted. The intention is to give a flavour of what is available and hopefully guidance and direction straight to particular information.

Organizations

The construction industry is supported by a wealth of professional organizations, trade and research associations and bodies representing standards and consumers interests. Many of these provide information and/or advisory services and the list below gives the reader some indication of the breadth and depth of the information provision in this field.

Of the many organizations concerned generally with the construction industry, of particular importance are CIRIA, the Construction Industry Research and Information Centre, whose activities are reflected in its name; the Building Research Establishment, important for its advisory services and publications; and the Building Centre Group, with its permanent exhibitions of building products, the Barbour Builder Building Centre Enquiry Service and the information services and provision of trade literature from London and the regional offices.

The National Council of Building Material Producers (BMP) unites companies, trade associations and federations. The Vitreous Enamel Development Council promotes the use of enamelled metals.

Technical expertise and/or commercial promotion of the use of steel in construction are in the hands of the British Constructional Steelwork Association; the Steel Construction Institute, previously CONSTRADO; the Steel Castings Research and Trade Association; and several services offered by British Steel – the Stainless Steel Advisory Centre, the Steel Sheet Information Centre (providing information on coated and uncoated products), a British Steel Sections Structural Advisory Service and British Steel Tubes, particularly the Structural Hollow Section advisory service.

Non-ferrous interests are upheld by the Zinc, Lead and Copper Development Associations, the National Brassfoundry Association, the Aluminium Federation and the Architectural Aluminium Association.

The multitude of metal building components is reflected by the variety of organizations concerned with components and particular structures. Major examples are the Farm Buildings Information Centre; Cold Rolled Sections Association; Building Services Research and Information Association, Copper Cylinder and Boiler Manufacturers; Metal Roof Deck Association (now incorporated in the Flat Roofing Contractors Advisory Board); National Federation of Roofing Contractors; Suspended Ceilings Association; Steel and Aluminium Window Associations; Door and Shutter Manufacturers Association; Partitioning Industry Association; International Truss Plate Association; Steel Lintel

Manufacturers Association; Construction Fixings Association; Guild of Architectural Ironmongers; and the Ductile Iron Pipe Committee.

Overseas organizations are best represented by the American and Canadian Institutes of Steel Construction; Canadian Sheet Steel Building Institute; and the (North American) Metal Building Manufacturers Association. In Europe there are the Office Technique pour l'Utilisation de l'Acier, Paris; Stichting Staalcentrum Nederland, Rotterdam; Centre Belgo-Luxembourgeois d'Information de l'Acier, Brussels; Beratungstelle für Stahlverwendung, Düsseldorf; Swedish Institute of Steel Construction, Stockholm; Centre Suisse de la Construction Métallique, Zurich; and the Osterreicher Stahlverband, Vienna.

The following are organizations concerned with construction (addresses will be found in the Appendix).

Aluminium Federation
Aluminium Window Association
American Institute of Steel Construction, Inc.
Architectural Aluminium Association
Beratungstelle für Stahlverwendung
British Constructional Steelwork Association Ltd
British Steel Sections and Commercial Steels Structural Advisory Service
British Steel Tubes. Structural Hollow Sections Advisory Service
Building Centre Group
Building Research Establishment (BRE)
Building Services Research and Information Association (BSIRA)
Canadian Institute of Steel Construction
Canadian Sheet Steel Building Institute
Centre Belgo-Luxembourgeois de l'Information de l'Acier
Centre Suisse de la Construction Métallique
Cold Rolled Sections Association
Construction Fixings Association
Construction Industry Research and Information Association (CIRIA)
Copper Cylinder and Boiler Manufacturers
Copper Development Association (CDA)
Door and Shutter Manufacturers' Association
Ductile Iron Pipe Committee
Farm Buildings Information Centre
Flat Roofing Contractors Advisory Board
Guild of Architectural Ironmongers
International Truss Plate Association
Lead Development Association (LDA)
Metal Building Manufacturers Association
National Brassfoundry Association
National Council of Building Materials Producers
National Federation of Roofing Contractors
Office Technique pour l'Utilisation de l'Acier
Osterreicher Stahlverband
Partitioning Industry Association
Stainless Steel Advisory Centre
Steel Castings Research and Trade Association (SCRATA)

Steel Construction Institute
Steel Lintel Manufacturers' Association
Steel Sheet Information Centre, British Steel Technical
Steel Window Association
Stichting Staalcentrum Nederland
Suspended Ceilings Association
Swedish Institute of Steel Constructions
Syndicat de la Construction Métallique de France
Vitreous Enamel Development Council Ltd
Zinc Development Association

Indexing and abstracting publications and online databases

Technical information

Although covering a broader field, sources with material on metals in construction include *Engineering Index* (Compendex as an online database); *Metals Abstracts* (Metadex online); *World Aluminum Abstracts* (section 8.1 deals with building and construction) which is also available online; *Zincscan* and *Leadscan Abstracts* (also available online); *Copper Abstracts*; and for non-ferrous metals generally, *BNF Metals Abstracts*, which is available as an online database up to 1983 but is not being added to. *Current Technology Index* is useful for its very specific indexing, while *Applied Science and Technology Index* covers a wider (English language) scene.

Those specifically in the construction sector include RIBA's *Architectural Periodicals Index*; *Current Information in the Construction Industry*, issued every fortnight by the Property Services Agency, cumulated every six months into Construction References and forming the basis for PICA, the PSA online database; IBSEDEX, the BSIRA database on building services from their bi-monthly *International Building Services Abstracts*. *International Building Science and Construction Abstracts* and *International Structural Engineering Abstracts* are extracted from *International Civil Engineering Abstracts*, available in published and compact-disc form. BRIX, 100,000 references strong and yet to include the pre-1970 material, is the Building Research Establishment's online database. British Standards on metal building components and metal designations used in them can be selected on BSI Standardline.

Commercial information

Predicast have various hard copy and database services allowing the selection of company, product, marketing and statistical information, only one of which, the monthly *PrediBrief on*

Building Products (containing a section on metal building products), has abstracts specific to the building sector, but, on the others, the use of standard industry classifications allows the isolation of industrial sectors or particular metals. The particular services are: *PROMPT* (Predicast's Overview of Markets and Technology); *F & S. Index* (United States, Europe, and International versions in published form); *Forecasts*; *Time Series*; and *New Product Announcements*.

In the UK, *Research Index*, now also available as a database, allows selection by company and industry. *Metals Abstracts' Steels Alert* carries some items on product applications including some in the construction industry.

Publications

Applied Science and Technology Index (monthly, quarterly and annual cumulations, H. W. Wilson).
Architectural Periodicals Index (quarterly, RIBA Publications).
BNF Metals Abstracts (monthly, BNF Metals Technology Centre).
Cadmium Abstracts (quarterly, Cadmium Association).
Construction References (six-monthly cumulation of Current Information in the Construction Industry, PSA Library).
Copper Abstracts (quarterly, Copper Development Association).
Current Technology Index (monthly, annual cumulation, Library Association).
Engineering Index (monthly, annual cumulation, Engineering Index).
International Building Science and Construction Abstracts (quarterly, CITIS, Dublin and New York).
International Building Services Abstracts (bi-monthly, Building Services Research and Information Ass.).
International Structural Engineering Abstracts (quarterly, CITIS, Dublin and New York).
Leadscan (quarterly, Lead Development Association).
Metals Abstracts (monthly, annual cumulation, Metals Information).
PrediBrief Building Products (monthly, Predicast).
Predicast F. & S. Index. Europe (monthly, quarterly and annual cumulations, Predicast).
Predicast F. & S. Index. International (monthly, quarterly and annual cumulations, Predicast).
Predicast F. & S. Index. United States (monthly, quarterly and annual cumulations, Predicast).
Predicast Prompt (monthly, Predicast).
Research Index (fortnightly, Business Surveys Ltd.).
World Aluminium Abstracts (monthly, Metals Information).
Zincscan (quarterly, Zinc Development Association).

Databases

BLAISE-Line – British Library.
BRIX (on ESA-IRS) – BRE.
Compendex (on ESA-IRS, DIALOG, Orbit, BRS, Data-star, PFDS (Pergamon Financial Data Services): likely to change to ORBIT, INKA, and FIZ-Technik) – Engineering Index.

IBSEDEX (on ESA/IRS) – BSIRA.
METADEX (on ESA/IRS, DIALOG, Pergamon, Orbit and INKA) – Metals Information.
Non-ferrous Metals Abstracts Data Base (on DIALOG (1961–83)) – BNF Metals Technology Centre.
PICA (on PFDS: likely to change to ORBIT) – PSA.
World Aluminium Abstracts Data Base (on ESA/IRS, DIALOG) – Metals Information.

Bibliographies and information guides

For tracking down books, Blaise-Line, the British Library online service is useful, but several publications are established as guides to construction information and relevant to our purpose. *Building Information*, published annually in association with the Building Industry Convention, covers organizations, journals, recent books and forthcoming exhibitions. *Architects' Journal*, as a supplement to its first issue of the year, issues its wide-ranging Information Guide. The Construction Industry Research and Information Association's *Guide to the Sources of Construction Information* covers relevant trade associations, research bodies and advisory services. The annual *International Directory of Published Market Research*, compiled by the British Overseas Trade Board in association with Arlington Management Publications, includes details of market reports on metal structures and building components. *Information Sources in Architecture* (Bradfield) has items relevant to seekers of metals information.

Publication lists are available from several organizations, notably the catalogue of the Building Bookshop, Steel Construction Institute (Publications), Property Services Agency (*PSA in Print*) and the Building Research Establishment (*Information Directory*). RIBA Data, the packaged information service, and *Specification*, the annual reference work, both cite references to further information.

Architects' Journal Information Guide (annually; supplement to the first issue each year to *Architects' Journal*, Architects Press Ltd).
Bradfield, V, *Information Sources in Architecture* (Butterworth, 1983).
British Overseas Trade Board, *International Directory of Published Market Research* (annual, BOTB: Arlington Management Publications Ltd).
Building Bookshop Catalogue (annually, Building Bookshop).
Building Information (annually, A4 Publications).
Building Research Establishment Information Directory (annually, BRE).
Construction Industry Research and Information Association, *Guide to Sources of Construction Information*, 4th edn (CIRIA, 1984).
PSA in Print (annually, PSA).
RIBA DATA (RIBA Services Ltd).
Specification (annually, Architectural Press).
Steel Construction Institute Publications (annually, Steel Construction Institute).

Directories and reference works

Price and costing information on metals and metal building components are published annually in the UK in three well-established price books – Spon's *Architects' and Builders' Price Book*; Laxton's *Building Price Book* and Griffith's *Building Pricebook* – and more recently in the *Comprehensive Building Price Book*, which claims the advantage of a greatly increased number of built-up and basic material prices.

For services, there is Spon's *Mechanical and Electrical Services Price Book*, while civil engineering is covered by Spon's *Civil Engineering Price Book*.

UBM includes actual prices in its annual catalogue (and on Builders' Viewdata).

Company and product information, including trade names, are found in *Where to Buy Building Construction and Maintenance Plant Supplies and Services*; *Sell's Building Index*; and their *Directory of Products and Services*; *Technical Indexes Construction and Civil Engineering* product data book; the *RIBA Product Selector* (both these two available separately from their parent package services); the *Barbour Compendium of Building Products*; and *Roofing, Cladding and Insulation Directory and Buyers Guide*. *Specification* contains such information but has more descriptive data on products and is valuable for its information on the whole range of building practices, especially those covering the use of metals.

All works indicated below are published annually unless stated otherwise:

Barbour Compendium of Building Products (Barbour-Builder).
Comprehensive Building Price Book (Wessex (Electronic) Publishing Ltd).
Griffiths Building Pricebook (Barton Publishers).
Laxton's Building Price Book (Thomas Skinner Directories).
RIBA Product Selector (RIBA Services).
Roofing, Cladding and Insulation Directory and Buyers Guide (Patey Doyle).
Sell's Building Index (Sell's Publications).
Sell's Directory of Products and Services (Sell's Publications).
Specification (Architectural Press).
Spon's Architects' and Builders' Price Book (updated quarterly, Associated Book Publishers).
Spon's Mechanical and Electrical Services Price Book (Associated Book Publishers).
Technical Indexes Construction and Civil Engineering Product Data Book (issued quarterly, Technical Indexes Ltd).
Where to Buy Building Construction and Maintenance Plant Supplies and Services (Where to Buy Ltd).

Packaged information services

Under this heading, sources are considered which go beyond the simple directory or yearbook to become miniature libraries or databases in their own right.

On microfiche are systems from *Barbour Index* and *Technical Indexes*. Barbour has two files, for products and for technical information. Barbour Product Microfile has about 30,000 pages of trade literature, fully indexed and updated annually. Because of more frequent changes in standards, regulations, technical reports and government publications, the Barbour Technical File is updated three times a year. Also on the large scale, covering more than 6000 manufacturers, is Technical Indexes' Construction and Civil Engineering Microfile. Access is by product group, trade name, or manufacturer. As with the other sections, the one on materials can be obtained separately, but, to the seeker of materials information, the sections on external and primary structural elements; secondary elements; services; and regulatory and guidance documents, are also relevant. The system is updated every three months. A joint RIBA-Technical Indexes Technical Information Microfile is also now available.

The long established Barbour Library is a bespoke collection of as-published catalogues to suit the subscriber's requirements, indexed, checked and updated annually. The RIBA Office Library Service is on similar lines. Of smaller scope is the annual Architects' Standard Catalogue Mini-File, which consists of four volumes of reproduced literature from some 600 manufacturers. There is some technical content and a list of British Standards. It is possible to obtain free copies.

RIBA Data offers A4 size trade literature in loose-leaf binders with suitable indexes, practice data sheets giving guidance to technical information, and information on legislation, standards and design.

Architects Standard Catalogue Mini-file (Standard Catalogue Information Services Ltd).
Barbour Library; Barbour Index Ltd; Barbour Technical Microfile; and Barbour Product Microfile (all from Barbour Microfiles).
RIBA-TI Technical Information Microfile; RIBA Data; and RIBA Office Library Services (RIBA Services Ltd).
Technical Indexes Construction and Civil Engineering Microfile (updated quarterly, Technical Indexes Ltd).

Viewdata

Architect's Viewdata uses the Mistel viewdata network to mount its company, product, and other information, including a structural design program. A description and example of the service can be seen on page 3511 of Prestel.

Builder's Viewdata, also from Context, is useful for price information for components as it incorporates prices of up to 25,000 UBM building products and the Spons price book updates.

Building Research Establishment Information is on page 50033 of Prestel. Tradelink gives access to product information by means of the product name or the subject index.

Architects' Viewdata (part of the CONTEXT service, AVS Intext).
Builders Viewdata (part of the CONTEXT service, AVS Intext).
Prestel (British Telecom).
Tradelink (video-based information service, Tradelink Communications Ltd).

Journals

Some journals are more or less central to the coverage of metals information in the construction industry. In this category fall *Metal Building Review*; *Steel Construction Today*; *CIDECT*; *Tubular Structures*; *Roofing, Cladding and Insulation*; and the *Journal of Constructional Steel Research*.

Of the general building or architectural journals, however, or even those covering specific sectors, most of the major publications are useful for the occasional article. Included here would be *Building*; *Architects' Journal*; *Architectural Design*; *Architectural Review*; *RIBA Journal*; *Building Today*; *Building Technical File*; *Roofing Contractor*; *Building Technology and Management*; *Civil Engineering*; and *Heating and Air Conditioning Journal*. *Aluminium* and *Modern Metals* are metallurgical journals featuring construction items amongst others.

Products

Journals specializing in new product information are *Building Products* and *What's New in Building*. Useful new product coverage is also included in *Building*, *Architects' Journal*, *RIBA Journal*, *Building Today*, *Building Design*, *Hardware Trades Journal*, and *Heating and Ventilating Review*. *Modern Metals*, *Metal Building Review*, *Building*, *Architects' Journal* and *RIBA Journal* also have articles or surveys on specific components or applications while *Building Technical File* includes reports of tests carried out on systems and products.

Of particular importance is the assessment and certification of products carried out by the British Board of Agrément and reported in the *BBA Publications Monthly*.

Prices

Building Today and *Building* publish regular pricing supplements. As well as its annual volume, the Building Cost Information Service issues the *Quarterly Review of Building Prices* based on an exchange of building cost information between members. The Property Services Agency is responsible for compiling the *Monthly Bulletin Construction Indices for Use with National Economic Development Office Price Adjustment Formulae*. Indices cover cast iron products, reinforcing steel and metal sections and structural steel.

Addresses for little-known publications are included below. Most will be readily found in Ulrich, however.

Aluminium: International Journal of Science, Technology, Economics (English language edn).
Architects' Journal.
Architectural Design.
Architectural Review.
British Board of Agrément, *BBA Publications Monthly* (monthly, BBA, PO Box 195, Bucknalls Lane, Garston, Watford, Herts WD2 7NG).
BSRIA Statistics Bulletin (quarterly, Building Seervices Research and Information Ass., Old Bracknell Lane, Bracknell, Berks. (0344 426511)).
Building.
Building Design.
Building Market Report.
Building Products.
Building Technical File.
Building Technology and Management.
Building Today.
CIDECT (Comité International pour le Développement et l'Etude de la Construction Tubulaire) (irregular, Bureau Technique CIDECT, 5 Ave Maurice Ravel, 92 300 Levallois-Perret, France. Available in the UK from British Steel, Tubes Sales Office, Corby NN17 1VA.).
Civil Engineering.
Hardware Trades Journal.
Heating and Air Conditioning Journal.
Heating and Ventilating Review.
Journal of Constructional Steel Research.
Metal Building Review.
Modern Metals.
Monthly Bulletin Construction Indices for Use with National Economic Development Office Price Adjustment Formula: Civil Engineering Works (HMSO).
RIBA Journal.
Roofing, Cladding and Insulation Journal (Wilmington House, Church Hill, Wilmington, Dartford, DA2 7EF).
Roofing Contractor (108 Epsom Lane South, Tadworth, Surrey, KT20 5TB).
Roofing/Siding/Insulation (757 Third Avenue, New York, New York 10017, USA).
Steel Construction Today.
Tubular Structures (irregular, British Steel, Tubes, Corby, Northamptonshire, NN17 1VA).
What's New in Building.

Standard specifications and regulations

Because of the large number of components made of or incorporating metal in some manner and the use of metal as a basic structural material, there are a large number of standards, codes of practice and guides to good practice relevant to this section. Concentrating, therefore, on the major ones we find *BSI Sectional List 16* covering the building standards and *BSI Handbook No. 3* as a useful compendium of abridgements.

The annual reference work, *Specification*, incorporates data and guidance from standards in the text and lists them in its bibliographies – around 150 of them for example in the Metalwork section. The Building Regulations make reference to British Standards and these are summarized in *BS Handbook 20*.

Ignoring the general metal specifications, such as BS 1470 on aluminium for engineering purposes or BS 2989 on galvanized steel sheet, which are, however, still relevant, there are standards covering the use of metals in structures. Foremost among these are BS 5950 and BS 449, structural steel in building, BS 4 and BS 4848, hot rolled structural, and BS 2994, cold rolled steel sections. BS 3083 is for corrugated galvanized and aluminium-zinc coated steel sheet.

CP 118 is on the structural use of aluminium, CP 1161 aluminium sections for structural purposes, BS 4868 profiled aluminium sheet. Milled lead sheet for building purposes is covered by BS 1178.

Sheet roof and wall coverings are dealt with by CP 143 – Aluminium – Parts 1 and 15, Galvanized corrugated steel – Part 10, Lead – Part 11, Copper – Part 12. Good practice in profiled sheet metal roofing and cladding has been expounded by the National Federation of Roofing Contractors.

It is when we come to individual components that problems of selection for a publication of this nature arise. Space rules out a comprehensive listing of metal building components. Use of the *BSI Yearbook* or the Standardline database is recommended to track down individual ones, but as way of example, and to show the variety of metal components in a building, the following are represented: conduit, cisterns, tanks, cylinders, flue pipes, water pipes, windows, cills, gutters and rainwater goods, wall ties, doors and door frames, lathing, timber connectors, flat, corrugated and profiled roofing and cladding sheets, drainage stack units, chimneys, partitioning, space heaters, ventilation and air-conditioning equipment, stairs, lintels, joist hangers, barriers, suspended ceilings, ducts, floors and lightning arresters. In addition, some standards give guidance on use, for example DD 24 on methods of protection against corrosion in light section steel used in building, and BS 5493 on protective coating of iron and steel structures against corrosion.

Codes of practice not yet mentioned cover agricultural buildings, air conditioning, central heating, hot water supply, non-loadbearing external vertical enclosures lightning protection, and pipework. Perhaps here also should be mentioned the HVCA specification on ductwork DW:142.

Listed below are some of the key standard specifications and Codes of Practice of interest to metals users in the UK construction industry.

BSI handbook 3: summaries of British Standards for building, including also Codes of Practice, Drafts for Development and other publications.
British Standards Institution. Manual of British Standards in building construction and specification. 1985. Hutchinson Education.
BSI handbook 20: Standards in the Building Regulations (England and Wales) summarized. Summaries of British Standards to which reference is made in the Building Regulations 1985 (SI 1065 and 1066). 1985.
BS 5950: Part 1: 1985.
Structural use of steelwork in building. Part 1. Code of practice for design in simple and continuous construction: hot rolled sections. 1985.
BS 5950: Part 2: 1985.
Structural use of steelwork in building. Part 2. Specification for materials, fabrication and erection: hot rolled sections. 1985.
BS 5950: Part 4: 1982.
Structural use of steelwork in building. Part 4. Code of practice for design of floors with profiled steel sheeting. 1982.
BS 5950: Part 5: 1987.
Code of practice for design of cold formed sections. 1987.
BS 449: Part 2: 1969.
Use of structural steel in building. Part 2. Metric units. 1969.
BS 449: Addendum No. 1 (1975) to BS 449: Part 2: 1969.
Use of cold formed steel sections in building. 1975.
BS 4: Part 1: 1980.
Structural steel sections. Part 1. Specification for hot rolled sections. 1980.
BS 4848: Part 2: 1975.
Hot rolled structural steel sections. Part 2. Hollow sections. 1975.
BS 4848: Part 4: 1972.
Hot rolled structural steel sections. Part 4. Equal and unequal angles. 1972.
BS 4848: Part 5: 1980.
Hot rolled structural steel sections. Part 5. Bulb flats. 1980.
BS 2994: 1976.
Specification for cold rolled steel sections. 1976.
BS 3083: 1988.
Specification for hot-dip zinc coated and hot-dip aluminium/zinc coated corrugated steel sheets for general purposes. 1988.
BS 1178: 1982.
Specification for milled lead sheet for building purposes. 1982.
CP 118: 1969.
The structural use of aluminium. 1969.

CP 143: Part 1: 1958.
Sheet roof and wall coverings. Part 1. Aluminium, corrugated and troughed. 1958.
CP 143: Part 15: 1973.
Sheet roof and wall coverings. Part 15. Aluminium. Metric units. 1973.
CP 143: Part 10: 1973.
Sheet roof and wall coverings. Part 10. Galvanized corrugated steel. Metric units. 1973.
CP 143: Part 11: 1970.
Sheet roof and wall coverings. Part 11. Lead. Metric units. 1970.
CP 143: Part 12: 1970.
Sheet roof and wall coverings. Part 12. Copper. Metric units. 1970.
BS 1161: 1977 (1984).
Specification for aluminium alloy sections for structural purposes. 1977.
Department of the Environment.
Building Regulations 1985. Part 7: Materials and workmanship. Approved document to support regulation 7. 1985. (HMSO).
Heating and Ventilating Contractors' Association.
Specification for sheet metal ductwork. Low, medium and high pressure/velocity air systems (DW:142). 1982 (HVCA).

Unless otherwise indicated, all are published by BSI.

Books

Over the years there have been many publications on building materials making reference to metals. Concentrating therefore on the more recent, we have *Concrete, Timber and Metals*, Illston *et al.* (Van Nostrand, 1977) which looks at their behaviour as structural materials; G. D. Taylor's *Materials of Construction* (Longman, 1983) and the *Building Materials Evaluation Handbook* by F. Wilson (Van Nostrand, 1984). A look at materials from a fundamental point of view is found in the interesting *Structures or Why Things Don't Fall Down*, J. E. Gordon (Plenum Press, 1978).

The prestigious *BRE Digests* have been gathered into book form and Vol. 2 covers building components and materials. Materials and workmanship as affected by the Building Regulations are reported in the Department of the Environment's Approved document to support regulation 7.

Composite construction can be represented by Robin Whittle's *Reinforcement Detailing Manual* (Viewpoint/Scholium Int., 1981); R. P. Johnson and R. J. Buckle's *Composite Structures of Steel and Concrete* (Granada, 1979); and Trevorrow's *Steel Reinforcement*.

In the design of metal structures it is essential to work from recognized authoritative sources and to have access to reliable data. Recognition comes from long-term use and/or from the prestige of the issuing body. There are several national publications aimed to promote good design – the *Manual of Steel Construction* of the American Institute of Steel Construction; the Canadian Institute's handbook, their code of standard practice, and steel joist recommended practice; Britain's *Steel Designers Manual* (latest edition by the Steel Construction Institute); the Kozai Club of Japan's Steel Construction Guidebooks, and the *Swedish Code for Light Gauge Metal Structures*. The American Iron and Steel Institute's *Cold Formed Steel Design Manual* has international recognition. The publications of the British Constructional Steelwork Association and of the Steel Construction Institute (CONSTRADO) carry some authority, although competitive materials such as concrete might dispute some of the economic findings. The European Convention for Constructional Steelwork has a useful series entitled *Recommendations for Steel Construction*, some of which are in the list of books at the end of this chapter.

The Metal Roof Deck Association's *Technical Requirements for the Design of Light Metal Roof Decks* and the National Federation of Roofing Contractors' guide to good practice *Profiled Sheet Metal Roofing and Cladding* have been joined by British Steel's *Roofing and Cladding in Steel: A Guide to Architectural Practice* and the Property Service Agency's summary of available guidance *Technical Guidance: Sheet Cladding Non-loadbearing Profiled Asbestos Cement, Steel and Aluminium*. Codes for suspended ceilings have been issued by the PSA and Suspended Ceiling Association. British Steel Sections have a *Construction Guide, Sections Book* and the *Piling Handbook* and British Steel Tubes advice on design in structural hollow sections. CIRIA in their reports and technical notes have covered, amongst others, profiled sheet as permanent formwork (Bryan) and composite beams and slabs (Lawson).

Help in the use of non-ferrous metals has been provided by the (North American) Aluminium Association in their guide to selecting building products, the *Aluminium Book*; by the recent *Aluminium Alloy Structures* by F. M. Mazzolani (Wiley, 1985); and the Lead Development Association's *Lead Sheet in Building*.

Data books supplying the raw materials for design include the BSCA:CONSTRADO *Structural Steelwork Handbook*; British Steel Sections *Structural Sections*; British Steel Tubes' data on RHS beams and hot finished structural hollow sections; and Fox and White's design tables for steel beams.

Being without too much vested interest, normally published books are of importance. For structural steelwork, recent examples are G. Ballio and S. M. Mozzolani, *Theory and Design of Steel Structures* (Chapman and Hall, 1983); A. B. Clarke and S. H.

Coverman, *Structural Steelwork* (Chapman and Hall, 1987); Lambert Tall *et al.*, *Structural Steelwork,* 2nd edn (Wiley, 1984); Knowles, *Design of Structural Steelwork* (Transatlantic Press, Philadelphia, 1977); T. J. Macginley, *Steel Structures: Practical Design Studies* (Spon, 1981); Trehair, *Behaviour and Design of Steel Structures* and, to show all is not necessarily well, the Michael R. Horne Conference on *Instability and Plastic Collapse* (Granada, 1983) and Boyd's *Brittle Fracture in Steel Structures* (Newnes/ Butterworths, 1970). Steel's continuing commitment to improvement was illustrated at the *International Conference on Construction: A Challenge to Steel*, in Luxembourg in 1980.

For an economic point of view, there is the NEDO guide to the use of modern fabrication equipment, *Structural Steelwork Productivity*. For cold formed sections, A. C. Walker (ed.), *Design and Analysis of Cold Formed Structures* (Halsted Press, 1975) and W. Yu, *Cold-formed Steel Design* (Krieger, 1973) complement the AISI design manual.

Based on research work at Salford University, E. R. Bryan and J. M. Davies produced a design guide to steel skin diaphragm roof decks and (J. M. Davies and E. R. Bryan) a *Manual of Stressed Skin Diaphragm Design* (Krieger, 1982).

The use of thin-walled structures has been promoted by the International conferences complemented by J. Rhodes and J. Spence (eds), *Behaviour of Thin-walled Structures* (Elsevier, 1984) and A. J. Brookes, *Cladding of Buildings* (Wiley, 1985).

Having concentrated on structures and building elements at the expense of components, just one work to cover fixings – P. Marshe and D. Beckett's *Mechanical Fixing Devices in the Building Industry* (Longman, 1978); and one on ironmongery – the Guild of Architectural Ironmongers' *Scheduling of Architectural Ironmongery*.

Allen, E., *Fundamentals of Building Construction Materials and Methods* (Wiley, 1986).
Aluminium Association, *Aluminium Book: A Guide to Selecting Building Products* (Aluminium Association, 1985).
American Institute of Steel Construction, *Manual of Steel Construction*, 8th edn (AISC, 1980).
American Iron and Steel Institute, *Cold Formed Steel Design Manual* (AISI, 1986).
Ballio, G. and Mazzolani, F., *Theory and Design of Steel Structures* (Chapman and Hall, 1983).
Bates, W., *Introduction to the Design of Industrial Buildings* (Constrado, 1978).
Boyd, G. M., *Brittle Fracture in Steel Structures* (Butterworths, 1970).
British Constructional Steelwork Association, *Metric Practice for Structural Steelwork* (BCSA, 1979).
British Constructional Steelwork Association and Constrado, *Structural Steelwork Handbook: Properties and Safe Load Tables* (Constrado, 1978).
British Steel Corporation: BSC Sections, *Piling Handbook*, 3rd edn (BSC, 1981).

British Steel Corporation: BSC Sections, *Structural Sections: Universal Beams, Columns, Joists, Channels and Angles and Related Design Data* (BSC, 1979).
British Steel Corporation: BSC Sections, *Construction Guide* (BSC, 1980).
British Steel Corporation: BSC Sections and Commercial Steels and British Constructional Steelwork Association Ltd., *Sections Book* (BSC:BCSA, 1982).
British Steel Corporation: BSC Strip Mill Products, *Roofing and Cladding in Steel: A Guide to Architectural Practice* (BSC, 1987).
British Steel Corporation: BSC Tubes, *RHS Beams: Buckling, Bearing and Safe Load Tables* (BSC, 1985).
British Steel Corporation: BSC Tubes, *Hot Finished Structural Hollow Sections: Technical Data* (BSC, 1986).
British Steel Corporation: BSC Tubes, *Hot Finished Structural Hollow Sections: Safe Load Tables* (BSC, 1986).
British Steel Corporation: BSC Tubes, *Design in SHS to BS 449* (BSC, 1985).
Brookes, A. J., *Cladding of Buildings* (Construction P., 1983).
Bryan, E. R. and Davies, J. M., *Steel Diaphragm Roof Decks: A Design Guide with Tables for Engineers and Architects* (Granada, 1981).
Bryan, E. R. and Leach, P., *Design of Profiled Sheeting as Permanent Formwork* (Construction Industry Research and Information. Technical Notes. 116. CIRIA, 1984).
Bucksch, H., *Dictionary of Architecture, Building Construction and Materials* (English and German, 2 volumes, French and European Publications, 1974–76).
Building Research Establishment, *BRE Digests* Vol. 2: *Building Components and Materials* (BRE, 1984).
Canadian Institute of Steel Construction, *CISC Code of Standard Practice for Structural Steel* (CISC, 1980).
Canadian Institute of Steel Construction, *Steel Joist Facts. . . . Recommended Practice*, 2nd edn (CISC, 1980).
Canadian Institute of Steel Construction, *Handbook of Steel Construction*, 4th edn (CISC, 1985).
Clarke, A. R. and Coverman, J., *Structural Steelwork* (Macmillan Education, 1983).
CONSTRADO, Durability of Steel Structures (Steel Construction Institute, 1984).
CONSTRADO, *Steel Designer's Manual*, 4th rev. edn (Collins, 1983).
CONSTRADO, *Steel Framed Multi-storey Building: Design Recommendations for Composite Floors and Beams Using Steel Decks*. Section 1: Structural. Section 2: Fire resistance.
CONSTRADO, *Steel Framed Multi-storey Building: the Economics of Construction in the UK*, 2nd edn (1985).
CONSTRADO, *Stressed Skin Construction: Principles and Practice*.
CONSTRADO, *Structural Steelwork Handbook for Standard Metric Angles: Properties and Safe Load Tables*.
CONSTRADO, *Use of Cold Rolled Purlins* (1978).
CONSTRADO, Elliott, D. A., *Fire and Steel Construction: Protection of Structural Steelwork*, 2nd edn (1981).
CONSTRADO, Hill, H. S., *Thermal Insulation of Profiled Steel Cladding* (1981).
CONSTRADO, Law, M. and O'Brien, T., *Fire and Steel Construction: Fire Safety of Bare External Steel*.
Corey, A. R., *The Development of Markets for New Materials: A Study of Building New End-product Markets for Aluminium, Fibrous Glass and the Plastics* (Books Demand UMI).
Davies, J. M. and Bryan, E. R., *Manual of Stressed Skin Diaphragm Design* (Granada, 1982).
Department of the Environment, *Building Regulations 1985*. Part 7: *Materials and Workmanship. Approved document to support regulation 7* (HMSO, 1985).

European Convention for Constructional Steelwork, *Design and Testing of Connections in Steel Sheeting and Sections* (ECCS, 1983).
European Convention for Constructional Steelwork, *Design of Composite Floors with Profiled Steel Sheet* (ECCS, 1975).
European Convention for Constructional Steelwork, *Design of Profiled Sheeting* (ECCS, 1984).
European Convention for Constructional Steelwork, *Stressed Skin Design of Steel Structures* (ECCS, 1984).
European Convention for Constructional Steelwork, *Composite Structures* (Construction Press, 1981).
European Convention for Constructional Steelwork, *Good Practice in Steel Cladding and Roofing* (ECCS, 1983).
European Convention for Constructional Steelwork, *Testing of Profiled Metal Sheets* (ECCS, 1978).
European Convention for Constructional Steelwork, *Mechanical Fasteners for Use in Steel Sheeting and Sections: Information and Testing* (ECCS, 1983).
Fox and White, *Design Tables 1: Steel Beams to BS 449* (1981).
Gordon, J., *The New Science of Strong Materials* (Pelican, 1979).
Guild of Architectural Ironmongers, *Scheduling of Architectural Ironmongery* (The Guild, 1982).
Ileston, J. M., *et al.*, *Concrete, Timber and Metals: The Nature and Behaviour of Structural Materials* (Van Nostrand, 1979).
International Conference on Construction: A Challenge for Steel, Luxembourg (1980), *Construction: A Challenge for Steel* (Westbury House, 1980).
International Conference on Thin-Walled Structures, University of Strathclyde (1979), *Thin Walled Structures: Recent Technical Advances and Trends in Design, Research and Construction* (Granada, 1980).
International Symposium on Thin Walled Steel Structures, 1st, University College of Swansea (1967), *Thin Walled Steel Structures: Their Design and Use in Building*, K. C. Rockey and H. V. Hill (eds) (Crosby Lockwood, 1969).
Johnson, R. P., *Composite Structures of Steel and Concrete*. Vol. 1: *Beams, Columns, Frames and Application in Building* (Granada, 1982).
Knowles, P., *Design of Structural Steelwork*, 2nd edn (Surrey Univ. Press, 1983).
Kozai Club, *Steel Construction Guidebook. Building Construction* (Kozai Club, Japan, 1984).
Kozai Club, *Steel Construction Guidebook. Civil Engineering* (Kozai Club, Japan, 1984).
Lambert, F. W., *Structural Steelwork*, 3rd edn (Godwin, 1982).
Lawson, F. M., *Composite Beams and Slabs with Profiled Steel Sheeting* (Construction Industry Research and Information Association. Reports. 99) (CIRIA, 1983).
Lead Development Association, *Lead Sheet in Building: A Guide to Good Practice* (LDA, 1978).
MacGinley, T. J., *Steel Structures: Practice Design Studies* (Spon, 1981).
Marsh, P. and Beckett, D., *Mechanical Fixing Devices in the Building Industry* (Construction P., 1975).
Mazzolani, F. M., *Aluminium Alloy Structures* (Longman, 1985).
Metal Roof Deck Association, *Technical Requirements for the Design of Light Metal Roof Decks*, 3rd edn (MRDA, 1970).
Michael R. Horne Conference on Instability and Plastic Collapse of Steel Structures, *Instability and Plastic Collapse of Steel Structures* (Granada, 1983).
National Economic Development Council: Civil Engineering EDC, *Price Adjustment Formulae for Civil Engineering Contracts. 1: Civil Engineering Works* (HMSO, 1973).

National Economic Development Council: Civil Engineering EDC, *Price Adjustment Formulae for Civil Engineering Contracts. 2: Structural Steelwork* (HMSO, 1974).
National Economic Development Office: Constructional Steelwork Sector Working Party, *Structural Steelwork Productivity: A Guide to the Use of Modern Fabrication Equipment* (NEDO, 1981).
National Federation of Roofing Contractors, *Profiled Sheet Metal Roofing and Cladding: A Guide to Good Practice* (The Federation, 1982).
National Swedish Committee on Regulations for Steel Structures and Swedish Institute of Steel Construction, *Swedish Code for Light-gauge Metal Structures* (Swedish Inst. of Steel Const., 1982).
Packham, G., *Structural Steelwork Detailing* (Orion Books, 1983).
Property Services Agency, *Technical Guidance: Sheet Cladding. Non-loadbearing Profiled Asbestos, Cement, Steel and Aluminium*, 2nd edn (PSA, 1979).
Property Services Agency, *Technical Guidance: Suspended Ceilings* (Methods of Building. 09.201) (PSA, 1978).
Rhodes, J. and Spence, J. (eds), *Behaviour of Thin-walled Structures* (Elsevier Applied Science, 1984).
Rosen, H. J., Construction Materials for Architecture (Wiley, 1985).
Suspended Ceiling Association, *Good Practice for the Installation of Suspended Ceilings* (Guides: 1) (SCA, 1982).
Suspended Ceiling Association, *Recommendations for Suspended Ceiling Grid Systems* (Guides: 2) (SCA, 1982).
Suspended Ceiling Association, *Introduction to Suspended Ceilings* (Data sheets No. 1) (SCA, 1982).
Taylor, G. D., *Materials of Construction*, 2nd edn (Construction P., 1983).
Trahair, N. S., *Behaviour and Design of Steel Structures* (Chapman and Hall, 1978).
Trevorrow, A., *Steel Reinforcement* (Construction P., 1984).
Walker, A. C., (ed.), *Design and Analysis of Cold Formed Structures* (Intertext, 1975).
Ward, J. W., *Construction, Information Source and Reference Guide*, 4th edn (Construction Publications, 1981).
Whittle, R., *Reinforcement Detailing Manual* (Viewpoint Publications, 1981).
Wilson, F., *Building Materials Evaluation Handbook* (Van Nostrand Reinhold (UK), 1984).
Yu, W. W., *Cold-formed Steel Design* (Wiley, 1984).

Conferences

The number of conferences relating to the use of building materials is growing and the following indicate the range of topics being addressed:

International Conference on Construction: A Challenge for Steel, Luxembourg (1980), *Construction: A Challenge for Steel* (Westbury House, 1980).
International Conference on Thin-Walled Structures, University of Strathclyde (1979), *Thin Walled Structures: Recent Technical Advances and Trends in Design, Research and Construction* (Granada, 1980).
International Symposium on Thin Walled Steel Structures, 1st, University College of Swansea (1967), *Thin Walled Steel Structures: Their Design and Use in Building*, K. C. Rockey and H. V. Hill (eds) (Crosby Lockwood, 1969).
Michael R. Horne Conference on Instability and Plastic Collapse of Steel Structures, *Instability and Plastic Collapse of Steel Structures* (Granada, 1983).

CHAPTER EIGHTEEN

Materials for the packaging industry

C. B. HUBBARD

The importance of packaging in our society has become of increasing significance in this century. It accounts for an ever greater share of the cost of the packaged product, and there is, in addition, an ever greater interdependence between the product, the packaging material and packaging design. The traditional packaging materials have always been metals, paper and glass, but they have been joined in modern times by plastics (both flexible film and rigid), which are taking an increasing share of the market by competing in most areas of packaging and container manufacture.

These four classes of materials are the ones which will be dealt with in this chapter, with the greatest emphasis on metals and plastics. Whereas metal container materials are either steel-based or aluminium-based, there are very many different polymers used which together constitute plastics packaging. A few, however, have emerged to dominate the industry, such as polyethylene terephthalate (PET), polyethylene (PE), polypropylene (PP), polystyrene (PS) and polyvinyl chloride (PVC). Laminated materials have been developed which consist of a blend of layers of different polymers. Many containers these days are composites of materials, e.g. metal and board container (for detergent powders), metal foil and plastics trays (for food), metal cans with plastics easy open ends and plastics containers with metal easy open ends. Paper/plastics laminates, metal foil/plastics laminates and metal foil/paper/plastics laminates have all been developed in recent years.

Unfortunately the developments in packaging technology have not been adequately documented for the serious student. Authoritative textbooks are produced very rarely and the greatest source of information is undoubtedly journal articles and conference papers, although even here the situation is complicated by the fact that packaging envelops such a wide area of technology. In order to keep abreast of developments with all packaging materials it is necessary to scan a large number of journals and periodicals, including the few that are concerned exclusively with packaging. Papers concerned with the development of a new material or new applications for existing materials, or concerned with developments in container manufacturing may appear in the most unexpected sources. Similarly the extraction of information from abstract journals and computer databases also usually requires a wide ranging search to be carried out.

Because of the absence of a comprehensive book reading list, it has been necessary to include a rather long list of journal articles and conference papers, but they do encompass all the modern technological developments concerned with packaging materials.

Choice of material

Selection

The choice of material for a particular packaging application depends on a number of factors. The most important can be summarized as follows:

The packaging material

(i) Formability – the ease with which it can be drawn, extruded, welded, deformed, stretched, folded, bent, etc.
(ii) Machinability – cutting, perforation, scoring.
(iii) Physical properties – weight, permeability to gases and water vapour.
(iv) Chemical properties – toxicity, resistance to corrosion, compatibility with product, degradability.
(v) Mechanical properties – resistance to fracture, crushing, denting, vibration, abrasion, temperature changes (sterilization, cooking, oven drying, freezing).
(vi) Printability – including decorating, coating, varnishing and lacquering, as well as printing.

The product to be packaged

(i) Physical properties – size, weight, shape, density.
(ii) Chemical properties – material composition, corrosiveness, compatibility, resistance to moisture, toxicity, volatility, degradability.

(iii) Mechanical properties – resistance to fracture, crushing, vibration, abrasion, temperature changes.

Transportation

(i) Type of transport – road, rail, sea, air.
(ii) Length of journey – both time and distance.
(iii) Form of containerization.
(iv) Degree of possible mechanical damage.
(v) Climatic conditions – high or low temperatures, humidity, dampness.

Market requirements

(i) Package design – shape, attractiveness, handling ability, stackability, desirable standards of printing and decoration.
(ii) Public acceptance.
(iii) Recycling considerations.

Cost of the material

(i) Availability – UK or foreign supplies.
(ii) Processing and refining.
(iii) Transportation.
(iv) Scrap value.
(v) Political situation – strikes, exchange rates, etc.

When the requirements of the container, with respect to both container material and product, have been established then the type of material that satisfies those requirements can be selected. The reasons for that choice may be very complex and may need periodical re-examination. There is a very wide range of materials to choose from – metals, plastics, paper and board, or glass. Table 18.1 shows the percentage share of the total market for each material for the years 1979, 1982 and 1986.

Table 18.1.

	% Total market		
	1979	*1982*	*1986*
Metals	23.0	23.0	21.0
Plastics	19.0	15.0	17.5
Paper and board	44.0	45.0	47.5
Glass	7.0	8.5	7.0
Other	7.0	8.5	7.0

Metal containers (cans, tins, aerosols, drums, trays)

Characteristics:

(i) Good tensile strength, good resistance to crushing, rigidity, high density.
(ii) A perfect barrier to permeability, when sealed.
(iii) Need joints and closures.
(iv) Can react with foods.
(v) Steel containers will eventually rust and degrade.

The market is divided between steel- and aluminium-based containers.

Steel-based

(i) Tinplate – low carbon, mild steel electrolytically coated with a layer of tin between 0.4 and 2.5 microns thick on each surface. Available in many specifications. Tinplate is the principal canmaking material in the UK. It combines the strength and formability of steel and the corrosion resistance, solderability and good appearance of tin. The mild steel base can be easily drawn and shaped whilst the tin coating protects the steel from product attack and vice versa. Tinplate can also be soldered or welded.
(ii) Tin-free steel (TFS) – low carbon, mild steel sheet electrolytically coated with a chromium/chromium oxide layer (0.04 microns thick) on each surface. The coating was developed as an alternative to tin, which is becoming increasingly more expensive. TFS has limited application in the UK.
(iii) Blackplate – low carbon, mild steel sheet, not tinned, but passivated. Its poor resistance to corrosion makes it unsuitable for food containers.
(iv) Alloyed tinplate (ATP) – similar to tinplate but with a very thin tin alloy layer to replace the normal expensive tin coating. This material has only been developed in recent years.
(v) Lightly tinned steel (LTS) – similar to ATP.
(vi) Nickel-plated steel – used only for primary battery cases.

With the exception of nickel-plated steel, the steel-based materials are collectively referred to as tin mill products.

Aluminium

Many different alloys are used, the specification often including manganese and magnesium to increase rigidity. Commercially pure aluminium is used for the production of foil and extruded containers (e.g. collapsible tubes). Aluminium is lighter and weaker than steel but more ductile and, therefore, has certain advantages and disadvantages by comparison. The containers will not rust but have limited resistance to chemical attack. They are always drawn or extruded, as soldering or welding is not possible.

Aluminium containers have a much larger share of the market in USA than the UK, because the price of aluminium is cheaper.

Plastics containers (bottles, cans, pots, tubes, aerosols, trays, drums) and film (flexible packaging)

Characteristics:

(i) Low density, usually low rigidity, tensile and tear strength variable.
(ii) Flexible.
(iii) Wide range of permeability properties.
(iv) Frequently transparent.

The following thermoplastic polymers are among the most commonly used:

Polyethylene terephthalate (PET)

PET was first developed as a fibre polymer and then introduced into packaging in the form of an oriented film. The market share of PET bottles in recent years has increased markedly and this polymer has gained a very substantial foothold in the plastics packaging industry. It has become a major challenger to glass for the soft drink bottle market. PET has the highest tensile strength and heat resistance of any commodity plastic and is also processable without further additives, making it the 'clearest' polymer available for food packing. The most significant property results from bi-axial orientation, when PET develops enhanced tensile, barrier and rigid properties combined with glass clarity. It is in this form that it has established uses in thin films and blow moulded containers. The stretch blow moulded bottles are clear, strong, impermeable to gas and well able to resist the pressures generated by normal carbonated soft drinks (up to 50–60 p.s.i.).

Table 18.2. Plastic bottle consumption (thousand tonnes)

	1978	*1982*	*1983*
LDPE	7.0	6.0	6.0
HDPE	40.5	42.0	44.0
PVC	16.0	22.0	25.0
PET	2.7	18.0	27.0

Polyethylene (PE)

Exists in two forms – low density and high density. Low density PE has high impact strength and can be blow-moulded into bottles and a wide range of shaped containers, where its flexibility is important. It is also used for flexible films, snap-on caps, collapsible tubes and various forms of spouts and dispensers. It is not suitable for packaging of oxygen-sensitive materials because of poor gas permeability properties. High-density PE is rigid and hard but has a lower impact strength. It has better resistance to permeability of water and gases, and is used for blow-moulded bottles.

Polypropylene (PP)

PP has special characteristics of rigidity, flexural resistance, heat resistance and compatibility with a wide range of foods and general products. It is harder than polyethylene with lower impact strength, but has similar permeability properties. It can be blow-moulded and injection-moulded and is widely used for closures and thin wall pots.

Polyvinyl chloride (PVC)

PVC is a hard, brittle material which can be softened by plasticization. It has good resistance to permeability of gases but not water vapour. It is used for rigid food packages and flexible packaging.

Polystyrene (PS)

Can be injection moulded or extruded into film and sheet, and blown into bottles. The sheet can subsequently be thermoformed to give a range of trays, tubs and blister packs. It is also used as a protective foam. Polystyrene is a hard brittle material with high tensile strength. Increase in impact strength is achieved at the loss of transparency.

Glass containers (bottles, aerosols)

Characteristics:

(i) Rigid, brittle, high density, transparent.
(ii) Perfect permeability barrier properties.
(iii) Inert towards foods.
(iv) Need separate closures.

Paper and board containers (cartons, tubs, tubes)

(i) Not brittle, lower tensile strength than metal, low density, good stiffness, tear easily.
(ii) Flexible and creasable.
(iii) No barrier properties without application of a wax coating.
(iv) Absorbent.

Sources of information

Journals and abstract journals

A number of journals are published in the UK, USA and Europe on the packaging scene in general, all of which are good sources of information on each type of packaging material. The foremost publications are:

Packaging News (UK)
Packaging Review (UK)
Packaging Today (UK)
Retail Packaging (UK)
Food Engineering (USA)
Packaging (USA)
Packaging Digest (USA)
Verpackungs Rundschau (West Germany)
Neue Verpackung (West Germany)
Emballages (France)
Emballages Digest (France)

Two UK publications contain specialized information on tinplate containers, i.e. *Tinplate Bulletin* (a bi-monthly survey of the open literature on all aspects of the production and applications of tinplate and tin mill products, published by British Steel) and *Tin International* (commercial information and new developments from the canmaking industry). *Tin and Its Uses* (International Tin Research Institute) occasionally contains review articles on steel canmaking techniques. The major USA journal for specialized information on canmaking and can materials is *Modern Metals*. It is primarily concerned with aluminium-based containers but also often contains comparative data on steel-based containers. Some issues of *Modern Metals* are almost entirely devoted to the canmaking industry, both in the USA and Europe.

Plastics packaging has specialist coverage in *FCD Packaging* (UK/Holland), whilst paper and board containers are covered by *Paperboard Packaging* (USA) and *Paper Trade Journal* (USA). Some information on glass containers appears in *Glass* (UK).

Those abstract journals and current awareness bulletins which cover the development and use of packaging materials to a lesser or greater extent are as follows:

All materials

PIRA International Packaging Abstracts (UK).
Packaging Science and Technology Abstracts (UK/W Germany)
Rutgers Current Packaging Abstracts (USA).
Food Science and Technology Abstracts (UK/W Germany).
Applied Science and Technology Index (USA).
Current Technology Index (UK).

Metals

Metals Abstracts (ASM/Metals Society).
Engineering Index (USA).
BSC Tinplate Bulletin (UK).
World Aluminium Abstracts (ASM, USA).
Corrosion Abstracts (NACE, USA) (Corrosive effects of food products on can materials).
Metal Finishing Abstracts (UK).
World Surface Coatings Abstracts (Paint RA, UK).

Plastics

RAPRA Abstracts (UK).

Paper and board

Abstract Bulletin of Institute of Paper Chemistry (USA).

Books and reports

Relatively few books have ever been written about packaging materials, but in recent years this situation has been remedied to a certain extent with the publication of four excellent texts which cover the subject in some depth. *A Handbook of Food Packaging*, F. A. Price and H. Y. Paine (eds) (Leonard Hill, 1983) reviews all of the packaging materials, discussing their chemical and physical properties, advantages and disadvantages, various applications and potentialities. *Developments in Food Packaging – 1*, S. J. Palling (ed.) (Applied Science Publishers, 1980) also reviews the full range of materials used. However, both of these books are only concerned with food packaging. Two recent publications, *Tinplate and Modern Canmaking Technology* by E. Morgan (Pergamon, 1984) and *Plastic Films for Packaging – Technology Applications and Process Economics*, by C. J. Benning (Technomic Publishing, 1983), are concerned only with their specialist areas but are the most up-to-date expert reviews. *Packaging – an*

Introduction by S. Sacharow and A. L. Brody (Harcourt Brace, 1987) fills the gap between introductory knowledge on packaging and detailed study.

Other specialist and general packaging books that have been published since the 1960s include:

Aluminium in Packaging, R. J. Sinia (Stamex, undated).
Basic Guide to Plastics in Packaging, S. Sacharow and R. C. Griffin (Cahners Books, 1973).
Cost and Availability of Packaging Media, R. Mills (PIRA Sponsored Survey, 1976).
A Guide to Thermoformed Plastic Packaging, S. E. Farnham (Cahners Books, 1972).
Guide to Tinplate (International Tin Research Institute Publication No. 622).
Handbook of Package Engineering, J. F. Hanlon (McGraw-Hill, 1971).
Know Your Packaging Materials. Part 1 – Foils, Paper and Boxboard. Part 2 – Films. Part 3 – Foam Plastics (American Management Association Packaging Report No. 5, 1958).
The Manufacture of Paper Containers, P. E. Verstone (Lewis Publications, 1960).
Modern Packaging Films, S. H. Pinner (ed.) (Butterworths, 1967).
Packaging in Glass, B. E. Moody (Hutchinson, 1963).
Packing Materials and Containers, F. A. Paine (Blackie & Son, 1967).
Paperboard Packaging Handbook, P. A. Toensmeier (Board Products Publications, 1961).
Plastic Containers for Pharmaceuticals: Testing and Control, J. Cooper (World Health Organisation Offset Publication No. 4, 1974).
Principles of Aseptic Processing and Packaging, P. E. Nelson *et al.* (Food Processors Institute, 1987).
The Technology of Tinplate, W. E. Hoare, E. S. Hedges and B. T. K. Barry (Edward Arnold, 1965).

The following are market research reports:

An Appraisal of Tinplate and Aluminium in the Packaging Industry (Rayner-Harwill Ltd., 1978).
Barrier Co-extruded Plastic Systems – A New Frontier in Food Packaging (Business Communications Co. Inc., Report P-071, 1983).
Material Substitution – Lessons from the Tin-Using Industries (Resources for the Future Inc., 1983).
Packaging (Plastics) (Keynote Publications, 1983).
Plastic Packaging 1979 (C. H. Kline & Co. Inc., 1979).

Key Papers

The following key papers have been selected from journals and conference proceedings. They mostly cover the recent advances made in the development of metals and plastics packaging materials:

'Aluminium alloys for rigid containers', G. C. Wood. Sheet Metal Industries, **52** (1) January 1975, pp. 40–3, 12 refs.
'Aluminium in packaging', K. A. Evans, S. M. Fenwick. Sheet Metal Industries, **52** (1) January 1975, pp. 28–33, 7 refs.

'Aluminium will play a stronger role', J. W. Davie. *Food Processing Industry*, **45** (539) October 1976, pp. 32–6.
'Cans: the looming metals battle (steel versus aluminium)', A. Serchuk. *Modern Packaging*, **51** (3) March 1978, pp. 29–32.
'Characteristics of lightly tin-coated steel sheet', H. Kuroda, I. Onoda, T. Inui and Y. Kondo. *2nd International Tinplate Conference*, International Tin Research Institute, October 1980, London. Paper 13, pp. 124–35, 10 refs.
'Chromium plated steel, a unique variety of tin free steel', H. Uchida, O. Yanabu, T. Hada and H. Sato. *Iron and Steel Engineer*, **46** (1), January 1969, pp. 75–9.
'Coated blackplate: Steel's salvation in beverage cans?'. *Modern Metals*, **32** (12) January 1977, pp. 60–4.
'Demands two-piece can technologies place on the properties of tinplate', G. A. Jenkins, G. Jefford and D. W. Evans. 1st International Tinplate Conference, International Tin Research Institute, October 1976, London. Paper 11, pp. 122–40, 60 refs.
'The development of a low tin coating mass tinplate: LTS', D. Salm, A. A. Towers, D. Kaan. *2nd International Tinplate Conference,* International Tin Research Institute, October 1980, London. Paper 12, pp. 112–23.
'Development of the use of tin free steel', R. Allouf. 5th International Meeting, Comité Internationale Permanent de la Conserve, Cannes, May 1980.
'Development of tinplate substitute for D. and I beverage cans', E. Nagel Soepenberg. *Proceedings International Conference on Production and Use of Coil – Coated Strip*, Birmingham, 1981.
'Developments in PET containers (various papers)'. *PET Bottles, 3rd International Seminar*, May 1982, Antwerp.
'Developments in rigid metal containers for food', B. J. McKernan. *Food Technology*, **37** (4) April 1983, pp. 134–7, 3 refs.
'Developments in the use of tinplate and tinplate containers', G. P. Clay. *Conference on Tin Consumption*. International Tin Council and Tin Research Institute, March 1972, London. Paper 21, pp. 385–95.
'Differentially coated tinplate for drawn and ironed cans', D. C. Shah. *2nd International Tinplate Conference*, International Tin Research Institute, October 1980, London. Paper 19, pp. 198–207, 2 refs.
'Future food packaging – plastic, convenient, aseptic', J. R. Duprey. *Packaging*, **29** (1) January 1984, pp. 43–6.
'Future trends and technical innovations in the glass industry', D. Whittaker. Food Manufacturers' Federation (FMF) Seminar on Packaging Materials, November 1982, London.
'Materials substitution and tin consumption in the beverage container industry', F. R. Demler, J. E. Tilton. *Materials and Society*, **4** (3) 1980, pp. 365–73, 7 refs.
'Metal used for beer cans', C. E. Scruggs, J. L. Krickl, N. J. Linde. *Master Brewers Association of America Technical Quarterly*, **8** (1) 1971, pp. 63–71, 8 refs.
'New materials for food packaging – the future', J. H. Briston. *Food Chemistry*, **8** (2) February 1982, pp. 147–55.
'A new product for the packaging industry: ATP (alloyed tinplate)', J. P. Servais, J. L.Empereur, L. Renard and V. Leroy. *CRM* (56) June 1980, pp. 43–52, 8 refs.
'New uses for PET containers'. *Food Engineering*, September 1979, pp. 123–6.
'Packaging in aluminium – a technology update'. *PIRA Seminar*, December 1975, Leatherhead, Surrey. 7 papers.
'Paper and board packaging materials', K. Bridge. *Food Manufacturers' Federation (FMF) Seminar on Packaging Materials*, November 1982, London.
'PET packaging – beverages and beyond'. *Packaging Institute (USA) Seminar*, July/August 1980, Wisconsin.
'PET – the packaging material of the Eighties', A. D. Campion. Neue Verpackung, **35** (2) February 1982, pp. 142–7 (in German).

'Polyolefins for food packaging', D. W. Shorten. *Food Chemistry*, **8** (2) February 1982, pp. 109–19.
'PVC as a food packaging material', R. B. Pearson. *Food Chemistry*, **8** (2) February 1982, pp. 85–96.
'Recent developments in steel-based materials for packaging', J. R. Bevan. *International Conference on Packaging Technology*, PIRA/IAPRI, London, March 1972, Paper 3. *also* with S. H. Melbourne, *Packaging Technology*, **15** (104) January 1969, pp. 10–12.
'The selection of plastics films for food packaging', C. R. Oswin. *Food Chemistry*, **8** (2) February 1982, pp. 121–7, 8 refs.
'The significance of developments in canmaking materials: aluminium-based and composites', J. W. Davie. *Developments in Packaging in Cans*, PIRA Seminar, April 1975, Leatherhead, Surrey.
'The significance of developments in canmaking materials: Steel-based', B. T. K. Barry, *Developments in Packaging with Cans*, PIRA Seminar, April 1975, Leatherhead, Surrey.
'Steel-based packaging materials', G. Jefford, *The Metallurgist and Materials Technologist*, **7** (10) October 1975, pp. 516–21, 34 refs.
'Styrene polymers and food packaging', C. A. Brighton. *Food Chemistry*, **8** (2) February 1982, pp. 97–107.
'Tinplate and the welded can', S. L. Jaques, J. Szczur. ATIC (Australian Tin Information Centre) Seminar The Welded Can, Melbourne, October 1982, 6 refs, 27 pp.
'Trends and innovations in plastics packaging', W. Cooper. *Food Manufacturers' Federation (FMF) Seminar on Packaging Materials*, November 1982, London.
'Welcco – a new light tin-coated steel for canmaking', R. Allouf, C. Mergey. *2nd International Tinplate Conference*, International Tin Research Institute, October 1980, London. Paper 23, pp. 241–53, 4 refs.

The UK journal *Packaging Review* presents an annual survey in its January edition of the UK packaging industry. The value of the UK market in the previous two years for each packaging material is given plus a review of sales, future prospects, packaging capacities and applications.

Standards

Steel-based materials

There are significant numbers of BSI, ISO, ASTM and EURONORM standards which are concerned with packaging and, more specifically, packaging materials. The reader should consult the yearbooks of these bodies and should also peruse the chapters concerned with tinplate and aluminium in particular.

BS 2920 (1973)	Cold-reduced tinplate and cold-reduced blackplate.
ISO R1111/1/2 (1983)	Cold-reduced tinplate and cold-reduced blackplate. Part 1. Sheet. Part 2. Coil.
Euronorm 145 (1978)	Tinplate and blackplate in sheet form. Qualities, dimensions and tolerances.
146 (1980)	Tinplate and blackplate in coil form for subsequent cutting into sheets. Qualities, dimensions and tolerances.

ASTM A599–84	Steel sheet, cold rolled, tin-coated by electro-deposition.
A623–87	General requirements for tin mill products.
A623M–87	General requirements for tin mill products (metric).
A624–86	Single reduced electrolytic tinplate.
A624M–86	Single reduced electrolytic tinplate (metric).
A625–85	Single reduced blackplate.
A626–86	Double reduced electrolytic tinplate.
A626M–87	Double reduced electrolytic tinplate (metric).
A650–83	Double reduced blackplate.
A650M–83	Double reduced blackplate (metric).
A657–87	Steel, cold rolled, single-and double-reduced tin mill blackplate, electrolytic chromium coated.
D3061–79	Tinplate fabricated aerosol cans.

Aluminium-based materials

BS 1470 (1987)	Wrought aluminium and aluminium alloys for general engineering purposes – plate, sheet and strip. Three grades of aluminium and six aluminium alloys.
1471 (1972)	Wrought aluminium and aluminium alloys for general engineering purposes – drawn tube. Two grades of aluminium and seven aluminium alloys.
1474 (1972)	Wrought aluminium and aluminium alloys for general engineering purposes – bars, extruded round tubes and sections. Two grades of aluminium and seven aluminium alloys.
1683 (1987)	Coated aluminium foil for wrapping processed cheese.
3313	Aluminium capping foil and strip for dairy product containers. Part 1 (1968) for glass containers. Part 2 (1968) for skirted closures for plastics containers.
3394 (1971)	Aluminium food storage canisters.
5313 (1976)	Aluminium catering containers and lids.
5439 (1977)	Specification for aluminium foil catering containers.
ISO R209 (1971)	Composition of wrought products of aluminium and aluminium alloys – chemical composition (per cent).
827 (1968)	Mechanical property limits for extruded products of aluminium alloys.
TR2136 (1977)	Wrought aluminium and aluminium alloys – rolled products – mechanical properties.
R2142 (1981)	Wrought aluminium and aluminium alloys – selection of specimens and test pieces for mechanical testing.
ASTM B209–86	Aluminium alloy – sheet and plate.
B209M–86	Aluminium alloy – sheet and plate (metric).
B210–86	Aluminium alloy – drawn seamless tubes.
B210M–86	Aluminium alloy – drawn seamless tubes (metric).
B221–85	Aluminium alloy – extruded bars, rods, shapes and tubes.
B221M–85	Aluminium alloy – extruded bars, rods, shapes and tubes (metric).
B479–85	Annealed aluminium alloy foil for flexible barrier applications.
B483–85	Aluminium alloy drawn tubes for general purpose applications.

B483M–85	Aluminium alloy drawn tubes for general purpose applications (metric).

Plastics materials

BS 1133	Packaging code. Section 22 (1967): Packaging in plastics containers.
1679	Containers for pharmaceutical dispensing. Part 4 (1969): Plastics containers for tablets and ointments.
1763 (1975)	Thin PVC sheeting.
4839	Blow moulded polyolefin containers. Part 1 (1972) Containers up to 5 litre capacity. Part 2 (1974) Containers over 5 litres and up to 60 litres capacity. Part 3 (1977) Closed head containers over 60 litres up to and including 210 litres.
5496 (1977)	Plastics catering containers and lids.
5597 (1978)	Non-refillable plastics aerosol containers up to 300 ml capacity.
ASTM D1201–81	Thermosetting polyester moulding compounds
D2463–74	Drop impact resistance of blow moulded thermoplastic containers.
D2561–70	Environmental stress crack resistance of blow moulded polyethylene containers.
D2569–84	Column crush properties of blown thermoplastic containers.
D2684–84	Permeability of thermoplastic containers.
D2741–84	Susceptibility of polyethylene bottles to soot accumulation.
D2911–82	Dimensions and tolerances of plastic bottles.

Paper and board materials

BS 1133	Packaging code Section 7 (1967) Paper and board wrappers, bags and containers. Section 7 Chapter 7.5 (1981) Fibreboard cases.
1679	Containers for pharmaceutical dispensing. Part 1 (1976) Paperboard containers for strip and blister packs.
5167 (1978)	Packages for washing and cleaning powders. Dimensions and volumes of cartons and drums from fibreboard.

Glass materials

BS 1133	Packaging code. Section 18 (1967) Glass containers and closures.
1679	Containers for pharmaceutical dispensing. Part 5 (1973) Eye-dropper bottles. Part 6 (1984) Glass medicine bottles. Part 7 (1968) Ribbed oval glass bottles.
1777 (1981)	Glass honey jars.
6106 (1981)	One pint (568 ml) multi-trip glass bottles for pasteurized milk.
6117 (1981)	Glass bottles for light wine.

6118 (1981)	Multi-trip glass bottles for beer and cider.
6119	Glass bottles for carbonated soft drinks. Part 1 (1981) 750 ml and 1 litre multi-trip bottles.
ASTM C147–86	Internal pressure test for glass containers.
C148–77	Polariscopic examination of glass containers.
C149–86	Thermal shock test on glass containers.
C224–78	Glass containers, sampling.
C225–85	Resistance of glass containers to chemical attack, tests for.
D3071–79	Drop testing of glass aerosol bottles.

General

BS 1133	Packaging code. Section 10 (1966) Metal containers. Section 21 (1976) Regenerated cellulose film, plastics film, aluminium foil and flexible laminate.

There are a number of BSI and ISO committees dealing with packaging and packaging materials listed in BSI and other standards organizations handbooks. Other important committees involved with the development and use of packaging materials are:

PIRA Packaging Division – all materials.
EEC SEFEL – steel-based and aluminium.
Sheet Metal Technology Group – steel-based and aluminium.

Conferences

In November 1982, the Food Manufacturers' Federation (UK) held a seminar on Packaging Materials in London. Future trends concerning all types of packaging material were discussed.

PIRA (UK) regularly organizes many seminars on the development and use of packaging materials.

The International Tinplate Conference is held in London every four years. It is organized by the International Tin Research Institute and previous conferences have included key papers on the use of tinplate in canmaking, and on the properties, production and performance of tinplate. Developments in lightly tin coated steel and tin free steel have also been revealed.

Other important conferences or seminars dealing with the metallic materials in the past have included the ASM Materials Conferences and International Deep Drawing Research Group Conferences. In May 1983, at London, the Metals Society, co-sponsored by the International Wire and Machinery Association and the Institution of Mechanical Engineers, organized 'Developments in the Drawing of Metals' which was partly concerned with the production of drawn tinplate and aluminium containers.

In 1981 the Metals Society organized an international conference on the 'Production and Use of Coil-Coated Strip' in Birmingham.

Packaging innovations are examined in Future-Pak Conferences.

Plastics materials have been dealt with exclusively in the following conferences:

International Seminars on PET Bottles, Antwerp.
International Conferences on Oriented Plastic Containers.
High Performance Plastics Containers.
SPE International Conferences on Bi-axial Oriented Bottles and Containers in PET and other Engineering and Plastic Resins.
The Ryder Conferences on Oriented Plastic Containers and Plastics Beverage Containers.
Packaging Institute (USA) Seminar on 'PET – Beverages and Beyond', 1980.
SPE Annual Technical Conferences.
TAPPI Paper Synthetics Conference 1983.
PI Seminar Hi-Tech Plastics for Food Packaging, Chicago 1983.

Organizations

The following UK organizations are able to provide information on packaging materials:

Can Makers Information Service,
25 North Row, London W1R 2BY
Tel: (01) 629 9621.

ITRI (International Tin Research Institute),
Fraser Road, Perivale, Greenford, Middlesex UB6 7AQ.
Tel: (01) 997 4254.

MPMA Ltd (Metal Packaging Manufacturers Association),
Castle Chambers, 3–9 Sheet Street, Windsor, Berks SL4 1BN.
Tel: (07535) 56012.

PIRA (Research Association for the Paper and Board, Printing and Packaging Industries),
Randalls Road, Leatherhead, Surrey KT22 7RU.
Tel: (0372) 376161

RAPRA (Rubber and Plastics Research Association),
Shawbury, Shrewsbury, Shropshire SY4 4NR
Tel: (0939) 250383.

Trade directories

Packaging – Buyers' Guide and Directory (Cahners Publishing, USA).
The Packaging Encyclopedia (Cahners Publishing, USA).
Packaging Digest – Machinery/Materials Guide (Delta Communications, USA).
The Institute of Packaging Directory and Packaging Review Buyers' Guide (Business Press International, UK).
International Container Directory (Paperboard Packaging Magazines for Industry, USA).

Plastics Technology – Manufacturing Handbook and Buyers' Guide (Bill Communications, USA).
Phillips Paper Trade Directory (Benn Business Information Services, UK).
BSC Tinplate Product Range (UK).

Computer databases

The Host System on which the database can be accessed is given in brackets.

General Packaging

PIRA Abstracts – Packaging Subfile (ORBIT)
PACKABS (Dialog, IRS Dialtech) – database version of Packaging Science and Technology Abstracts.
FSTA (Dialog, IRS Dialtech, ORBIT) – database version of Food Science and Technology Abstracts.

Metals

METADEX (Dialog, ORBIT, IRS Dialtech) – database version of Metals Abstracts.
MDF/1 (ORBIT) – Metals Datafile.
COMPENDEX (Dialog, IRS Dialtech, ORBIT, Datastar) – database version of Engineering Index.
WAA (Dialog, IRS Dialtech) – database version of World Aluminum Abstracts.

Plastics

RAPRA (ORBIT) – database version of RAPRA Abstracts.
CAS (STN, ORBIT, IRS Dialtech, Datastar) – database version of Chemical Abstracts.

Paper and board

PAPERCHEM (Dialog) – database version of Abstract Bulletin of Institute of Paper Chemistry.

All materials

PROMT (Dialog, Datastar) – commercial and market information from Predicasts Inc.
WSCA (ORBIT) – database version of World Surface Coatings Abstracts.
WPI, WPIL (Dialog, ORBIT) – worldwide patents databases marketed by Derwent Publications Ltd.
INPADOC (ORBIT) – worldwide patents database.
CLAIMS (Dialog, ORBIT) – USA patents database.
MATERIALS BUSINESS FILE (Dialog, ORBIT) – technical and commercial developments.

APPENDIX I

Abbreviations of organizations

AASHTO	American Association of State Highway and Transportation Officials
ABBF	Association of Bronze and Brassfounders
ACGIH	American Conference of Government Industrial Hygienists
ACS	Association of Consulting Scientists
AFNOR	Association française de normalisation
AGA	American Gas Association
AIM	Associazione Italiana de Metallurgia
AIME	American Institute of Mining, Metallurgical and Petroleum Engineers
AINDT	Australian Institute for Non-Destructive Testing
AISC	American Institute of Steel, Inc.
AISE	Association of Iron and Steel Engineers
ALFED	Aluminium Federation
AMTRI	Advanced Manufacturing Technology Research Institute (was MTIRA)
ANSI	American National Standards Institute
AOTC	Associated Offices Technical Committee
API	American Petroleum Institute
ASM	American Society for Metals
ASME	American Society of Mechanical Engineers
ASNT	American Society for Nondestructive Testing
ASTM	American Society for Testing and Materials
ATP	American Technical Publishers
AWA	Aluminium Window Association
AWD	Association of Welding Distributors
AWI	Australian Welding Institute
AWRA	Australian Welding Research Association
AWS	American Welding Society
AWTAC	American Welding Technology Application Center
BABS	British Association for Brazing and Soldering
BACO	British Aluminium Co. Ltd (part of TI)
BBA	British Board of Agrément
BCG	Building Centre Group
BCIRA	British Cast Iron Research Association

BCSA	British Constructional Steelwork Association Ltd
BEAMA	Federation of British Electrotechnical & Allied Mfrs. Assns.
BICTA	British Investment Casters Technical Association
BInstN-DT	British Institute of Non-Destructive Testing, The
BISPA	British Independent Steel Producers' Association
BL	British Library
BLDSC	British Library Document and Supply Centre
BLLD	British Library Lending Division (now BLDSC)
BMI	Battelle Memorial Institute
BMT	British Maritime Technology
BNF	British Non-Ferrous Metals Association (now BNF Metals Technology Centre)
BNFMF	British Non-Ferrous Metals Federation
BRA	British Robot Association
BRE	Building Research Establishment
BRS	Building Research Station (now part of DoE Building Research Establishment)
BSC	British Steel Corporation
BSI	British Standards Institution
BSRA	British Ship Research Association (now BMT Ltd)
BSRIA	Building Services Research and Information Association
CA	Cadmium Association
CDA	Copper Development Association
CFA	Construction Fixings Association
CIDEC	Conseil International pour la Développement du Cuivre
CIRIA	Construction Industry Research and Information Association
CISC	Canadian Institute of Steel Construction
COFREND	Comité Française d'Étude des Essais Non Destructifs
CONSTRADO	Constructional Steel Research and Development Organization
CSNDT	Canadian Society for Nondestructive Testing
CSWIP	Certification Scheme of Weldment Inspection Personnel
DELTA	DELTA Materials Research Ltd.
DIN	Deutsches Institüt fur Normung
DTI	Department of Trade and Industry
DVS	Deutscher Verband Schweisstecnik (German Welding Association)
EC	Engineering Council
EITB	Engineering Industry Training Board
ERA	Electrical Research Association (now ERA Technology Ltd.)
ESDU	Engineering Sciences Data Unit Ltd.
ESDU	Engineering Sciences Data Unit (now ESDU International Ltd.)
FRI	Fulmer Research Institute
GDMB	GDMB Gesellschaft Deutscher Metallhuetten-und Bergleute
GEC	General Electric Company
GKN	GKN Techology Ltd.
HSE	Health and Safety Executive
ICE	Institution of Civil Engineers
IEC	International Electrotechnical Commission
IEE	Institution of Electrical Engineers
IEEE	Institute of Electrical and Electronic Engineers
IIW	International Institute of Welding
ILZRO	International Lead Zinc Research Association
IMarE	Institute of Marine Engineers
IMechE	Institution of Mechanical Engineers

IMM	Institution of Mining and Metallurgy
INCO	INCO Europe Ltd.
INCRA	International Copper Research Association Inc.
IP	Institute of Petroleum
IQA	Institute of Quality Assurance
IS	Institut de Soudure
ISO	International Standards Organisation
ITRI	International Tin Research Institute
JIM	Japan Institute of Metals
JWRI	Japan Welding Research Institute
JWS	Japan Welding Society
LDA	Lead Development Association
LRQA	Lloyds Register Quality Assurance Certification Association
MATSU	Marine Technology Support Group – Department of Energy, Petroleum Division
MBMA	Metal Builders Manufacturers Association
MIRA	Motor Industry Research Association
MIT	Massachusetts Institute of Technology
MS	Metals Society (now Institute of Metals)
MTIRA	Machine Tool Industry Research Association (now AMTRI)
NACE	National Association of Corrosion Engineers
NBA	National Brassfoundry Association
NBS	National Bureau of Standards
NEL	National Engineering Laboratory
NORDTEST	Norsk NDT-Forening
NPL	National Physical Laboratory
NTIAC	Nondestructive Testing Information Analysis Centre
NWSA	National Welding Supply Association
OTUA	Office Technique pour l'Utilisation de l'Acier
PERA	Production Engineering Research Association
PVQAB	Pressure Vessel Quality Assurance Board
RAE	Royal Aircraft Establishment
RoSPA	Royal Society for the Prevention of Accidents
RWMA	Resistance Welder Manufacturer's Association
SAE	Society of Automotive Engineers
SCRATA	Steel Castings Research and Trade Association
SFM	Société Française de Metallurgie
SIRA	Scientific Instrument Research Association (now SIRA Institute Ltd)
SRI	Stanford Research International
SSFA	Stainless Steel Fabricator's Association
THE	Technical Help to Exporters
TI	TI Research Laboratories (formerly Tube Investments)
TRRL	Transport and Road Research Laboratory
UEG	Underwater Engineering Group – Department of Energy, Petroleum Division
UKAEA	UK Atomic Energy Authority
WI	Welding Institute, The
WIC	Welding Institute of Canada
WMA	Welding Manufacturer's Association
WRC	Welding Research Society
ZDA	Zinc Development Association

APPENDIX II

Addresses of Organizations

Admiralty Research Establishment
Holton Heath
Poole
Dorset BH16 6JU
UK

Advanced Manufacturing Technology Research Institute (was MTIRA)
(AMTRI)
Hulley Road
Hurdsfield
Macclesfield
Cheshire SK10 2NE
UK

Aluminium Federation
(ALFED)
Broadway House
Calthorpe Road
Birmingham B15 1TN
UK

Aluminium Window Association
(AWA)
323/324 Golden House
28–31 Great Pulteney Street
London W1R 3DD
UK

American Association of State Highway and Transportation Officials
(AASHTO)
444 North Capitol
Washington DC 20001
USA

American Conference of Government Industrial Hygienists
(ACGIH)
Bldg D-7
6500 Glenway Avenue
Cincinnati
OH 45211
USA

American Gas Association
(AGA)
1515 Wilson Blvd.
Arlington
Vancouver 22209
USA

American Institute of Mining, Metallurgical and Petroleum Engineers
(AIME)
345 East 47th Street
New York
NY 10017
USA

American Institute of Steel, Inc.
(AISC)
Room 1580
1221 Avenue of the Americas
New York
NY 10020
USA

American National Standards Institute
(ANSI)
1403 Broadway
New York
NY 10018
USA

American Petroleum Institute
API)
2101 'L' Street
NW Washington DC 20037
USA

American Society for Testing and Materials
(ASTM)
1916 Race Street
Philadelphia
PA 19103
USA

American Society for Metals
(ASM)
Metals Park
OH 44073
USA

American Society for Nondestructive Testing
(ASNT)
4153 Arlington Plaza
Caller no. 28518
Columbus
Ohio 43228–0518
USA

American Society of Mechanical Engineers
(ASME)
United Engineering Center
345 East 47th Street
New York
NY 10017
USA

American Technical Publishers
(ATP)
68a Wilbury Way
Hitchin
Hertfordshire SG4 0TP
UK

American Welding Society
(AWS)
550 N.W. Le Jeune Road
PO Box 351040
Miami
Florida 33135
USA

Anglian Welding School
New Road
Acle
Norwich NR13 3BD
UK

Architectural Aluminium Association
193 Forest Road
Tunbridge Wells
Kent TN2 5JA
UK

Asociacion de Industrias Metalurgicas y Metalmecanicas
Agustinas 785
Piso 4
Santiago
Chile

Asociacion Mexicano del Cobre AC
Avenue Sonora
No. 166–1er Piso
Colonia Hippodromo Delagacion
Cuauhtemoc
06100 Mexico D.F.
Mexico

Associated Offices Technical Committee
(AOTC)
St Mary's Parsonage
Manchester M60 9AP
UK

Association française de normalisation
(AFNOR)
Tour Europe
Cedex 7
92080 Paris la Défense
France

Association of Iron and Steel Engineers
(AISE)
Three Gateway Center
Suite 2350
Pittsburgh
PA 15222
USA

Association of Bronze and Brassfounders
(ABBF)
c/o Heathcote and Coleman
136 Hagley Road
Edgbaston
Birmingham B16 9PN
UK

Association of Consulting Scientists
(ACS)
Owles Hall
Buntingford
Hertfordshire
UK

Association of Light Alloy Refiners
635 Grand Buildings
Trafalgar Square
London WC2N 5HN
UK

Associazione Italiana de Metallurgia
(AIM)
Piazza Velasca 8
20122 Milan
Italy

Australasian Institute of Mining and Metallurgy
Clunies Ross House
191 Royal Parade
Parkville
Victoria 3052
Australia

Australian Institute for Non-Destructive Testing
(AINDT)
National Science Centre
191 Royal Parade
Parkville
Victoria 3052
Australia

Australian Lead Development Association
95 Collins Street
Melbourne
Victoria 3000
Australia

Babcock & Wilcox Ltd.
PO Box 8
Birmingham New Road
Tipton
West Midlands DY4 8YY
UK

Battelle Memorial Institute – Metals and Ceramics Information Centre
(BMI)
505 King Avenue
Columbus
Ohio 43201–2693
USA

Benelux Metallurgie
3 rue Ravenstein
B-1000 Brussels
Belgium

Beratungstelle für Stahlverwendung
Kasernstrasse 36
Postfach 1611
4000 Düsseldorf 1
West Germany

BNF Metals Technology Centre
Wantage Business Park
Denchworth Road
Wantage
Oxfordshire OX12 9BJ
UK

Bowfort Engineering Services Ltd.
Welding Technology Centre
Mobet Trading Estate
Workington
Cumbria CA14 5AE
UK

British Aluminium Co. Ltd. (part of TI)
BACO
Chalfont Technological Centre
Chalfont Park
Gerrards Cross
Bucks S29 0QB
UK

British Association for Brazing and Soldering
(BABS)
BNF Metals Technology Centre
Denchworth Road
Wantage
Oxfordshire OX12 9BJ
UK

British Board of Agrément
(BBA)
PO Box 195
Bucknalls Lane
Garston
Watford
Hertfordshire WD2 7NG
UK

British Cast Iron Research Association
(BCIRA)
Alvechurch
Birmingham B48 7QB
UK

British Constructional Steelwork Association Ltd.
(BCSA)
92–96 Vauxhall Bridge Road
London SW1V 2RL
UK

British Independent Steel Producers Association
(BISPA)
5 Cromwell Road
London SW7 2HX
UK

British Institute of Non-Destructive Testing, The
(BInstN-DT)
1 Spencer Parade
Northampton NN1 5AA
UK

British Investment Casters Technical Association
(BICTA)
2nd Floor
Royston House
George Road
Edgbaston
Birmingham B15 1NU
UK

British Library Bibliographic Services Division
British Library
2 Sheraton House
London W1V 4BM
UK

British Library Document and Supply Centre
(BLDSC)
Boston Spa
Wetherby
West Yorkshire LS23 7BQ
UK

British Library Lending Division (now BLDSC)

British Library Science Reference and Information Service
25 Southampton Buildings
Chancery Lane
London WC2A 1AW
UK

British Maritime Technology
(BMT)
Wallsend Research Station
Wallsend
Tyne & Wear NE28 6UY
UK

British Non-Ferrous Metals Association (now BNF Metals Technology Centre)

British Non-Ferrous Metals Federation
(BNFMF)
10 Greenfield Crescent
Edgbaston
Birmingham B15 3AU
UK

British Powder Metal Federation
124A Compton Road
Wolverhampton WV3 9QZ
UK

British Robot Association
(BRA)
39 High Street
Kempston
Bedford MK42 7BT
UK

British Ship Research Association (now BMT Ltd.)

British Shipbuilders Ltd.
Scottish Regional Centre
Holmfauld Road
Linthouse
Glasgow G51 4RY
UK

British Shipbuilders Ltd.
Ellison Street
Hebburn
Tyne & Wear NE31 1YN
UK

British Standards Institution
(BSI)
2 Park Street
London W1A 2BS
UK

British Steel Corporation (BSC)
Swinden Laboratories
Information Library and Research Services
Moorgate
Rotherham
South Yorkshire S60 3AR
UK

British Steel Corporation (BSC)
9 Albert Embankment
London SE1 7SN
UK

BSC Sections and Commercial Steels Structural Advisory Service (BSC)
PO Box 24
Steel House
Redcar
Cleveland TS10 5QL
UK

BSC Stainless Steel Advisory Centre
PO Box 161
Shepcote Lane
Shefield S9 1TR
UK

BSC Steel Sheet Information Centre
PO Box 32
Port Talbot SA13 2NG
UK

BSC Tubes Structural Hollow Sections Advisory Section
(BSC)
Market Development & Technical Sales (SHS)
Corby
Northamptonshire NN1 1UA
UK

Building Centre Group
(BCG)
26 Store Street
London WC1E 7BT
UK

Building Research Establishment
(BRE)
Building Research Station
Garston
Watford WD2 7JR
UK

Building Research Station (now part of DoE Building Research Establishment)
(BRS)
Bucknalls Lane
Garston
Watford
Hertfordshire WD2 7NG
UK

Building Services Research and Information Association
(BSRIA)
Old Bracknell Lane
Berkshire
UK

Bundesanstalt für Materialprufung
The Federal Institute for Materials Testing
1000 Berlin
East Germany

Cadmium Association
(CA)
34 Berkeley Square
London W1X 6AJ
UK

Canadian Institute of Mining and Metallurgy
Suite 400
1130 Sherbrooke Street West
Montreal
Quebec H3A 2M8
Canada

Canadian Institute of Steel Construction
(CISC)
Suite 300
201 Consumers Road
Willowdale
Ontario M2J 4G8
Canada

Canadian Sheet Steel Building Institute
Consumers Road
Willowdale
Ontario M2J 4G8
Canada

Canadian Society for Nondestructive Testing
(CSNDT)
c/o Mohawk College of Applied Arts and Technology
135 Fennell Avenue
West Hamilton
Ontario L8N 3T2
Canada

Carnegie-Mellon University
5000 Forbes Avenue
Pittsburgh
PA 15213
USA

Centre Belgo-Luxembourgeois de l'Information de l'Acier
Rue Montoyer 47
1040 Bruxelles
Belgium

Centre d'Information du Plomb
79 Avenue Denfert-Rochereau
75014 Paris
France

Centre Suisse de la Construction Metallique
Seefeldstrasse 25
CH-8034 Zurich
Switzerland

Certification Scheme of Weldment Inspection Personnel
(CSWIP)
The Welding Institute
Abington Hall
Abington
Cambridge CB1 6AL
UK

Chinese Society of Metals
46 Dongsix Dajie
Beijing
China

Cold Rolled Sections Association
Centre City Tower
7 Hill Street
Birmingham B5 4UU
UK

Comité Française d'Étude des Essais Non Destructifs
(COFREND)
Secretariat de COFREND
Institut de Soudure
32 Boulevard de la Chapelle
75880 Paris Cedex 18
France

Confederacion Espanola de Organicaciones Empresariales dei Metal
Principe de Vergara 74
28006 Madrid
Spain

Conseil International pour la Développement du Cuivre
(CIDEC)
100 rue du Rhone
1211 Geneva
Switzerland

Construction Fixings Association
(CFA)
Light Trades House
3 Melbourne Avenue
Sheffield S10 2QJ
UK

Construction Industry Research and Information Association (CIRIA)
6 Storeys Gate
London SW1P 3AU
UK

Constructional Steel Research and Development Organization
(CONSTRADO)
NLA Tower
12 Addiscombe Road
Croydon CR9 3JH
UK

Copper Cylinder and Boiler Manufacturers
56 Oxford Street
Manchester M1 6EV
UK

Copper Development Association (CDA)
Orchard House
Mutton Lane
Potters Bar
Hertfordshire EN6 3AP
UK

DELTA Materials Research Ltd.
(DELTA)
PO Box 22
Hadleigh Road
Ipswich
Suffolk
UK

Department of Trade and Industry
(DTI)
1 Victoria Street
London SW1H 0ET
UK

Deutsche Gesellschaft für Zerstorungsfreie Prufung EV
Unter den Eichen 87
D-1000 Berlin 45
West Germany

Deutscher Verband Schweisstecnik (German Welding Association)
(DVS)
4000 Düsseldorf
Postfach 2725
West Germany

Deutsches Institüt für Normung
(DIN)
Burggrafenstrasse 4–10
Postfach 1107
D-1000 Berlin 30
West Germany

Door and Shutter Manufacturer's Association
5 Greenfield Crescent
Edgbaston
Birmingham B15 3BE
UK

Ductile Iron Pipe Committee
Bridge House
Small Brooke
Queensway
Birmingham B5 4JP
UK

EAGIT (Engineering) Ltd.
Hurricane Way
Norwich Airport
Norwich NR6 6EY
UK

École Nationale Superieure des Arts et Metiers
151 Blvd de L'hôpital
75640 Paris
France

Edison Welding Institute
Columbus
Ohio
USA

Electrical Research Association (now ERA Technology Ltd.)
(ERA)
Cleeve Road
Leatherhead
Surrey KT22 7SA

Engineering Council
(EC)
10 Maltravers Street
London WC2R 3ER
UK

Engineering Industry Training Board
EITB
54 Clarendon Road
Watford WD1 1LB
UK

Engineering Sciences Data Unit (now ESDU International Ltd.)
(ESDU)
251–259 Regent Street
London W1R 7AD
UK

ERA Technology Limited (formerly Electrical Research Association)
Cleeve Road
Leatherhead
Surrey KT22 7SA
UK

European Patent Office
Erhardstrasse 27
D-8000 München 2
West Germany

European Powder Metallurgy Federation
Institute of Metals
1 Carlton House Terrace
London SW1Y 5DB
UK

Eutectic Co. Ltd.
Central Way
North Feltham Trading Estate
Feltham
Middlesex TW14 0UE
UK

Fachverband Pulvermetallurgie
Goldene Pfortel
5800 Hagen-Ernst
Postfach 921
West Germany

Farm Buildings Information Centre
National Agricultural Centre
Stoneleigh
Kenilworth
Warwickshire CV8 2LG
UK

Federation of British Electrotechnical & Allied Mfrs. Assns.
(BEAMA)
8 Leicester Street
London WC2H 7BN
UK

Federation des Chambres des Minerais et des Metaux Non Ferreux
30 avenue de Messine
75008 Paris
France

Flat Roofing Contractors Advisory Board
Maxwelton House
Boltro Road
Haywards Heath
West Sussex RH16 1BJ
UK

Foster Wheeler Automated Welding Ltd
17 Stadium Way
Tilehurst
Reading
Berkshire RG3 6BX
UK

Fulmer Research Institute
(FRI)
Stoke Poges
Slough
Berkshire SL2 4QD
UK

GDMB Gesellschaft Deutscher Metallheuten- und Bergleute
(GDMB)
Paul Ernst Strasse 10
PO Box 210
D-3392 Clausthal-Zellerfeld
West Germany

General Dynamics, Convair Division
Convair School for NDT
PO Box 80847
San Diego
California 92138
USA

General Electric Company (Materials Information Services)
(GEC)
120 Erie Boulevard
Schenectady
New York
NY 12305
USA

GKN Technology Ltd.
(GKN)
Birmingham New Road
Wolverhampton WV4 6BW
UK

Guild of Architectural Ironmongers
8 Stepney Green
London E1 3JU
UK

Health and Safety Executive
(HSE)
Baynards House
1 Chepstow Place
London W2 4TY
UK

Herriot Watt University – Institute of Offshore Engineering
Chambers Street
Edinburgh EH1 1HX
UK

INCO Europe Ltd.
(INCO)
1/3 Grosvenor Place
London SW1X 7EA
UK

Indian Lead Zinc Information Centre
B-6/7 Shopping Centre
Safdarjung Enclave
New Delhi 110029
India

Institut de Soudure
(IS)
32 Boulevard de la Chapelle
75880 Paris Cedex 18
France

Institute of Electrical and Electronic Engineers
(IEEE)
345 East 47th Street
New York
NY 10017
USA

Institute of Marine Engineers
(IMarE)
76 Mark Lane
London EC3R 7JN
UK

Institute of Metals
1 Carlton House Terrace
London SW1Y 5DB
UK

Institute of Petroleum
(IP)
61 New Cavendish Street
London W1M 8AR
UK

Institute of Polymer Technology and Materials Engineering
University of Technology
Loughborough
Leicestershire LE11 3TU
UK

Institute of Quality Assurance
(IQA)
54 Princes Gate
Exhibition Road
London SW7 2PG
UK

Institute of Offshore Engineering (see Herriot Watt University)

Institution of Civil Engineers
(ICE)
Great George Street
Westminster
London SW1P 3AA
UK

Institution of Electrical Engineers
(IEE)
Savoy Place
London WC2R 0BL
UK

Institution of Mechanical Engineers
(IMechE)
1 Birdcage Walk
Westminster
London SW1H 9JJ
UK

Institution of Mining and Metallurgy
(IMM)
44 Portland Place
London W1N 4BR
UK

Instituto Brasileiro de Informacao du Chumbo, Niquel e Zinco
Avenida Nova de Julho 4015
0417 Sao Paulo
Brazil

Instituto Brasileiro de Mineracao
Avenida Alfonso Pena 3880
4° e 5° Andares
Belo Horizonte
Minas Gerais
Brazil

Instituto Mexicano del Zinc, Plomo y Coproductos AC
Avenue Sonora
No. 166–1er Piso
Colonia Hippodromo Delagacion
Cuauhtemoc
06100 Mexico DF
Mexico

International Copper Research Association Inc.
(INCRA)
Brosnan House
Darkes Lane
Potters Bar
Hertfordshire EN6 1BW
UK

International Electrotechnical Commission
(IEC)
3 rue de Varembe
1211 Geneva
Switzerland

International Institute of Welding
(IIW)
54 Princes Gate
Exhibition Road
London SW7 2PG
UK

International Lead Zinc Research Association
(ILZRO)
292 Madison Avenue
New York
NY 10017
USA

International Society for Powder Metallurgy
PO Box 74
A-6600 Reulte
Tirol
Austria

International Standards Organisation
(ISO)
1 rue de Varembe
Case Postale 56
CH-1211 Geneva
Switzerland

International Tin Research Institute
(ITRI)
Brunel Science Park
Kingston Lane
Uxbridge
Middlesex UB8 3PJ
UK

International Translations Centre
Doelenstraat 101
2611 NS Delft
Netherlands

International Truss Plate Association
PO Box 44
Halesowen
West Midlands
UK

Iron and Steel Institute of Japan, The
Keidanren Kaikan (3rd floor)
1–9–4 Otemachi
Chiyoda-ku
Tokyo 100
Japan

ITM-Head Wrightson Teesdale Ltd.
PO Box 10
Stockton-on-Tees
Cleveland TS17 6AZ
UK

Japan Institute of Metals
(JIM)
Nihon Kinzoku Gakkai
Shiro Banya
Sendai 980
Japan

Japan Lead Zinc Development Association
New Hibiya Building
3–6 Uchisaiwaicho
1-chome
Chiyoda-Ku
Tokyo
Japan

Japan Welding Society
(JWS)
1–11 Kanda Sakuma-cho
Chiyoda-ku
Tokyo 101
Japan

Jernkontoret
Jungstradgardsgatan 10
PO Box 1721
S-11187 Stockholm
Sweden

Johnson and Matthey
Blounts Court
Sonning Common
Reading RG4 6NH
UK

Lead Development Association
(LDA)
42 Weymouth Street
London W1N 3LQ
UK

Lead Industries Association Inc.
292 Madison Avenue
New York
NY 10017
USA

Lloyds British School of Welding
Arcon Works
Walsall Road
Bridgtown
Cannock
Staffordshire WS11 3HG
UK

Lloyds Register Quality Assurance Certification Association (LRQA)
Norfolk House
Croydon
Surrey
UK

Machine Tool Industry Research Association (now AMTRI)
(MTIRA)
Hulley Road
Hurdsfield Industrial Estate
Macclesfield
Cheshire SK10 2NE
UK

Marine Technology Support Group – Department of Energy, Petroleum Division
(MATSU)
Atomic Energy Research Establishment
Harwell
Didcot
Oxfordshire OX11 0RA
UK

Massachusetts Institute of Technology
(MIT)
Cambridge
Massachusetts 02139
USA

Metal Builders Manufacturers Association
(MBMA)
1230 Keith Building
Cleveland
Ohio 4415
USA

Metalbox plc.
Denchworth Road
Wantage
Oxfordshire OX12 9BP
UK

Metallurgical Society of AIME
420 Commonwealth Drive
Warrendale
Pennsylvania
PA 13086
USA

Metals Information (see ASM/Institute of Metals entries)

Metals Information Service (in conjunction with Sheffield City Libraries)
Sheffield University Advisory Centre
Sheffield
UK

Metals Society (now Institute of Metals)

Mining, Geological and Metallurgical Institute of India
29 Chowringhee Road
Calcutta 700016
India

Mining and Materials Processing Institute of Japan
Nogizaka Bldg
9–6–41 Akasaka
Minato-ku
Tokyo 107
Japan

Motor Industry Research Association
(MIRA)
Watling Street
Nuneaton
Warwickshire CV10 0TU
UK

Murex Training Centre
Plume Street
Aston
Birmingham B6 7RU
UK

National Board of Boiler and Pressure Vessel Inspectors
1055 Crupper Avenue
Columbus
Ohio 43229
USA

National Association of Corrosion Engineers
(NACE)
PO Box 218340
Houston
TX 77218
USA

National Brassfoundry Association
(NBA)
5 Greenfield Crescent
Edgbaston
Birmingham B15 3BE
UK

National Bureau of Standards
(NBS)
A323 Physics Building
Washington DC 20234
USA

National Council of Building Materials Producers
33 Alfred Place
London WC1 7EN
UK

National Engineering Laboratory
(NEL)
East Kilbride
Glasgow G75 0DU
UK

National Engineering Training Association Ltd
Pennine Avenue
North Tees Industrial Estate
Portrack Lane
Stockton-on-Tees
Cleveland TS18 2RG
UK

National Federation of Roofing Contractors
15 Soho Square
London W1V 5FB
UK

National Physical Laboratory
(NPL)
Teddington
Middlesex TW11 0LW
UK

National Research Institute for Metals (Japan)
2–3–12 Nakemeguro
Meguro-ku
Tokyo 153
Japan

National Welding Supply Association
(NWSA)
1900 Arch Street
Philadelphia
PA 19103
USA

Nondestructive Testing Information Analysis Centre
(NTIAC)
6220 Culebra Road
San Antonio
TX 7824
USA

Nordic Galvanisers Association
Kungsgatan 37
4tr S-111
56 Stockholm
Sweden

Norsk NDT-Forening
(NORDTEST)
Kronprinsensgt 9
Oslo 2
Norway

North West Welder Training
11 Westbrook Road
Trafford Park
Manchester M17 1AY
UK

Office Technique pour l'Utilisation de l'Acier
(OTUA)
5 bis rue de Madrid
75008 Paris
France

Oilfab Group Ltd.
Hareness Circle
Althens Industrial Estate
Aberdeen AB1 4LY
UK

Osterreicher Stahlverband
Larochegasse 28
A-1130 Vienna
Austria

Partitioning Industry Association
1 Lansdale Avenue
Solihull
West Midlands B92 0PP
UK

Patent Office
66–71 High Holborn
London WC1R 4TP
UK

Powder Metallurgy Group
Institute of Metals
1 Carlton House Terrace
London SW1Y 5DB
UK

Powder Metallurgy Research Group Institute of Polymer Technology
University of Technology
Loughborough
Leicestershire LE11 3TU
UK

Pressure Vessel Quality Assurance Board
(PVQAB)
c/o Institution of Mechanical Engineers
1 Birdcage Walk
London SW1H 9JJ
UK

Production Engineering Research Association
(PERA)
Melton Mowbray
Leicestershire LE13 0PB
UK

Royal Aircraft Establishment
(RAE)
Farnborough
Hampshire GU14 6TD
UK

Royal Society for the Prevention of Accidents
(RoSPA)
Cannon House
The Priory
Queensway
Birmingham B4 6BS
UK

Scandinavian Lead Zinc Association
Sturegatan 22
S-114 36 Stockholm
Sweden

Scientific Instrument Research Association (now SIRA Institute Ltd)
(SIRA)
South Hill
Chislehurst
Kent BR7 5CH
UK

The Scottish School of NDT
Paisley College of Technology
High Street
Paisley
Renfrew PA1 2BE
UK

Sheffield University – Department of Metallurgy
Sheffield
S10 2TN
UK

Société Française de Metallurgie
(SFM)
1–3 rue Cezanne
F75008 Paris
France

Society of Mining Engineers of AIME
8307 Shaffer Parkway Caller no.D
Littleton
Colorado 80127
USA

Society of Automotive Engineers
(SAE)
400 Commonwealth Drive
Warrendale
PA 15096
USA

South African Institute of Mining and Metallurgy
PO Box 61019
Marshalltown 2107
South Africa

Stainless Steel Advisory Centre (see BSC entries)

Stainless Steel Fabricator's Association
(SSFA)
14 Knott Road
Dorking
Surrey RH4 3EW
UK

Stanford Research International
(SRI)
333 Ravenswood Avenue
Menlo Park
California 94025
USA

Steel Castings Research and Trade Association
(SCRATA)
5 East Bank Road
Sheffield S2 3PT
UK

Steel Construction Institute
NLA Tower
12 Addiscombe Road
Croydon CR9 3JH
UK

Steel Lintel Manufacturers Association
PO Box 10
Newport
Gwent NPT 0XN
UK

Steel Sheet Information Centre (see BSC entries)

Steel Window Association
Building Centre
26 Store Street
London WC1E 7BT
UK

Svejsecentralen (Danish Welding Institute)
Park Alle 345
DK-2600 Glostrup
Denmark

Technical Help to Exporters
(THE)
British Standards Institution
Linford Wood
Milton Keynes MK14 6LE
UK

TI Research Laboratories
(TI)
Hinxton Hall
Hinxton
Saffron Walden
Essex CB10 1RH
UK

Transport and Road Research Laboratory
(TRRL)
Crowthorne
Berkshire RG11 6AU
UK

UK Atomic Energy Authority
(UKAEA)
Harwell
Didcot
Oxfordshire OX11 0RA
UK

The Welder Training Centre (Lancaster) Ltd
Unit 28
Ladies Walk
Caton Road
Lancaster LA1 3NX
UK

Welding Institute, The
(WI)
Abington Hall
Cambridge CB1 6AL
UK

Welding Institute of Canada
(WIC)
391 Burnhamthorpe Road East
Oakville
Ontario L6J 6C9
Canada

Welding Manufacturer's Association
(WMA)
8 Leicester Street
London WC2H 7BN
UK

ZCCM Ltd
PO Box 48
74 Independence Avenue
Lusaka
Zambia

Zinc and Lead Asian Service
95 Collins Street
Melbourne
Victoria 3000
Australia

Zinc Development Association
(ZDA)
34 Berkeley Square
London W1X 6AJ
UK

Index